VISIT US AT

Syngress is committed to publishing high-quality books for IT Professionals and delivering those books in media and formats that fit the demands of our customers. We are also committed to extending the utility of the book you purchase via additional materials available from our Web site.

SOLUTIONS WEB SITE

To register your book, visit www.syngress.com/solutions. Once registered, you can access our solutions@syngress.com Web pages. There you will find an assortment of value-added features such as free e-booklets related to the topic of this book, URLs of related Web site, FAQs from the book, corrections, and any updates from the author(s).

ULTIMATE CDs

Our Ultimate CD product line offers our readers budget-conscious compilations of some of our best-selling backlist titles in Adobe PDF form. These CDs are the perfect way to extend your reference library on key topics pertaining to your area of expertise, including Cisco Engineering, Microsoft Windows System Administration, CyberCrime Investigation, Open Source Security, and Firewall Configuration, to name a few.

DOWNLOADABLE EBOOKS

For readers who can't wait for hard copy, we offer most of our titles in downloadable Adobe PDF form. These eBooks are often available weeks before hard copies, and are priced affordably.

SYNGRESS OUTLET

Our outlet store at syngress.com features overstocked, out-of-print, or slightly hurt books at significant savings.

SITE LICENSING

Syngress has a well-established program for site licensing our ebooks onto servers in corporations, educational institutions, and large organizations. Contact us at sales@syngress.com for more information.

CUSTOM PUBLISHING

Many organizations welcome the ability to combine parts of multiple Syngress books, as well as their own content, into a single volume for their own internal use. Contact us at sales@syngress.com for more information.

SYNGRESS®

How to Cheat at

Securing a Wireless Network

Chris Hurley
Brian Baker
Christian Barnes
Tony Bautts
Darren Bonawitz
Randy Hiser
Jan Kanclirz Jr.
Andy McCullough
Jeffrey A. Wheat

KEY	SERIAL NUMBER
001	HJIRTCV764
002	PO9873D5FG
003	829KM8NJH2
004	HJPOOLL783
005	CVPLQ6WQ23
006	VBP965T5T5
007	HJJJ863WD3E
008	2987GVTWMK
009	629MP5SDJT
010	IMWQ295T6T

PUBLISHED BY
Syngress Publishing, Inc.
800 Hingham Street
Rockland, MA 02370

How to Cheat at Securing a Wireless Network

1 2 3 4 5 6 7 8 9 0
ISBN: 1597490873
Printed in Canada

Publisher: Andrew Williams
Acquisitions Editor: Erin Heffernan
Technical Editor: Chris Hurley
Cover Designer: Michael Kavish

Page Layout and Art: Patricia Lupien
Copy Editor: Darlene Bordwell
Indexer: Nara Wood

Distributed by O'Reilly Media, Inc. in the United States and Canada.
For information on rights, translations, and bulk sales, contact Matt Pedersen, Director of Sales and Rights,
at Syngress Publishing; email matt@syngress.com or fax to 781-681-3585.

Acknowledgments

Syngress would like to acknowledge the following people for their kindness and support in making this book possible.

Syngress books are now distributed in the United States and Canada by O'Reilly Media, Inc. The enthusiasm and work ethic at O'Reilly are incredible, and we would like to thank everyone there for their time and efforts to bring Syngress books to market: Tim O'Reilly, Laura Baldwin, Mark Brokering, Mike Leonard, Donna Selenko, Bonnie Sheehan, Cindy Davis, Grant Kikkert, Opol Matsutaro, Steve Hazelwood, Mark Wilson, Rick Brown, Tim Hinton, Kyle Hart, Sara Winge, Peter Pardo, Leslie Crandell, Regina Aggio Wilkinson, Pascal Honscher, Preston Paull, Susan Thompson, Bruce Stewart, Laura Schmier, Sue Willing, Mark Jacobsen, Betsy Waliszewski, Kathryn Barrett, John Chodacki, Rob Bullington, Kerry Beck, Karen Montgomery, and Patrick Dirden.

The incredibly hardworking team at Elsevier Science, including Jonathan Bunkell, Ian Seager, Duncan Enright, David Burton, Rosanna Ramacciotti, Robert Fairbrother, Miguel Sanchez, Klaus Beran, Emma Wyatt, Krista Leppiko, Marcel Koppes, Judy Chappell, Radek Janousek, Rosie Moss, David Lockley, Nicola Haden, Bill Kennedy, Martina Morris, Kai Wuerfl-Davidek, Christiane Leipersberger, Yvonne Grueneklee, Nadia Balavoine, and Chris Reinders for making certain that our vision remains worldwide in scope.

David Buckland, Marie Chieng, Lucy Chong, Leslie Lim, Audrey Gan, Pang Ai Hua, Joseph Chan, June Lim, and Siti Zuraidah Ahmad of Pansing Distributors for the enthusiasm with which they receive our books.

David Scott, Tricia Wilden, Marilla Burgess, Annette Scott, Andrew Swaffer, Stephen O'Donoghue, Bec Lowe, Mark Langley, and Anyo Geddes of Woodslane for distributing our books throughout Australia, New Zealand, Papua New Guinea, Fiji, Tonga, Solomon Islands, and the Cook Islands.

Technical Editor
and Contributor

Chris Hurley (Roamer) is a Senior Penetration Tester working in the Washington, DC area. He is the founder of the WorldWide WarDrive, a four-year effort by INFOSEC professionals and hobbyists to generate awareness of the insecurities associated with wireless networks and is the lead organizer of the DEF CON WarDriving Contest.

Although he primarily focuses on penetration testing these days, Chris also has extensive experience performing vulnerability assessments, forensics, and incident response. Chris has spoken at several security conferences and published numerous whitepapers on a wide range of INFOSEC topics. Chris is the lead author of *WarDriving: Drive, Detect, Defend*, and a contributor to *Aggressive Network Self-Defense, InfoSec Career Hacking, OS X for Hackers at Heart,* and *Stealing the Network: How to Own an Identity*. Chris holds a bachelor's degree in computer science. He lives in Maryland with his wife Jennifer and their daughter Ashley.

Contributing Authors

Brian Baker is a computer security penetration tester for the U.S. government, located in the Washington, D.C., area. Brian has worked in almost every aspect of computing, from server administration to network infrastructure support and now security. Brian has been focusing his work on wireless technologies and current security technologies.

I'd like to thank my wife, Yancy, and children, Preston, Patrick, Ashly, Blake and Zakary. A quick shout out to the GTN lab dudes, Chris, Mike, and Dan.

Chapter 2 is dedicated to my mother, Harriet Ann Baker, for the love, dedication, and inspiration she gave her three kids, raising us as a single parent. Rest in peace, and we'll see you soon...

Christian Barnes (CCNA, CCDA, MCSE, CNA, A+) is a Network Consultant for Lucent Technologies in Overland Park, KS. His career in the IT industry began with supporting NT and NetWare servers and NT workstations for a large banking company in Western New York. It quickly evolved into support of high-level engineers and LAN and WAN administrators as they attempted to troubleshoot and design their networks, and then on to consulting. Chris has a wife and four sons.

Tony Bautts is a Senior Security Consultant with Astech Consulting. He currently provides security advice and architecture for clients in the San Francisco Bay area. His specialties include intrusion detection systems, firewall design and integration, post-intrusion forensics, bastion hosting, and secure infrastructure design. Tony's security experience has led him to work with Fortune 500 companies in the United States as well as two years of security consulting in Japan. He is also involved with the BerkeleyWireless.net project, which is working to build neighborhood wireless networks for residents of Berkeley, CA.

Darren Bonawitz is a Network Systems Engineer with Lucent Worldwide Service. Darren started his career pursuing entrepreneurial endeavors in electronic commerce. In January 2001, he joined Lucent Worldwide Service as a Network Systems Engineer, bringing his knowledge of the desktop platform and a general understanding of a broad range of technologies in areas such as remote access, ATM, frame relay, and wireless. In addition, his background includes consulting with universities and corporate clients on a pre- and post-sales basis, business/technology planning, and a proven dedication to customer service. He studied Electrical

Engineering with an emphasis in Communication Systems at Kansas State University. In 2000, Darren was nominated for Kansas Young Entrepreneur of the Year, and he was also recently recognized by *The Los Angeles Times* for commitment to online customer service.

Anthony Bruno (CCIE #2738, CCDP, CCNA-WAN, MCSE, NNCSS, CNX-Ethernet) is a Principal Consultant with Lucent Worldwide Services. As a consultant, he has worked with many customers in the design, implementation, and optimization of large-scale, multiprotocol networks. Anthony has worked on the design of wireless networks, voice over technologies, and Internet access. Formerly, he worked as an Air Force Captain in network operations and management. While in this role, he implemented wireless LANs on the base network. Anthony received his master's degree in Electrical Engineering from the University of Missouri-Rolla in 1994 and his B.S. in Electrical Engineering from the University of Puerto Rico-Mayaguez in 1990. He is the coauthor of *CCDA Exam Certification Guide* and has performed technical reviews for several Cisco professional books.

Dan Connelly (MSIA, GSNA) is a Senior Penetration Tester for a Federal Agency in the Washington, D.C., area. He has a wide range of information technology experience, including Web applications and database development, system administration, and network engineering. For the last five years he has been dedicated to the information security industry, providing penetration testing, wireless audits, vulnerability assessments, and network security engineering for many federal agencies. Dan holds a Bachelor of Science degree in Information Systems from Radford University and a Master of Science degree in Information Assurance from Norwich University.

I would like to thank Chris Hurley, Mike Petruzzi, Brian Baker, and everyone at GTN and CMH for creating such an enjoyable work environment. Thanks to everyone at ERG for letting me do what I love to do and still paying me for it.

I would also like to thank my mom and dad for their unconditional support, wisdom, and guidance; my brother for his positive influence; and my sister for always being there. I would particularly like to thank my beautiful wife, Alecia, for all her love and support throughout the years and for blessing our family with our son, Matthew Joseph. He is truly a gift from God and I couldn't imagine life without him.

Chuck Fite is a Consultant currently working for Iconixx Systems Engineering on Sprint ION. He has been a technical writer, a test technician, and a business analyst in the computer and telecommunications industries for the past eight years. Chuck received a B.S. in Physics and an M.A. in Rhetoric and Professional Communication from Iowa State University.

Randy Hiser is a Senior Network Engineer for Sprint's Research, Architecture & Design Group, with design responsibilities for home distribution and DSL self-installation services for Sprint's Integrated On Demand Network. He is knowledgeable in the areas of multimedia services and emerging technologies, has installed and operated fixed wireless MMDS facilities in the Middle East, and has patented network communication device identification in a communications network for Sprint. Randy lives in Overland Park, KS, with his wife, Deborah, and their children, Erin, Ryan, Megan, Jesse, and Emily.

Jan Kanclirz Jr. (CCIE #12136-Security, CCSP, CCNP, CCIP, CCNA, CCDA, INFOSEC Professional) is a Senior Network Information Security Engineer working for IBM Global Services. Currently, he is responsible for strategic and technical evolution of a large multicustomer/multidata center networks and their security environment. Jan specializes in multivendor, hands-on implementations and architectures of network technologies such as routers, switches, firewalls, intrusion sensors, content networking, and wireless networks. Beyond network design and engineering, Jan's background includes extensive experience with Linux and BSD administration and security implementations.

Andy McCullough (BSEE, CCNA, CCDA) has been in network consulting for over seven years. He is currently working at Lucent Enhanced Services and Sales as a Distinguished Member of the Consulting Staff. Andy has done architecture and design work for several global customers of Lucent Technologies, including Level 3 Communications, Sprint, MCI/WorldCom, the London Stock Exchange, and British Telecom. His areas of expertise include network architecture and design, IP routing and switching, and IP Multicast. Prior to working for Lucent, Andy ran a consulting company and a regional ISP.

Andy is coauthor of *Building Cisco Remote Access Networks* (Syngress Publishing, ISBN: 1-928994-13-X). He is also an assistant professor teaching networking classes at a community college in Overland Park, KS.

Mike Petruzzi is a senior penetration tester in the Washington, D.C., area. Mike has performed a variety of tasks and assumed multiple responsibilities in the information systems arena. He has been responsible for performing the role of Program Manager and InfoSec Engineer, System Administrator and Help Desk Technician, and Technical Lead for companies such as IKON and SAIC. Mike also has extensive experience performing risk assessments, vulnerability assessments, and certification and accreditation. Mike's background includes positions as a brewery representative, liquor salesman, and cook at a greasy spoon diner.

Jackie Tucker is a Kansas-based Technical Consultant with over 14 years' experience in technical writing, interface design, and Web development. She has participated in all phases of software design at several software companies, including a long tenure at Informix Software, Inc., worked extensively on Sprint ION, and is currently consulting in the network division of Sprint Corporation. She graduated with honors from St. Mary College with a B.S. in Computer Science and from Baker University with a M.S. in Management. After work, Jackie spends as much time as possible with her husband, Bob, and her two little girls, Sarah and Jessie, in a sports-filled household.

Jeffrey A. Wheat (Lucent WaveLAN Wireless Certification, FORE ATM Certification) is a Principal Member of the Consulting Staff at Lucent Worldwide Services. He currently provides strategic direction and architectural design to Lucent Service Provider and Large Enterprise customers. His specialties include convergence and wireless architectures, and he is an ATM and Testing Methodology Subject Matter Expert within Lucent. Jeff's background with Lucent includes design engagements with Metricom, Sprint ION, Sprint PCS, Raytheon, and Marathon Oil. Prior to Lucent, he spent 11 years working for the U.S. Intelligence Agencies as a Network Architect and Systems Engineer. Jeff graduated from the University of Kansas in 1986 with a B.S. in Computer Science and currently resides in Kansas City with his wife, Gabrielle, and their two children, Madison and Brandon.

Mark Wolfgang (RHCE) is a Senior Information Security Engineer based out of Columbus, OH. He has over five years of practical experience in penetration testing and over 10 years in the information technology field. Since June 2002, he has worked for the U.S. Department of Energy, leading and performing penetration testing and vulnerability assessments at DOE facilities nationwide. He has published several articles and white papers and has twice spoken at the U.S. Department of Energy Computer Security Conference.

Prior to his job as a contractor for the U.S. DOE, he worked as a Senior Information Security Consultant for several companies in the Washington, DC, area, performing penetration testing and vulnerability assessments for a wide variety of organizations in numerous industries. He spent eight years as an Operations Specialist in the U.S. Navy, of which, four years, two months, and nine days were spent aboard the USS DeWert, a guided missile frigate. After an honorable discharge from the Navy, Mark designed and taught the Red Hat Certified Engineer (RHCE) curriculum for Red Hat, the industry leader in Linux and open source technology.

He holds a bachelor of science in computer information systems from Saint Leo University and is a member of the *Delta Epsilon Sigma* National Scholastic Honor Society.

Contents

Introduction to Wireless: From Past to Present

Solutions in this chapter:

- **Exploring Past Discoveries That Led to Wireless**
- **Exploring Present Applications for Wireless**
- **Exploring This Book on Wireless**
- **Summary**
- **Solutions Fast Track**
- **Frequently Asked Questions**

☑ **Summary**

☑ **Solutions Fast Track**

☑ **Frequently Asked Questions**

Introduction

You've been on an extended business trip and have spent the long hours of the flight drafting follow-up notes from your trip while connected to the airline's onboard server. After deplaning, you walk through the gate and continue into the designated public access area. Instantly, your personal area network (PAN) device, which is clipped to your belt, beeps twice announcing that it automatically has retrieved your e-mail, voice mail, and video mail. You stop to view the video mail—a finance meeting—and also excerpts from your children's school play.

Meanwhile, when you first walked into the public access area, your personal area network device contacted home via the Web pad on your refrigerator and posted a message to alert the family of your arrival. Your spouse will know you'll be home from the airport shortly.

You check the shuttle bus schedule from your PAN device and catch the next convenient ride to long-term parking. You also see an e-mail from your MP3 group showing the latest selections, so you download the latest MP3 play list to listen to on the way home.

As you pass through another public access area, an e-mail comes in from your spouse. The Web pad for the refrigerator inventory has noted that you're out of milk, so could you pick some up on the way home? You write your spouse back and say you will stop at the store. When you get to the car, you plug your PAN device into the car stereo input port. With new music playing from your car stereo's MP3 player, you drive home, with a slight detour to buy milk at the nearest store that the car's navigation system can find.

The minute you arrive home, your PAN device is at work, downloading information to various devices. The data stored on your PAN device is sent to your personal computer (PC) and your voicemail is sent to the Bluetooth playback unit on the telephone-answering device. The PAN device sends all video to the television, stored as personal files for playback. As you place the milk in the refrigerator, the Web pad updates to show that milk is currently in inventory and is no longer needed. The kids bring you the television remote and you check out possible movies together to download later that night.

A few weeks later, you are en route back to the airport, but this time with the family for vacation. During the drive, you try to figure out what you could have forgotten. You check the status of the home security system on the PAN device, which was indeed activated—for alarm zones, interior light, and stereo activation to make it look and sound as if people are home—and check the lawn sprinkler mode as well, which was *not* on, so you set it to activate in the evenings while you're away. Your spouse had already ordered the grocery and dry cleaning delivery for the day you return, from the Web pad on your refrigerator.

As you head toward the check-in lines at the airport, you walk through the designated public access area. Your PAN device beeps, showing that there is an update on your flight plans–the flight is delayed. Your teenagers, unflappable as always, make the best of the situation, flopping in the nearest chair and pulling from the carry-on bag the Web pad that they had been sure to remove from the refrigerator just before they left. They initiate a game session to pass the time. You head over to the Bluetooth kiosk to print out a new temporary message about the flight. The message you print out also includes a coupon for the new coffee shop around the corner from the gate; your spouse takes the coupon to go get some much needed caffeine for the two of you. You sit down with the kids and they relinquish the game so you can watch a movie together with a video-on-demand session on the Web pad. This helps settle the family's nerves and makes the time go by quickly.

Do these scenarios sound familiar? If they are not familiar now, they will be soon. All of the personal wireless technologies are available for purchase now or in the very near future. Innovative service providers plan to create new services and value-added services for public access areas around the world. Soon, we will be able to communicate wirelessly throughout the world. This technology is the leading edge of future technology, a new revolution in communication.

Now that you've seen a glimpse of the current and future applications of this technology, this chapter will explain some of the history behind this technology; explore how some of the modern trends in wireless communications have developed and how business and the private sector utilize wireless networks; and discuss how that service is delivered.

Exploring Past Discoveries That Led to Wireless

Wireless technology is the method of delivering data from one point to another without using physical wires, and includes radio, cellular, infrared, and satellite. A historic perspective will provide you with a general understanding of the substantial evolution that has taken place in this area. The common wireless networks of today originated from many evolutionary stages of wireless communications and telegraph and radio applications. Although some discoveries occurred in the early 1800s, much of the evolution of wireless communication began with the emergence of the electrical age and was affected by modern economics as much as by discoveries in physics.

Because the current demand of wireless technology is a direct outgrowth of traditional wired 10-Base-T Ethernet networks, we will also briefly cover the advent of the computer and the evolution of computer networks. Physical networks, and their limitations, significantly impacted wireless technology. This section presents some of the aspects of traditional computer networks and how they relate to wireless networks. Another significant impact to wireless is the invention of the cell phone. This section will briefly explain significant strides in the area of cellular communication.

Discovering Electromagnetism

Early writings show that people were aware of magnetism for several centuries before the middle 1600s; however, people did not become aware of the correlation between magnetism and electricity until the 1800s. In 1820, Hans Christian Oersted, a Danish physicist and philosopher working at that time as a professor at the University of Copenhagen, attached a wire to a battery during a lecture; coincidentally, he just happened to do this near a compass and he noticed that the compass needle swung around. This is how he discovered that there was a relationship between electricity and magnetism. Oersted continued to explore this relationship, influencing the works of contemporaries Michael Faraday and Joseph Henry.

Michael Faraday, an English scientific lecturer and scholar, was engrossed in magnets and magnetic effects. In 1831, Michael Faraday theorized that a changing magnetic field is necessary to induce a current in a nearby circuit. This theory is actually the definition of *induction*. To test his theory, he made a coil by wrapping a paper cylinder with wire. He connected the coil to a device called a *galvanometer*, and then moved a magnet back and forth inside the cylinder. When the magnet was moved, the galvanometer needle moved, indicating that a current was induced in the coil. This proved that you must have a moving magnetic field for electromagnetic induction to occur. During this experiment, Faraday had not only discovered induction but also had created the world's first electric generator. Faraday's initial findings still serve as the basis of modern electromagnetic technology.

Around the same time that Faraday worked with electromagnetism, an American professor named Joseph Henry became the first person to transmit a practical electrical signal. As a watchmaker, he constructed batteries and experimented with magnets. Henry was the first to wind insulated wires around an iron core to make electromagnets. Henry worked on a theory known as *self-inductance*, the inertial characteristic of an electric current. If a current is flowing, it is kept flowing by the property of self-inductance. Henry found that the property of self-inductance is affected by how the circuit is configured, especially by the coiling of wire. Part of his experimentation involved simple signaling.

It turns out that Henry had also derived many of the same conclusions that Faraday had. Though Faraday won the race to publish those findings, Henry still is remembered for actually finding a way to communicate with electromagnetic waves. Although Henry never developed his work on electrical signaling on his own, he did help a man by the name of Samuel Morse. In 1832, Morse read about Faraday's findings regarding inductance, which inspired him to develop his ideas about an emerging technology called the telegraph. Henry actually helped Morse construct a repeater that allowed telegraphy to span long distances, eventually making his Morse code a worldwide language in which to communicate. Morse introduced the repeater technology with his 1838 patent for a Morse code telegraph. Like so many great inventions, the telegraph revolutionized the communications world by replacing nearly every other means of communication—including services such as the Pony Express.

Exploring Conduction

Samuel Morse spent a fair amount of time working on wireless technology, but he also chose to use mediums such as earth and water to pass signals. In 1842, he performed a spectacular demonstration for the public in which he attempted to pass electric current through a cable that was underwater. The ultimate result of the demonstration was wireless communication by *conduction*, although it was not what he first intended. Morse submerged a mile of insulated cable between Governor's Island and Castle Garden in New York to prove that a current could pass through wire laid in water. He transmitted a few characters successfully, but, much to his dismay, the communication suddenly halted—sailors on a ship between the islands, unseen to the spectators, raised their ship's anchor and accidentally pulled up the cable, and not knowing what it was for, proceeded to cut it. Morse faced considerable heckling from the spectators and immediately began modification to the experiment. He successfully retested his idea by transmitting a wireless signal between copper plates he placed in the Susquehanna River, spanning a distance of approximately one mile. In doing so, he became the first person to demonstrate wireless by conduction. Conduction is the flow of electricity charges through a substance (in this case, the water in the river) resulting from a difference in electric potential based on the substance.

Inventing the Radio

After the significant discoveries of induction and conduction, scientists began to test conduction with different mediums and apply electricity to machinery. The scholars and scientists of the day worked to apply these discoveries and explore the parameters of the properties. After the theory of conduction in water was proven, new theories were derived about conduction in the air. In 1887, a German named Heinrich Hertz became the first person to prove electricity travels in waves through the atmosphere. Hertz went on to show that electrical conductors reflect waves, whereas nonconductors simply let the waves pass through the medium. In addition, Hertz also proved that the velocity of light and radio waves are equal, as well as the fact that it is possible to detach electrical and magnetic waves from wires and radiate. Hertz served as inspiration to other researchers who scrambled to duplicate his results and further develop his findings. Inventors from all across the world easily validated Hertz's experiments, and the world prepared for a new era in *radio*, the wireless transmission of electromagnetic waves.

An Italian inventor called Guglielmo Marconi was particularly intrigued by Hertz's published results. Marconi was able to send wireless messages over a distance of ten miles with his patented radio equipment, and eventually across the English Channel. In late 1901, Marconi and his assistants built a wireless receiver in Newfoundland and intercepted the faint Morse code signaling of the letter "S" that had been sent across the Atlantic Ocean from a colleague in England. It was astounding proof that the wireless signal literally curved around the earth, past the horizon line—even Marconi could not explain *how* it happened, but he had successfully completed the world's first truly long-distance communication, and the communication world would never be the same.

Today we know that the sun's radiation forms a layer of ionized gas particles approximately one hundred miles above the earth's surface. This layer, the ionosphere, reflects radio waves back to the earth's surface, and the waves subsequently bounce back up to the ionosphere again. This process continues until the energy of the waves dissipates.

Another researcher by the name of Reginald Fessenden proceeded to further develop Marconi's achievements, and he became the first person to create a radio band wave of human speech. The importance of his results was felt worldwide, as radio was no longer limited to telegraph codes.

Mounting Radio-Telephones in Cars

In 1921, mobile radios began operating in the 2MHz range, which is just above the amplitude modulation (AM) frequency range of current radios. These mobile radios were generally used for law enforcement activities only. They were not integrated with the existing wireline phone systems that were much more common at that time—since the technology was still so new, the equipment was considered experimental and not practical for mass distribution. In fact, people originally did not consider mobile radio as a technology for the public sector. Instead, the technology was developed for police and emergency services personnel, who really served as the pioneers in mobile radio.

It was not until 1924 that the voice-based wireless telephone had the ability to be bi-directional, or two-way. Bell Laboratories invented this breakthrough telephone. Not only could people now receive messages wirelessly, they could also respond to the message immediately, greatly increasing convenience and efficiency. This improved system was still not connected to landline telephone systems, but the evolution of wireless communication had taken one more major step. One issue that still plagued this early mobile radio system was the sheer size of the radio; it took up an entire trunk. Add to the size restriction, the cost of the radio system that was almost as expensive as the vehicle.

In 1935, Edwin Howard Armstrong introduced frequency modulation (FM). This technology not only increased the overall transmission quality of wireless radio but also drastically reduced the size of the equipment. The timing could not have been any better. World War II had begun, and the military quickly embraced FM technology to provide two-way mobile radio communication. Due to the war, companies immediately sensed the urgency to develop the FM technology rapidly, and companies such as Motorola and AT&T immediately began designing considerably smaller equipment. Many of these new inventions became possible due to the invention of the circuit board, which changed the world of electronic equipment of all types.

Inventing Computers and Networks

Though the beginning of the computer age is widely discussed, computer discoveries can be attributed to a long line of inventors throughout the 1800s, beginning with the Englishman Charles Babbage, who in 1822 created the first calculator called the "Difference Engine." Then came Herman Hollerith, who in 1887 produced a punch card reader to tabulate the American census for 1890. Later developments led to the creation of different punch card technologies, binary representation, and the use of vacuum tubes.

The war effort in the 1940s produced the first decoding machine, the Colossus, used in England to break German codes. This machine was slow, taking about 3 to 5 seconds per calculation. The next significant breakthrough was the creation of the Electronic Numerical Integrator and Computer (ENIAC) by Americans John Presper Eckert and John W. Mauchley. The ENIAC was the first general-purpose computer that computed at speeds 1000 times greater than the Colossus. However, this machine was a behemoth, consuming over 160 kilowatts of power—when it ran, it dimmed lights in an entire section of Philadelphia. The main reason these machines were so huge was the vacuum tube technology. The invention of the transistor in 1948 changed the computer's development and began shrinking the machinery. In the next thirty years, the computers got significantly faster and smaller.

In 1981, IBM introduced the personal computer for the home, school, and business. The number of PCs more than doubled from 2 million in 1981 to 5.5 million in 1982; more than 65 million PCs were being used ten years later. With the surge of computer use in the workplace, more emphasis was being placed on how to harness their power and make them work together. As smaller computers became more powerful, it became necessary to find a way to link them together to share memory, software, and information, and to find a way for them to communicate together. Network technology to this point consisted of a mainframe that stored the information and performed the processes hooked to several "dumb terminals" that provided the input.

Ethernet was developed in the early 1970s and was used to link multiple PCs within a physical area to form what is known as a local area network (LAN). A LAN connects network devices over a short distance. Common applications include offices, schools, and the home. Sometimes businesses are composed of several LANs that are connected together. Besides spanning a short distance, LANs have other distinctive attributes. LANs typically are controlled, owned, and operated by a single person or department. LANs also use specific technologies, including Ethernet and Token Ring for connectivity. There are typically two basic components to the LAN configuration: a client and a server. The client is the node that makes a request, and the server is the node that fulfills that request. The client computer contains the client software that allows for access to shared resources on the server. Without the client software, the computer will not actively participate in either of the two network models.

Wide area networks (WANs) span a much wider physical distance. Usually a WAN is a widely dispersed collection of LANs. The WAN uses a router to connect the LANs physically. For example, a company may have LANs in New York, Los Angeles, Tokyo, and Sydney; this company would then implement a WAN to span the LANs and to enable communication throughout the company. WANs use different connectivity technology than LANs—typically, T1 or T3 lines, Asynchronous Transfer Mode (ATM) or Frame Relay circuits, microwave links, or higher speed Synchronous Optical Network (SONET) connections.

The largest WAN is the Internet. The Internet is basically a WAN that spans the entire globe. Home networks often implement LANs and WANs through cable modems and digital subscriber line (DSL) service. In these systems, a cable or DSL router links the home network to the provider's WAN and the provider's central gateway to reach the Internet.

A wireless local area network transmits over the air by means of base stations, or access points, that transmit a radio frequency; the base stations are connected to an Ethernet hub or server. Mobile end-users can be handed off between access points, as in the cellular phone system, though their range generally is limited to a couple hundred feet.

Inventing Cell Phones

Wireless technology is based on the car-mounted police radios of the 1920s. Mobile telephone service became available to private customers in the 1940s. In 1947, Southwestern Bell and AT&T launched the first commercial mobile phone service in St. Louis, Missouri, but the Federal Communications Commission (FCC) limited the amount of frequencies available, which made possible only 23 simultaneous phone conversations available within a service area (the mobile phones offered only six channels with a 60kHz spacing between them). Unfortunately, that spacing schema led to very poor sound quality due to cross-channel interference, much like the cross talk on wireline phones. The original public wireless systems generally used single high-powered transmitters to cover the entire coverage area. In order to utilize the precious frequencies allotted to them, AT&T developed an idea to replace the single high-powered transmitter approach with several smaller and lower-powered transmitters strategically placed throughout the metropolitan area; calls would switch between transmitters as they needed a stronger signal. Although this method of handling calls certainly eased some of the problems, it did not eliminate the problem altogether. In fact, the problem of too few voice channels plagued the wireless phone industry for several years.

The problem was that demand always seemed to exceed supply. Since the FCC refused to allocate more frequencies for mobile wireless use, waiting lists became AT&T's temporary solution as the company strove for the technological advances necessary to accommodate everyone. For example, in 1976, there were less than 600 mobile phone customers in New York City, but there were over 3,500 people on waiting lists. Across the United States at that time, there were nearly 45,000 subscribers, but there were still another 20,000 people on waiting lists as much as ten years long. Compare this situation to today's, in which providers give away free phones and thousands of minutes just to gain a subscriber.

Cellular technology has come a long way. The term *cellular* describes how each geographic region of coverage is broken up into *cells*. Each cell includes a radio transmitter and control equipment. Early cellular transmission operated at 800 MHz on analog signals, which are sent on a continuous wave. When a customer makes a call, the first signal sent identifies the caller as a customer, verifies that he or she is a customer of the service, and finds a free channel for the call. The mobile phone user has a wireless phone that in connection with the cellular tower and base station, handles the calls, their connection and handoff, and the control functions of the wireless phone.

Personal communications services (PCS), which operates at 1850 MHz, followed years later. PCS refers to the services that a given carrier has available to be bundled together for the user. Services like messaging, paging, and voicemail are all part of the PCS environment. Sprint is the major carrier that typically is associated with PCS. Some cellular providers began looking into digital technology (digital signals are basically encoded voice delivered by bit streams). Some providers are using digital signals to send not only voice, but also data. Other advantages include more power of the frequency or bandwidth, and less chance of corruption per call. Coverage is based on three technologies: Code Division Multiple Access (CDMA), Time Division Multiple Access (TDMA), and Global System for Mobile Communication (GSM).

Exploring Present Applications for Wireless

Many corporations and industries are jumping into the wireless arena. Two of the industries most committed to deploying wireless technologies are airports and hotels, for business travelers' communications needs. If they are traveling in a car, they use their wireless phones. When they are at work or home, they are able to use their computers and resources to again be productive. But when staying in a hotel for the night or even a week, there are few choices—a business traveler can look for the RJ-11 jack and connect to the Internet via 56-kilobit modem, not connect at all, or connect wirelessly. When a hotel provides the correct configuration information based on the provider, and a software configuration, a business traveler with wireless capabilities can connect to their network without worrying about connection speed or out-of-date modems.

Airports offer such services to increase travelers' productivity at a time when they would otherwise be isolated from business resources. The same configuration applies: set the configuration in the wireless client software and, voilà, you are connected. This wireless technology allows users to get access to the Internet, e-mail, and even the corporate intranet sites utilizing a virtual private network (VPN) solution. Now, the work (or in some cases, gaming) can be done during what used to be known as idle time. This increase in productivity is very attractive to corporations who need their increasingly mobile workforce to stay connected. This scenario is accomplished using the following scheme:

- A wireless Internet service provider contracts with the airport or hotel to set up wireless access servers and access points.

- Access points are located in specific locations to provide wireless coverage throughout the hotel or airport.

Using this scenario, anyone with an account to that service provider can get access to the Internet by walking into the location where the service is offered with their laptop, Personal Digital Assistant (PDA), or other wireless device, such as a mobile phone with 802.11 capability. This access includes such applications as e-mail, Intranet connection via VPN solution, push content such as stock updates or email, and Web browsing. Not that this is not all work and no play–you can also set up online gaming and video-on-demand sessions. In fact, nonwork scenarios open up the possible user base to children and families, multiplying the use and demand of this technology.

Applying Wireless Technology to Vertical Markets

There are several vertical markets in addition to airports and hotels that are realizing the benefits of utilizing wireless networks. Many of these markets, including delivery services, public safety, finance, retail, and monitoring applications, are still at the beginning of incorporating wireless networks, but as time passes and the demand and popularity grows, they will integrate wireless networking more deeply.

Using Wireless in Delivery Services

Delivery and courier services, which depend on mobility and speed, employ a wireless technology called Enhanced Specialized Mobile Radio (ESMR) for voice communication between the delivery vehicle and the office. This technology consists of a dispatcher in an office plotting out the day's events for a driver. When the driver arrives at his location, he radios the dispatcher and lets them know his location. The benefit of ESMR is its ability to act like a CB radio, allowing all users on one channel to listen, while still allowing two users to personally communicate. This arrangement allows the dispatcher to coordinate schedules for both pick-ups and deliveries and track the drivers' progress. Drivers with empty loads can be routed to assist backlogged drivers. Drivers that are on the road can be radioed if a customer cancels a delivery. This type of communication benefits delivery services in two major areas, saving time and increasing efficiency.

United Parcel Service (UPS) utilizes a similar wireless system for their business needs. Each driver carries a device that looks like a clipboard with a digital readout and an attached penlike instrument. The driver uses this instrument to record each delivery digitally. The driver also uses it to record digitally the signature of the person who accepts the package. This information is transmitted wirelessly back to a central location so that someone awaiting a delivery can log into the Web site and get accurate information regarding the status of a package.

Using Wireless for Public Safety

Public safety applications got their start with radio communications for maritime endeavors and other potentially hazardous activities in remote areas. Through the use of satellite communications and the coordination of the International Maritime Satellite Organization (INMARSAT), these communications provided the ships with information in harsh weather or provide them a mechanism to call for help. This type of application led to Global Positioning Systems (GPS), which are now standard on naval vessels. In many cases, a captain can use the 24 satellites circling the globe in conjunction with his ship's navigational system to determine his exact location and plot his course. GPS is also used for military applications, aviation, or for personal use when tracking or pinpointing the user's location could save his or her life.

Today, there are medical applications that use wireless technology such as ambulance and hospital monitoring links. Remote ambulatory units remain in contact with the hospital to improve medical care in the critical early moments. An emergency medical technician can provide care under a doctor's instruction during transport prior to arriving in the hospital's emergency room. Standard monitoring of critical statistics are transmitted wirelessly to the hospital.

Using Wireless in the Financial World

Wireless applications can keep an investor informed real-time of the ticker in the stock market, allowing trades and updates to be made on the go. No longer is the investor tied to his desk, forced to call into his broker to buy and sell. Now, an online investor has the opportunity to get real-time stock quotes from the Internet pushed to his wireless device. He can then make the needed transactions online and make decisions instantaneously in response to the market.

There are also services that allow you to sign up and get critical information about earmarked stocks. In this scenario, you can set an alarm threshold on a particular stock you are following. When the threshold is met, the service sends a page to you instantly. Again, this improves the efficiency of the investor.

Using Wireless in the Retail World

Wireless point-of-sale (POS) applications are extremely useful for both merchant and customer, and will revolutionize the way retail business transactions occur. Registers and printers are no longer fixed in place and can be used at remote locations. Wireless scanners can further assist checkout systems. Wireless technology is used for connecting multiple cash registers through an access point to a host computer that is connected to the WAN. This WAN link is used to send real-time data back to a corporate headquarters for accounting information.

Another type of wireless point-of-sale application is inventory control. A hand-held scanner is used for multiple purposes. The operator can check inventory on a given product throughout the day and wirelessly transfer the data back to the main computer system. This increases efficiency in that the device is mobile and small, and the data is recorded without manually having to enter the information.

Using Wireless in Monitoring Applications

We have been using wireless technologies for monitoring for years. There are typically two types of monitoring: passive and active. Active monitoring is conducted by use of radio signals being transmitted, and any of a number of expected signals received. An example of this implementation is the use of radar guns in traffic control. In this case, the patrolman points the gun and pulls the trigger, and a specific reading of a specific target is displayed on the radar unit. Passive monitoring is a long-term implementation whereby a device listens to a transmitter and records the data. An example of this is when an animal is tagged with a transmitter and the signal is collected and data is gathered over a period of time to be interpreted at a later date.

Monitoring applications in use today include NASA listening to space for radio signals, and receiving pictures and data relayed from probes; weather satellites monitoring the weather patterns; and geologists using radio waves to gather information on earthquakes.

Applying Wireless Technology to Horizontal Applications

Along with the many vertical markets and applications, you can apply wireless technologies to horizontal applications, meaning that delivery services, public safety, finance, retail, and monitoring can all use and benefit from them. The next section gives an overview of some of the more popular horizontal trends in wireless technology.

Using Wireless in Messaging

The new wave of messaging is the culmination of wireless phones and the Wireless Application Protocol (WAP) and Short Message Service (SMS). This service is similar to the America Online Instant Messaging service. The ability for two-way messaging, multiservice calling, and Web browsing in one device creates a powerful tool for consumers, while providing the vendors the ability to generate higher revenues. Look for wireless messaging services to be introduced in local applications, particularly within restaurants, to replace conventional wait lists.

Using Wireless for Mapping

Mapping in a wireless environment, of course, relates back to the GPS system; GPS not only assists the maritime industry with navigation, but also commercial vehicles and private cars for safety. In a few cars out today, a GPS receiver is placed on board to prevent drivers from becoming lost. It will also display a map of the surrounding area. The signal from the GPS satellites is fed into an onboard computer, which contains an application with software that contains a topographical map. The more current the software is, the more accurate the map will be. The coordinates of the receiver are placed on the topographical map in the program, usually in the form of a dot, and a display screen provides a visible picture of where in relation to the map someone is at that moment. This is updated live as the receiver moves.

Using Wireless for Web Surfing

In addition to the standard laptop computer connected to a wireless LAN with Internet connectivity, there has been an explosion of other wireless units that offer multiple voice and data applications integrated in one piece of equipment. Typically, personal organizer functionality and other standard calculation-type services are offered, but now, these devices are used with appropriate software to get access to the Internet. This brings the power of the Internet and the vast repository of information to the palm of the hand.

PDAs, Palm, Inc.'s handheld devices, and wireless phones with the appropriate hardware and software are now being used for Internet access at speeds of up to 56 Kbps. With new technologies such as Evolution Data Only (EVDO), some wireless phones now even offer speeds up 400-700 Kbps with maximum speeds of 2.4 Mbps. This is moving wireless into the realm of not only browsing the Internet, which is a big accomplishment in and of itself, but Internet gaming. As the interface of the wireless devices gets better and better, the gaming community will be able to offer high quality online games played on your PDA.

Using Bluetooth Wireless Devices

In recent years Bluetooth devices that also transmit in the 2.4GHz frequency range have become increasingly popular. With the convenience of Bluetooth, it is now possible to wirelessly sync devices such as PDAs or smartphones with laptop computers. Bluetooth headsets that allow hands-free, wireless communication with wireless phones can be seen almost everywhere. In fact, many new cars now come with Bluetooth capability so that wireless phones can be paired with the car stereo, allowing hands free calls to be made and received without even requiring a headset.

As more organizations and corporations realize the convenience that Bluetooth devices offer, the popularity of these devices will only continue to increase. In addition to headsets and syncing capabilities, some wireless phones that have Internet access allow tethering via Bluetooth. Tethering allows you to connect your phone to the Internet through your wireless phone and access the Internet through your laptop computer.

Exploring This Book on Wireless

This book provides network administrators that are unfamiliar with the possibilities and the pitfalls of using wireless networking in its many forms with the fundamental information they need to successfully and securely deploy various wireless technologies in their environment. This book is also valuable to managers who are considering the benefits of wireless networking in their workplace or for their employees. Unlike with many other technologies, wireless networking effectively 'extends' the boundaries of your network beyond the physical area of your facility. Because of this, and because it is a best practice when deploying any new technology, the first thing to look at is the security implications. Chapter 2 of this book discusses secure implementations of several popular wireless networking devices. Chapter 3 continues in this vein, detailing the specific dangers that wireless networking introduces into a corporate environment.

Perhaps there is already wireless in your environment, and you don't even realize it. This is often because an employee or group of employees have set up their own wireless network connected to your existing network. These are called Rogue Access points, and the dangers of Rogues and how to find them are detailed in Chapter 4.

One of the most important concepts to understand when integrating wireless into an existing environment is how to protect the rest of the network from the potential hazards of the wireless network. Chapters 5 and 6 prepare you for this task by providing information on both specific VLANs for the wireless network and on architecture and design strategies that are most effective with wireless networking.

Once a good architecture is in place, an effective monitoring and intrusion detection strategy needs to be implemented. Chapter 8 of this book details some effective ways to accomplish these tasks.

Since it is often easier to understand a concept by visualizing a scenario of how these technologies have actually been implemented, Chapters 9 and 10 provide case studies for deploying wireless networks securely in different types of corporate environments.

Understanding the concepts presented in this book and realizing the potential productivity benefits of wireless, as well as the potential problems associated with these benefits, will allow you to develop a plan to successfully deploy your wireless network. This book gives you the foundation and building blocks you need to successfully implement wireless networking and devices into your environment without requiring you to learn every technical detail of how wireless networking functions, essentially teaching you How To Cheat at Securing a Wireless Network.

Summary

In this chapter we have explored some of the history of how wireless technology evolved into what it is today. Wireless technology has been around a long time, considering the decades of development in radio and cellular telephone technology. These technologies have been quietly developing in the background while PDAs, Palm Pilots, and other handheld wireless devices have been gaining notoriety. Other uses such as GPS and satellite communications to the home have also been developing for mainstream applications. These applications offer consumers many advantages over wireline counterparts, including flexibility, mobility, and increased efficiency and timeliness.

In surveying all that wireless has achieved, one problem still exists: needing many devices to access the various services offered. These devices can be cumbersome to carry, and the options and selection make the situation more complicated. Without question it is time for professionals to be free from having to carry a pager, a cell phone, and a PDA to get services they need. The future of wireless technology is to integrate voice and data services into one system that will allow end users to get all their requirements met in one device. The entire network infrastructure in wired technology is moving toward a converged network where voice and data are integrated into one network—the move is monumental for end-user quality and services offered, it makes sense for carriers and service providers from a revenue standpoint, and once completed, will allow services to be distributed by request. Wireless fits into that by extending the last mile of communications to the mobile user. As wireless experiences the cost reduction and benefits enhancement that it has over the previous few years, there is little to oppose the notion of a globally connected wireless world.

Solutions Fast Track

Exploring Past Discoveries That Led to Wireless

- ☑ *Wireless technology* is the method of delivering data from one point to another without using physical wires, and includes radio, cellular, infrared, and satellite.

- ☑ The discovery of electromagnetism, induction, and conduction provided the basis for developing communication techniques that manipulated the flow of electric current through the mediums of air and water.

☑ Guglielmo Marconi was the first person to prove that electricity traveled in waves through the air, when he was able to transmit a message beyond the horizon line.

☑ The limitations on frequency usage that hindered demand for mobile telephone service were relieved by the development of the geographically structured cellular system.

Exploring Present Applications for Wireless

☑ Vertical markets are beginning to realize the use of wireless networks. Wireless technology can be used for business travelers needing airport and hotel access, gaming and video, for delivery services, public safety, finance, retail, and monitoring.

☑ Horizontal applications for wireless include new technology for messaging services, mapping (GPS) and location-based tracking systems, and Internet browsing.

Frequently Asked Questions

The following Frequently Asked Questions, answered by the authors of this book, are designed to both measure your understanding of the concepts presented in this chapter and to assist you with real-life implementation of these concepts. To have your questions about this chapter answered by the author, browse to **www.syngress.com/solutions** and click on the **"Ask the Author"** form.

Q: Why did it take so long to develop wireless technologies?

A: The scientific principles behind wireless technologies have been developing at the same time as wireline technologies, and include major advances for military and industrial needs—but in comparison, the potential mainstream consumer applications for wireless have not been embraced. One reason is that the FCC has strictly regulated service providers' access to the necessary frequencies.

Q: What is the difference between wireless voice and wireless networking?

A: Technologically, they are very similar. In short, wireless voice is a traditional conversation between two or more people with at least one person not connected to wires, whereas wireless networking often implies that data is transmitted rather than voice.

Chapter 2

Wireless Security

Solutions in this chapter:

- **Enabling Security Features on a Linksys WRT54G 802.11g Access Point**

- **Enabling Security Features on a D-Link DI-624 AirPlus 2.4GHz Xtreme G Wireless Router with Four-Port Switch**

- **Enabling Security Features on Apple's Airport Extreme 802.11g Access Point**

- **Enabling Security Features on a Cisco 1100 Series Access Point**

- **Configuring Security Features on Wireless Clients**

- **Understanding and Configuring 802.1X RADIUS Authentication**

- ☑ **Summary**
- ☑ **Solutions Fast Track**
- ☑ **Frequently Asked Questions**

Introduction

By the end of this chapter, the reader will be able to correctly, and securely, configure several different types of access points. The access points chosen are industry-leading brands and the most popular devices on the market today. The reader will also have a clear understanding of how to connect a wireless-capable workstation to any encrypted wireless network.

Enabling Security Features on a Linksys WRT54G 802.11g Access Point

The most popular and best-selling 802.11g device on the market today is the Linksys WRT54G 802.11b/g access point/router. The WRT54G gained popularity in 2003 as 802.11g devices became more common and affordable. The 802.11g devices operate on the 2.4GHz band, like 802.11b, but offer speeds up to 54 megabits per second (Mbps). Additionally, 802.11g devices are compatible with 802.11b cards. Because this is such a popular product for Linksys, the company has released a Compact Version (WRT54GC) and a version with speed-boosting features (WRT54GS).

Linksys has since released version 5 of its WRT54G device, with the current firmware version being 1.00.9. Still popular among Linksys devices for its ability to be reflashed with a small Linux distribution is version 4 of the WRT54G device. This section focuses on a WRT54G version 4 device, with firmware version 4.30.5.

This section details the minimum steps you should take to securely configure the WRT54G. All the steps outlined in this section should be done from a computer that is connected to your wired network. Configuration of the device from a wirelessly connected workstation should be disabled. If by chance your wireless network becomes compromised, you will want to limit the intruder's ability to administer the device.

Setting a Unique SSID

The first security measure you should enable on the Linksys WRT54G is setting a unique SSID. When you log in to the WRT54G, by default the username is *admin* and the password is *admin*. Logging in brings up the initial setup screen (see Figure 2.1).

Figure 2.1 The Linksys WRT54G Initial Setup Screen

Click the **Wireless** tab. In the **SSID** text box, enter a unique SSID, as shown in Figure 2.2.

Figure 2.2 Setting a Unique SSID on the WRT54G

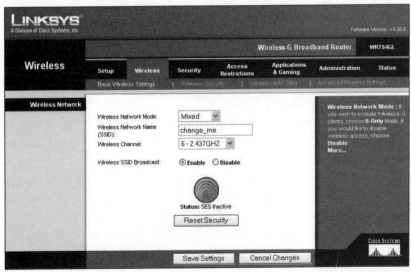

Then click the **Save Settings** button to save your settings.

Disabling SSID Broadcast

After you have set a unique SSID, disable the SSID broadcast. From the Wireless setup screen, select the **Disable** radio button from the **SSID Broadcast** option, as shown in Figure 2.3.

Figure 2.3 Disable SSID Broadcast

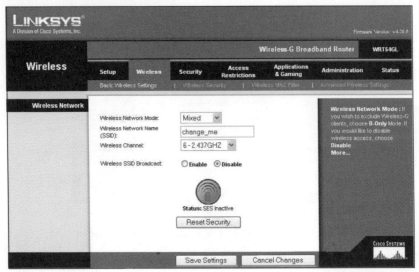

Then click the **Save Settings** button to save your settings and disable SSID broadcast.

Enabling Wired Equivalent Privacy

Once you have set a unique SSID and have disabled SSID broadcast, you need to require the use of 128-bit Wired Equivalent Privacy (WEP) encryption. From the Wireless setup screen, choose the **Wireless Security** tab. From the drop-down list, choose the **WEP** option, to enable and configure WEP encryption, as shown in Figure 2.4.

Next select **128 bits 26 hex digits** from the **WEP Encryption** drop-down box, to require 128-bit WEP. Type a strong passphrase in the **Passphrase** text box. This is the passphrase that will be used as the basis for generating WEP keys. Click the **Generate** button to generate four WEP keys, as shown in Figure 2.5.

Figure 2.4 Enable WEP on the WRT54G

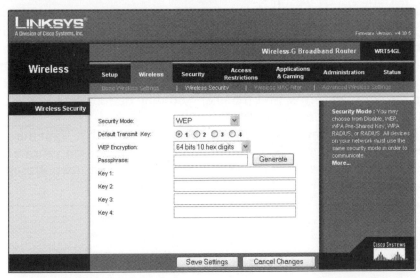

Figure 2.5 The WEP Keys Window

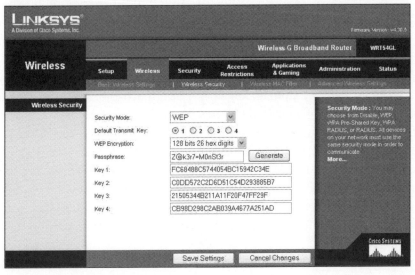

Next, select the key (1–4) that you will initially use by choosing the appropriate radio button next to **Default Transmit Key**. Finally, click **Save Settings** in the Wireless Security tab to save your settings.

SOME INDEPENDENT ADVICE

Some people will argue that WEP is a "broken" standard and should not be used. Yes, WEP is an easy protocol to hack and allows intruders to gain the encryption key to your wireless network using tools included in the Aircrack suite. However, due to wireless connections by other devices (game consoles, PDAs, and the like), you may be forced to use WEP instead of the more secure WPA.

Remember that *no security is bad security,* and that something is always better than nothing. Enabling WEP encryption on your network may be the difference between your network or your unencrypted neighbor's being hacked.

Enabling Wi-Fi Protected Access

An alternative and more secure approach to wireless security on an access point is to use *Wi-Fi Protected Access*, or WPA. WPA uses an improved encryption process based on the Temporal Key Integrity Protocol (TKIP). TKIP jumbles the keys and incorporates an integrity-checking feature to ensure that the keys have not been tampered with.

WPA also includes client authentication via the Extensible Authentication Protocol (EAP). EAP uses a public key encryption mechanism to ensure that only authorized systems have access to the access point.

In late 2004, the Institute of Electrical and Electronics Engineers (IEEE) ratified the 802.11i specification, more commonly referred to as WPA2. WPA2 uses AES as the encryption standard, whereas WPA uses the TKIP standard. This is not to say that WPA is not secure but to acknowledge that wireless security is ever changing. WPA2 also supports a personal authentication implementation (PSK) and an enterprise authentication implementation (RADIUS). This chapter focuses on the WPA standard.

Log in to the WRT54G and click the **Wireless** tab. Click the **Wireless security** subtab to enable WPA. From the drop-down list, choose **WPA-Personal**, as shown in Figure 2.6.

Figure 2.6 The WRT54G WPA Setup Screen

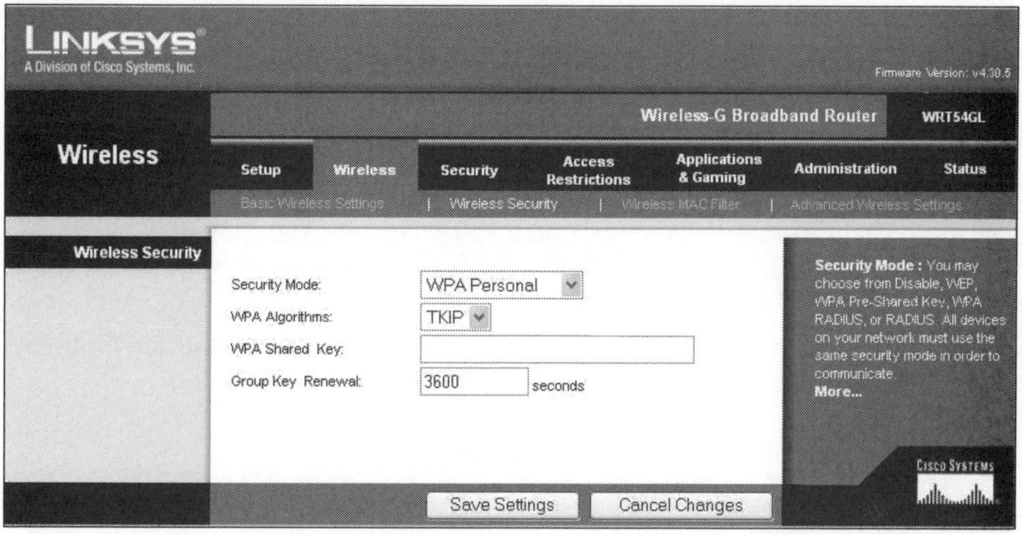

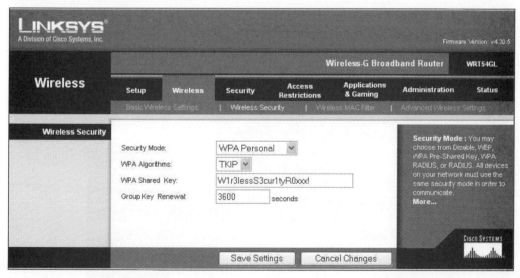

Leave the WPA algorithm as **TKIP**. Enter a shared key of between 21 and 63 characters in the **WPA Shared Key**: text box. Leave the **Group Key Renewal** at its default of **3600 seconds** (see Figure 2.7).

Figure 2.7 WPA Shared Key

Click **Save Settings** to save the WPA settings on the WRT54G. It is still a good idea to follow the previous security steps to enable wireless MAC filters and disable the SSID broadcast. Be careful not to set the SSID to anything personal to you, such as your phone number, home address, or name.

Filtering by Media Access Control (MAC) Address

After you have set a unique SSID, disabled SSID broadcast, and enabled WEP encryption, you need to filter access to the WRT54G by MAC address. Filtering access to the access point allows only those MAC addresses specified in the list the ability to access the wireless network.

First, from the main **Wireless** tab, click the **Wireless MAC Filter** tab to display the option to enable or disable Wireless MAC filtering (see Figure 2.8).

Figure 2.8 The Wireless MAC Filter screen

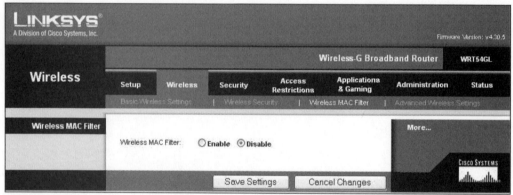

Next select **Enable** from the **Wireless MAC Filter** radio buttons. This will reveal the MAC filter options, as shown in Figure 2.9.

Figure 2.9 The Wireless MAC Filter Options

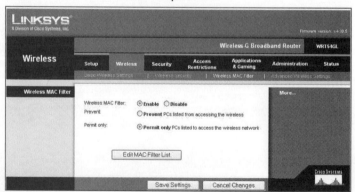

Choose the **Permit Only PCs listed to access the wireless network** radio button, and click the **Edit MAC Filter List** button to display the **MAC Address Filter List** window (see Figure 2.10).

Figure 2.10 The MAC Address Filter List Window

In the provided text boxes, enter the MAC addresses of wireless clients that are allowed to access your wireless network, and then click **Apply**, as shown in Figure 2.11.

Figure 2.11 Enter Allowed MAC Addresses

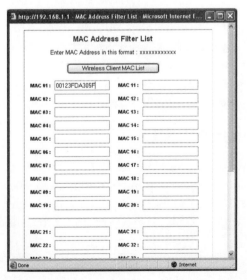

Finally, click **Save Settings** in the **Advanced Wireless** window to save your settings and enable filtering by MAC address. Keep in mind that this should not be the only security measure implemented. Using various tools in Windows and/or Linux, it is easy for an attacker to spoof his or her local MAC address to gain access to your wireless network.

> ### Some Independent Advice
>
> Finding your MAC address is a simple process with any operating system. Using Windows XP, from a command line, you can type:
> *ipconfig /all*
> to show the MAC address of the installed network devices.
> Linux makes the process just as simple. From a terminal window, type:
> *ifconfig –a*
> And find the *HWaddr* for the requested network interface. This is the MAC address.

Enabling Security Features on a D-Link DI-624 AirPlus 2.4GHz Xtreme G Wireless Router with Four-Port Switch

Although Linksys has a sizable share of the home access point market, D-Link also has a large market share. D-Link products are sold at most big computer and electronics stores such as Best Buy and CompUSA. This section details the steps you need to take to enable the security features on the D-Link 624 AirPlus 2.4GHz Xtreme G Wireless Router with Four-Port Switch. The DI-624 is an 802.11g access point with a built-in router and switch, similar in function to the Linksys WRT54G.

Setting a Unique SSID

The first security measure to enable on the D-Link DI-624 is setting a unique SSID. First you need to log into the access point. Configure your local workstation with a static IP in the 192.168.0.0/24 subnet and point your browser to 192.168.0.1. Use the username **admin** with a blank password to access the initial setup screen (see Figure 2.12).

Figure 2.12 The D-Link DI-624 Initial Setup Screen

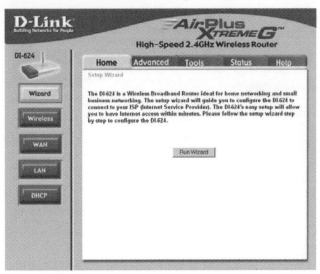

Next click the **Wireless** button on the left side of the screen to bring up the **Wireless Settings** screen, as shown in Figure 2.13.

Figure 2.13 The Wireless Settings Screen

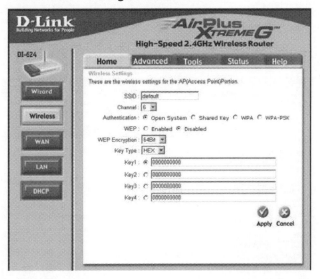

In the **SSID** textbox, enter a unique SSID, as shown in Figure 2.14, and click **Apply** to save and enable the new SSID.

Figure 2.14 Set a Unique SSID

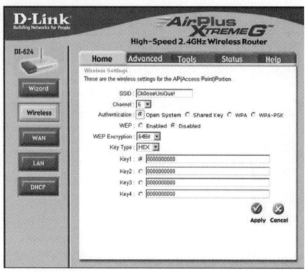

Disabling SSID Broadcast

After you have set a unique SSID, enabled 128-bit WEP, and filtered access by MAC address, you need to disable SSID broadcast.

From the **Advanced Features** screen, click the **Performance** button, as shown in Figure 2.15.

Figure 2.15 The Advanced Performance Options

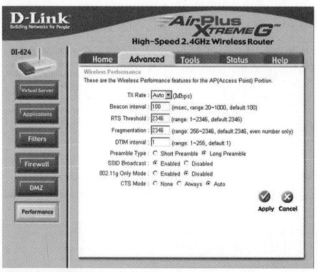

Select the **Disabled** radio button next to **SSID Broadcast**, and click **Apply** to save your settings, as shown in Figure 2.16.

Figure 2.16 Disabling SSID Broadcast

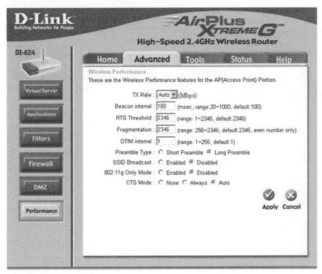

Enabling Wired Equivalent Privacy

After you have set a unique SSID, you will need to enable 128-bit WEP encryption. First, choose the **Enabled** radio button next to WEP, as shown in Figure 2.17.

Figure 2.17 Enable WEP

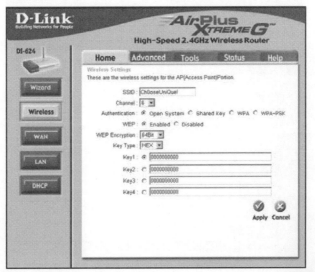

Next choose **128Bit** from the **WEP Encryption** drop-down box, as shown in Figure 2.18.

Figure 2.18 Require 128-Bit WEP Encryption

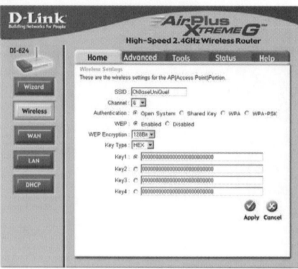

Then you need to assign a 26-character hexadecimal number to at least Key1 (see Figure 2.19). A 26-digit hexadecimal number can contain the letters A–F and the numbers 0–9.

Figure 2.19 Assign WEP Keys

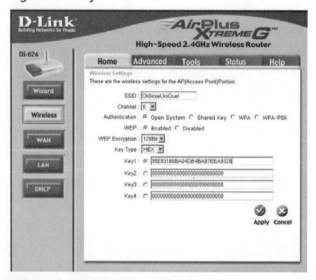

Finally, after you have assigned your WEP keys, click **Apply** to save your settings. Any wireless clients that connect to the DI-624 must be configured to use this WEP key.

Enable Wi-Fi Protected Access

To enable WPA on the access point, on the left side of the screen click the **Wireless** button. To enable WPA, click the radio button labeled **WPA-PSK** next to the **Authentication** option (see Figure 2.20).

Figure 2.20 Enabling WPA

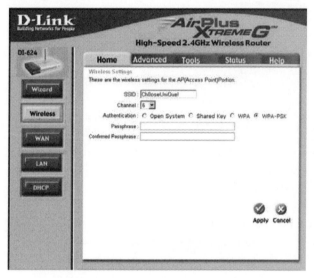

Enter a passphrase into the Passphrase text box, and retype the passphrase in the Confirmed Passphrase text box to verify it, as shown in Figure 2.21.

Click **Apply** to confirm the settings and enjoy added wireless security protection!

Figure 2.21 WPA Passphrase

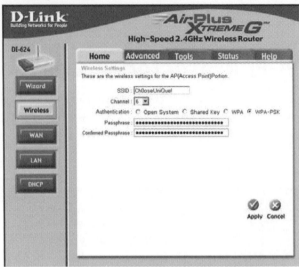

Filtering by Media Access Control Address

After you have set a unique SSID and enabled 128-bit WEP encryption, you should filter access to the wireless network by Media Access Control (MAC) address.

First click the **Advanced** tab, as shown in Figure 2.22.

Figure 2.22 The Advanced Options Screen

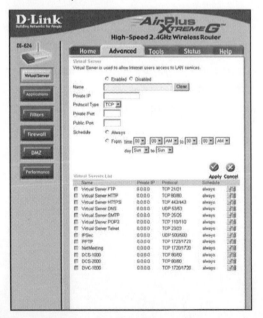

Next click the **Filters** button on the left side of the screen, as shown in Figure 2.23.

Figure 2.23 The Advanced Filters Options

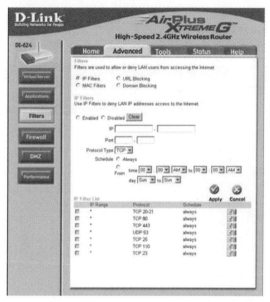

Then choose the **MAC Filters** radio button. This makes the MAC filtering options visible, as shown in Figure 2.24.

Figure 2.24 The MAC Filtering Options

Finally, select the **Only allow computers with MAC address listed below to access the network** radio button and enter the MAC address of each client card that is allowed to access the network. You must also enter a descriptive name of your choice for each client in the **Name** text box (see Figure 2.25). Note that you must click **Apply** after each MAC address entered.

Figure 2.25 Filter by MAC Address

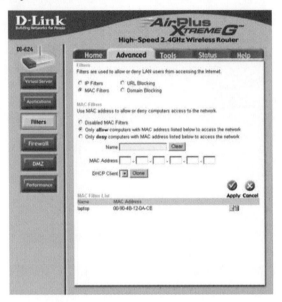

Enabling Security Features on Apple's Airport Extreme 802.11g Access Point

In early 2003, Apple released the Airport Extreme base station to the masses, supporting the 802.11b and 802.11g protocols. Even though this access point was released as an Apple product, it fully supports Apple, Windows, and Linux clients running WEP or WPA encryption.

Configuring the Airport Extreme is usually done from an Apple, whether a Powerbook, iBook, or MacBook. Apple provided applications for configuring the Airport for Windows-based operating systems, but it is a much easier process from an Apple workstation. This section focuses on configuring the Airport Extreme from a Apple Powerbook G4.

Connecting to the AirPort Extreme and Setting a Unique SSID

The easiest way to connect to the Airport is via the wireless connection. Ensure that your wireless card is enabled by clicking the wireless symbol at the top right of the screen and clicking **Turn AirPort On**, as shown in Figure 2.26.

Figure 2.26 Enabling the AirPort Card on the Apple PowerBook

Once you enable the Airport card, you can reclick the wireless symbol and see any access points broadcasting in your area. We want to click the **Apple Network ######** listing to connect to our AirPort (see Figure 2.27).

NOTE

To ensure that you are connecting to the correct access point, verify that the network number listed in the drop-down list matches the last six characters of your Airport ID, located on the access point itself.

Figure 2.27 Connect to the Appropriate Airport Access Point

Once you have connected to the Airport, you will use the AirPort Admin Utility in Mac OS X to configure the Airport. Launch the AirPort Admin Utility by clicking the **Finder**, then **Applications | Utilities | AirPort Admin Utility** (see Figure 2.28). This series of clicks will open the **AirPort Admin Utility**. Click **Rescan** to locate the Airport if it does not automatically populate the window after a few seconds.

Figure 2.28 Launching the Admin Utility and Finding the Airport Base Station

Click the appropriate base station, and click **Configure** to enter the base station properties (see Figure 2.29).

Setting a Unique SSID

At the main properties screen, we will set the SSID by changing the **Name** text box, under the **AirPort Network** heading. Type in the SSID, remembering not to include any personal information such as address as part of the SSID. At this point, it would also be a good idea to change the **Name** of the Airport under the **Base Station** heading, to obfuscate the fact that this is an Apple Airport product (see Figure 2.30). Click **Update** to save the SSID.

Figure 2.29 Airport Default Properties

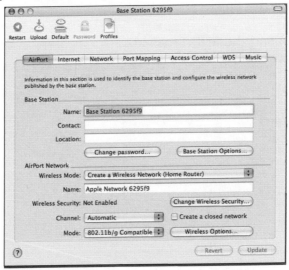

Figure 2.30 Setting the SSID

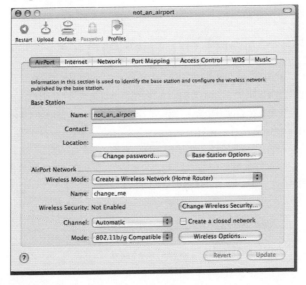

Disabling SSID Broadcast

To disable the broadcast of the Airport's SSID, click the **Create a closed network** check box. This will not allow the SSID to be broadcast to clients. You will be prompted on whether or not to disable the broadcast. Click **OK.** However, any client authorized to connect to the Airport must know the SSID beforehand to make the connection (see Figure 2.31).

Figure 2.31 Disabling the SSID Broadcast

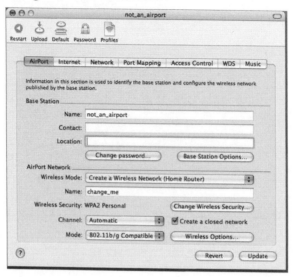

Setting a Password on the Airport

Because the Airport is in a default configuration, it is wise to set a password on the Airport to disable the ability of anyone making unauthorized changes. From the main base station properties windows, click the **Change Password...** button and enter and confirm a password for the Airport. Click **OK** to set the password. Click **Update** to save the changes to the Airport (see Figure 2.32).

Figure 2.32 Setting a Password on the Airport

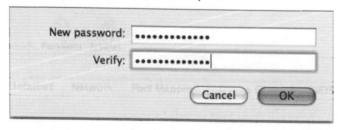

Enabling Wired Equivalent Privacy

To enable WEP on the Airport, click the **Change Wireless Security...** button to open the Properties dialog box (see Figure 2.33).

Figure 2.33 WEP Default Setting

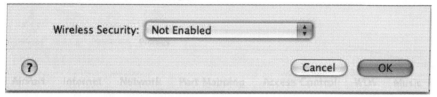

Click **WEP** from the drop-down menu. You will be presented with the options to add your encryption key. Type in an encryption key that is not easily guessable, and retype the key to confirm. Ensure that the **Encryption Type:** is set to **128 bit WEP**, and click **OK** to enable WEP encryption (see Figure 2.34).

Figure 2.34 Configuring a WEP Encryption Key

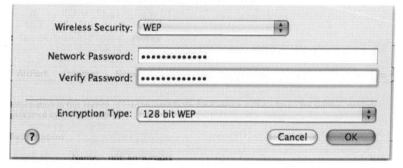

Anyone who attempts to this access point will now be required to enter the encryption key to make the connection.

Enabling Wi-Fi Protected Access

Enabling WPA on the Airport is just as simple as enabling WEP encryption. From the main setup screen, click the **Change Wireless Security...** button to open the Wireless Security dialog box. Change the **Wireless Security:** drop-down list to **WPA2 Personal** (see Figure 2.35).

Figure 2.35 WPA Settings

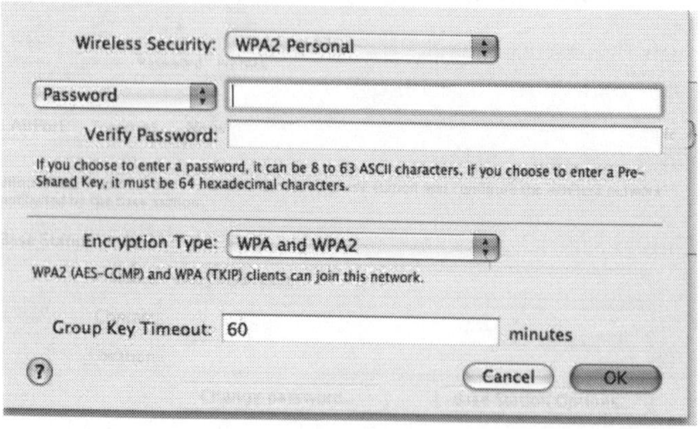

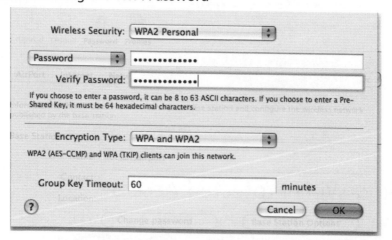

Ensure that the **Password** option is set, and enter a password or passphrase of between 8 and 63 ASCII characters. The **Encryption Type:** may be left at the default WPA and WPA2 option to allow both WPA and WPA2 connections. If only WPA clients or only WPA2 clients will be connecting, you may change this option to reflect that fact. Leave the **Group Key Timeout:** at its default of 60 minutes. Click **OK** to save the settings and enable WPA (see Figure 2.36).

Figure 2.36 Entering the WPA Password

Filtering by Media Access Control Address

To prevent connections to the Airport by workstations not authorized to do so, enable filtering by the MAC address. The MAC address of the connecting wireless network card will need to be entered manually. From the main options screen, click **Access Control** to view the settings (see Figure 2.37).

Figure 2.37 The Access Control Options

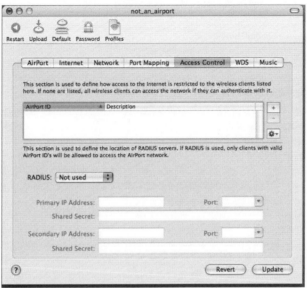

Click the **+ (plus)** sign next to the main dialog box to enter the MAC address of the client. A dialog box will open, requesting the Airport ID (MAC address) and the Description (see Figure 2.38).

Figure 2.38 Default MAC Address Filter Window

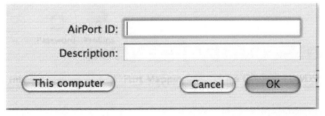

Enter the 12-character MAC address and provide a description if needed. Click **OK** to add the MAC address to the list (see Figures 2.39 and 2.40).

Figure 2.39 Entering the MAC Address

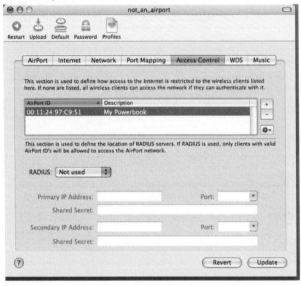

Figure 2.40 Confirming the List

Click **Update** to save the settings to the Airport.

Enabling Security Features on a Cisco 1100 Series Access Point

The Cisco Aironet series of access points are used largely by businesses and local hotspots that need the robustness of a Cisco product and the ease of use of a small office/home office (SOHO) product. The Cisco 1100 Series Access Point provides 802.11b/g services, operating on the 2.4GHz band. Unlike most SOHO router/AP products, the Cisco 1100 does not include a built-in switch and can only be used as a standalone wireless access point.

The easiest way to configure the Cisco 1100 is to connect via the Web interface. You will need to assign your local host a static IP between 10.0.0.2 and 10.0.0.10, with a subnet mask of 255.255.255.0 and a default gateway of 10.0.0.1. You may use either a straight-through Ethernet cable or a cross-over cable from your host computer to the Ethernet port on the access point.

When you power up the access point, it will attempt to connect to a DHCP server. If none exists, after a few moments the access point will default to the static IP of 10.0.0.1. If no connection is made within five minutes, it will default back to searching for a DHCP server indefinitely. To restart the process, unplug the access point for a few seconds, and retry the connection.

Setting a Unique SSID

The first step to configuring the Cisco 1100 is to set a unique SSID. Upon initial connection to the access point, you will be greeted with the initial setup screen (see Figure 2.41).

Figure 2.41 The Cisco 1100 Initial Setup Screen

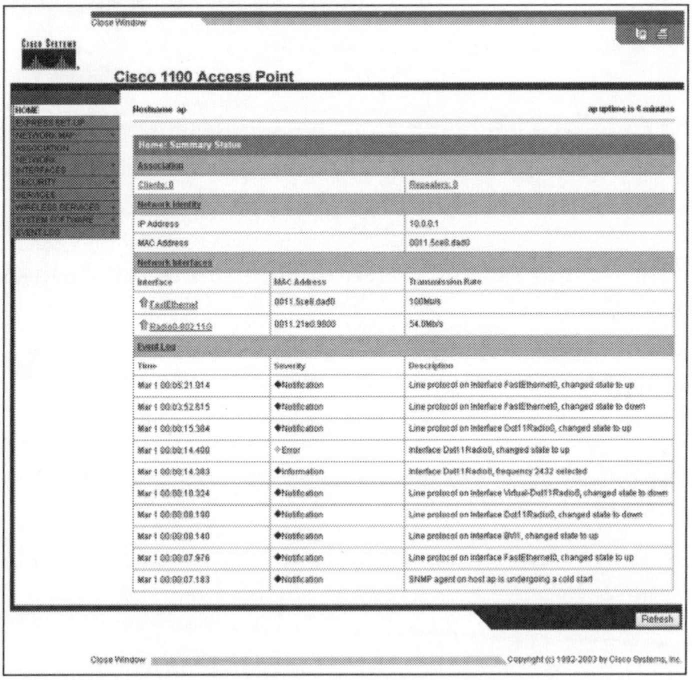

On the left-hand menu, click the **Security** option. On this screen you will have direct access to the current setup of the access point (see Figure 2.42).

Figure 2.42 Cisco 1100 Security Settings

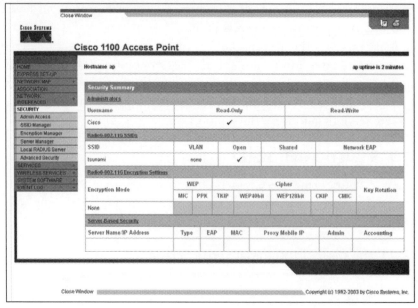

Because the Cisco 1100 does not come by default with an administrator pass-word, it would be wise to set one now. Click the **Admin Access** option. Enter and confirm a **Default Authentication Password** (see Figure 2.43).

Figure 2.43 The Admin Access Screen to Enter a Default Authentication Password

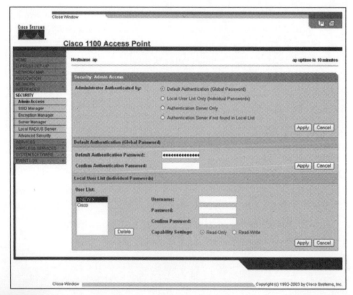

Once you click **Apply**, the password will be saved and you will now be required to authenticate back to the access point. Leave the **Username:** blank, and enter your new password in the **Password:** field (see Figure 2.44). You will be returned to the Admin Access screen.

Figure 2.44 The Authentication Request

Once you are back to the **Admin Access** screen, click the **SSID Manager** option on the left side of the screen (see Figure 2.45).

Figure 2.45 The SSID Manager

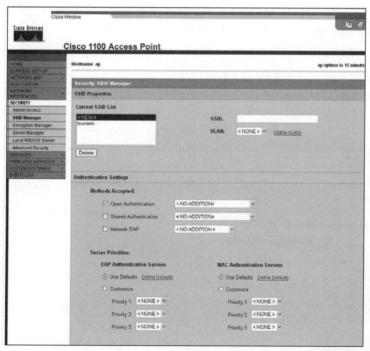

To set the SSID, you will first have to remove the default SSID of Tsunami and replace it with your own. From the Current SSID List, select **Tsunami** and click **Delete**. You will be asked to confirm the deletion, and then the Current SSID List will contain nothing but <NEW> (see Figure 2.46).

Figure 2.46 The SSID List

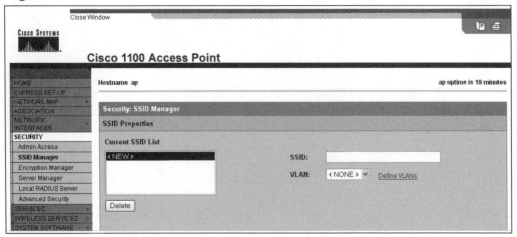

Assign a new SSID by typing it into the **SSID:** text box. Scroll down and click **Apply** to save the new SSID. Keep in mind that when setting the SSID, you should not use any revealing personal information such as address, home phone number, or surname (see Figure 2.47).

Figure 2.47 Setting the New SSID

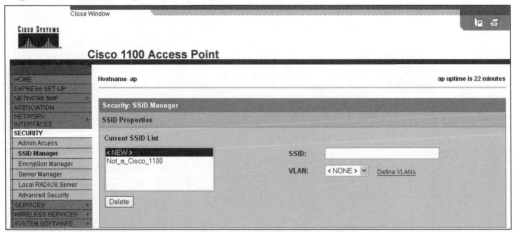

Disabling SSID Broadcast

At this point, we want to disable the broadcast of our SSID to the general public. The easiest way to ensure that the SSID is not being broadcast is from the Express Set-up menu. You can access the Express Set-up menu from the left-hand options. Click the **Express Set-up** option, and ensure that the **Broadcast SSID in Beacon** radio button is set to **No** (see Figure 2.48).

Figure 2.48 Disabling the SSID Broadcast

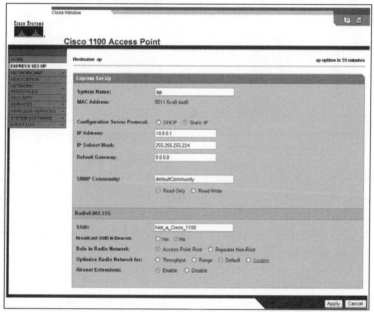

Click **Apply** to disable the SSID broadcast. From this menu, you also have the option of setting the SSID, the default SNMP Community string, and whether you want the access point to have a static or DHCP address.

Enabling Wired Equivalent Privacy

The option to enable WEP encryption on the access point is found in the Encryption Manager settings under the main Security options. From the left-hand menu, choose **Security** then **Encryption Manager** (see Figure 2.49).

Figure 2.49 Encryption Manager Settings

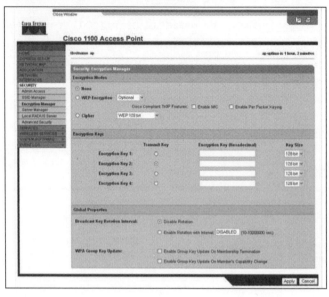

This screen offers several options to enable encryption on the access point. For simplicity's sake, we will stick to the Encryption Modes and Encryption Keys headings. From the **Encryption Modes** options, select the **WEP Encryption** radio button and choose **Mandatory** from the drop-down menu, to force clients to use WEP encryption, as shown in Figure 2.50.

Figure 2.50 Enabling WEP Encryption

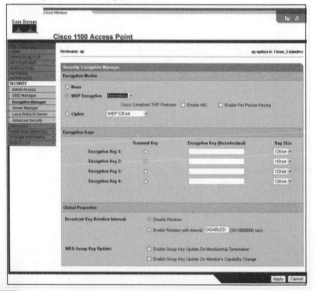

Ensure that the **Cipher** option is set to **WEP 128 bit** to force strong encryption.

The Cisco 1100 Access Point does not provide an option to enter a 13-character ASCII string and convert it automatically to hexadecimal format. Several Web sites will convert ASCII to hex. A Google search will reveal many converters; choose one that suits your taste.

In this case, we have decoded the 13-character ASCII string *CiscoRules!!!* to its hexadecimal equivalent of 436973636f52756c6573212121, which is 26 characters. Make sure that there are no hyphens, spaces, or percent symbols in the key and that all characters are either numerals 0–9 or letters a–f. When you enter the key into the access point, it will not show up as characters but instead is obfuscated by black dots.

Enter the encryption key into the **Encryption Key 1:** text box. Ensure that the **Transmit Key** radio button is selected for **Encryption Key 1:** and that the **Key Size** is set to **128 bit** (see Figure 2.51).

Figure 2.51 Entering the 26-Character Hexadecimal Encryption Key

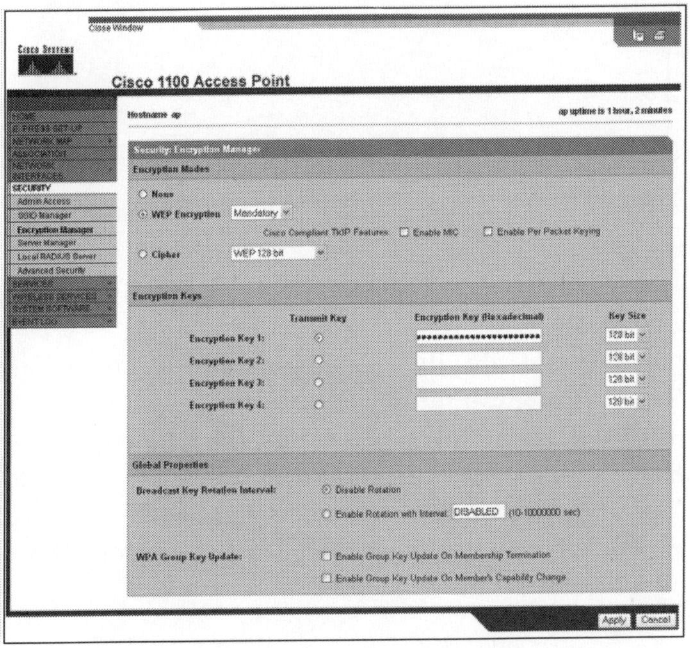

You may enter more then one encryption key if you plan to rotate keys on a prescheduled basis. Keep in mind that if you use a rotating key schedule, each of the clients must have the required keys to keep his or her existing connection. Click **Apply** to save the changes and enable 128-bit WEP encryption.

Enabling Wi-Fi Protected Access

Enabling WPA on the Cisco 1100 access point is just as easy as enabling WEP encryption. Return to the **Encryption Manager** settings from the **Security** menu on the left side of the screen (see Figure 2.52).

Figure 2.52 The Encryption Manager Settings

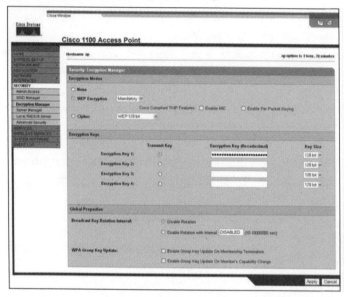

Click the radio button next to **Cipher** and choose **TKIP** for the encryption cipher (see Figure 2.53).

The Cisco 1100 requires that the encryption key for WPA be entered into the **Encryption Key 2:** text box. If the device was previously set up for WEP encryption, you will need to clear the **Encryption Key (Hexadecimal)** text box for the **Encryption Key 1:** option. Select the **Transmit Key** radio button for the **Encryption Key 2:** row and enter your password or passphrase for your WPA-enabled devices, remembering that the password/passphrase for WPA needs to be between 21 and 63 characters (see Figure 2.54).

Figure 2.53 Choose TKIP for the Encryption Cipher

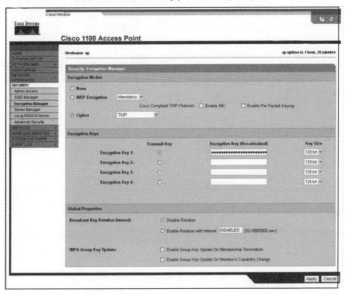

Figure 2.54 Enabling the WPA Password/Passphrase

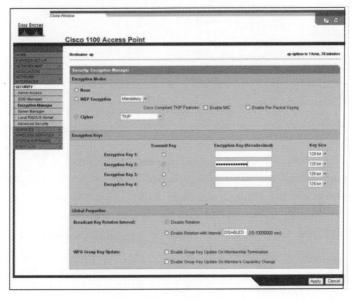

Click **Apply** to save the settings on the device and enable WPA encryption.

Filtering by Media Access Control Address

You can ensure that only approved devices are allowed to connect to the access
point by filtering the MAC address of only those approved devices. MAC filtering
on the Cisco 1100 access point is done in the **Advanced Security** options of the
AP. You may choose to use a local list served from the access point or from an
authentication server. This section focuses on a locally stored list on the access point.

Click **Security**, then click **Advanced Security** on the left menu. You will see
the **Advanced Security** options, as shown in Figure 2.55.

Figure 2.55 The Advanced Security Options

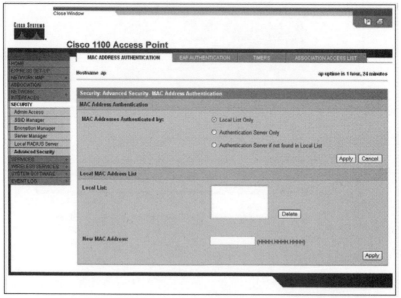

Ensure that **Local List Only** radio button is selected under the **MAC Address
Authenticated by**: option. From the **Local MAC Address List** section, enter a
valid MAC address in the **New MAC Address:** text box. Note that the format in
which the address should be input is *HHHH.HHHH.HHHH* (see Figure 2.56).

Figure 2.56 Entering the Approved MAC Address

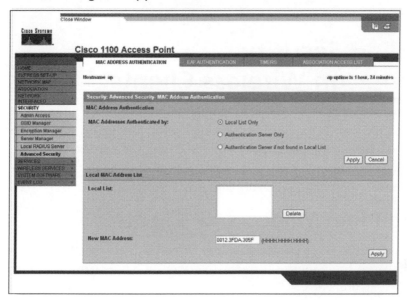

Click **Apply** to save the MAC address in the **Local List:** text box (see Figure 2.57).

Figure 2.57 Adding the MAC Address to the List

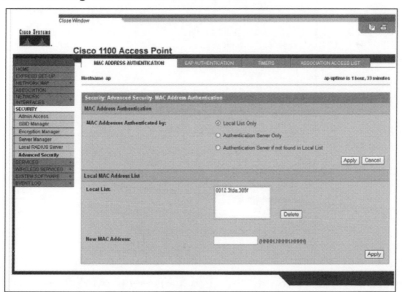

Only clients added to this list will have access to authenticate and use the access point.

Enabling Security Features on Wireless Clients

After you have configured your access points to utilize security measures, you will need to configure each wireless client to work with your access points. This means that each wireless client needs to be configured with the correct SSID and have the appropriate WEP key entered and selected. We will also look at enabling WPA on certain clients for use in a WPA-enabled environment.

Configuring Windows XP Clients

For Windows XP Clients, double-click the **Wireless Network Connections** icon on the taskbar to bring up the **Wireless Network Connection Properties** window (see Figure 2.58).

Figure 2.58 The Windows XP Wireless Network Connection Properties Window

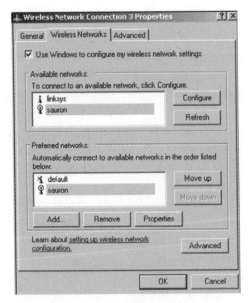

Next click the **Add** button to add a new preferred network. Enter the SSID of your network in the **SSID** text box, and uncheck **The key is provided for me**

automatically. Then enter the WEP key for your wireless network in the **Network Key** and **Confirm Network Key** text boxes, as shown in Figure 2.59.

Figure 2.59 Configuring Windows XP Clients

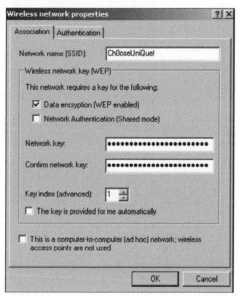

Finally, choose the appropriate key index (1–4) and click **OK**. If your settings are correct, you will be connected to your access point.

Configuring Windows XP Clients (WPA)

Configuring Windows XP clients for WPA encryption is just as easy as configuring them for WEP. Double-click the **Wireless Network Connections** icon on the taskbar. You should now have the familiar **Wireless Network Connection Properties** window open (see Figure 2.58). Click the **Add...** button to add a new Preferred Network. Type the **Network name (SSID)**. Choose **WPA-PSK** from the **Network Authentication** drop-down box and **TKIP** from the **Data Encryption** drop-down box (see Figure 2.60).

Type the WPA password in the **Network key:** text box, and verify the password in the **Confirm network key:** text box (see Figure 2.61).

Figure 2.60 WPA Settings

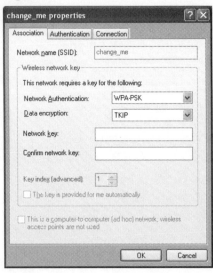

Figure 2.61 WPA Network Key

Click **OK** to save the settings and make the WPA connection (see Figure 2.62).

Figure 2.62 Making the WPA Connection

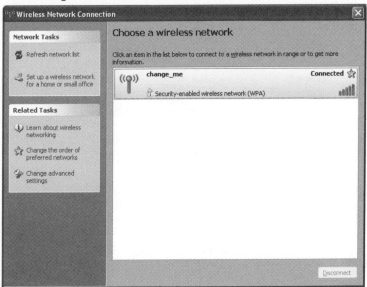

Configuring Windows 2000 Clients

Unlike Windows XP, with Windows 2000 you need to configure the wireless client software that came with your wireless card. The examples shown in this section are for the ORiNOCO client manager. The exact steps for other client managers might differ, but the basic idea is the same:

1. Create a wireless profile for your access point.

2. Enter the SSID of your access point.

3. Enter the WEP key for your wireless network.

To create a wireless profile for your access point, double-click the **Client Manager** icon on the taskbar to open the Client Manager. Choose **Actions | Add/Edit Configuration Profile** to open the **Add/Edit Configuration Profile** window, as shown in Figure 2.63. Select the radio button next to a blank profile and enter the name of the profile.

Next click the **Edit Profile** button. On the **Basic** screen, enter the SSID of your access point in the **Network Name** text box. Then click the **Encryption** tab, as shown in Figure 2.64. Choose the **Use Hexadecimal** radio button and enter your WEP keys (1–4).

Figure 2.63 The Configuration Profiles

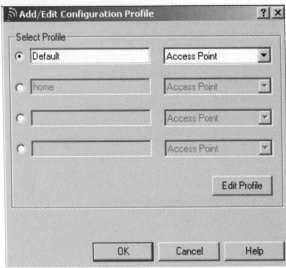

Figure 2.64 Entering the WEP Key

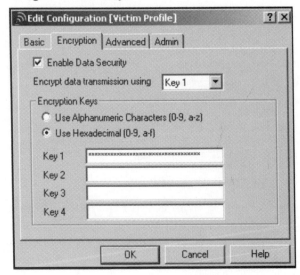

Finally, click **OK** to save your settings. Your wireless client is now configured for use with your access point.

Configuring Windows 2000 Clients

Configuring WPA in Windows 2000 can only be done using third-party software applications, either from a software provider or from the wireless card manufacturer. Microsoft has decided not to release WPA drivers for Windows 2000, and hence the company suggests that the user download provided client software from the wireless network card vendor. WPA is only natively supported in Windows XP SP2, or in pre-SP2 versions of Windows XP after a patch is applied.

Configuring MAC Clients

The Apple operating system makes it very simple to configure and make a connection to a WEP encrypted network. From the desktop, click the **wireless Airport** icon at the top right of the screen to view the wireless connections. Because the Airport we are connecting to does not have its SSID broadcast, we will have to manually enter the SSID and the WEP encryption key. Click the **Other...** option to manually enter the SSID and key (see Figures 2.65 and 2.66).

Figure 2.65 Wireless List

Figure 2.66 Closed Network Configuration

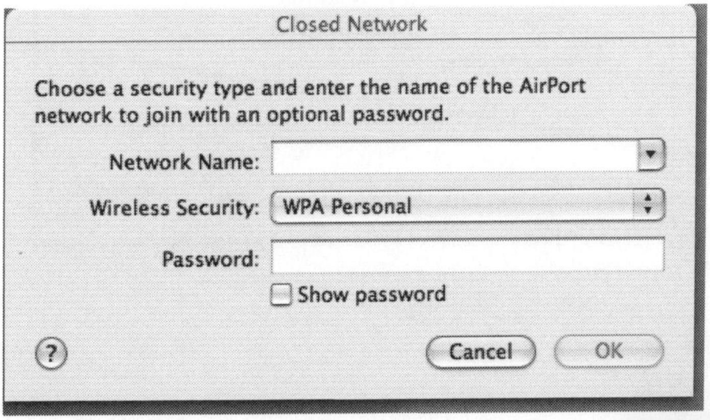

Enter the SSID in the **Network Name:** text box and choose either **WEP 40/128-bit hex** or **WEP 40/128-bit ASCII**, depending on whether the encryption key is in hex or ASCII characters. Click in the **Password:** text box and enter the specified encryption key. Click **OK** to make the connection (see Figure 2.67).

Figure 2.67 Making the WEP Connection

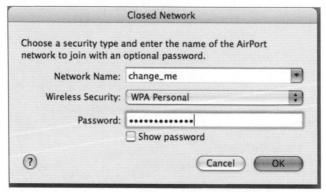

Configuring MAC Clients

Again, making a WPA connection from an Apple workstation is just as easy as making a WEP connection.

From the **Closed Network** dialog box, enter the SSID in the **Network Name:** text box and choose **WPA Personal** from the **Wireless Security:** drop-down menu. Enter the specified encryption key in the **Password:** text box and click **OK** to make the connection (See Figure 2.68).

Figure 2.68 Making the WPA Connection

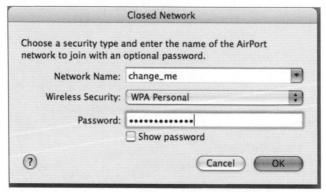

Configuring Linux Clients

Configuring wireless clients for use with your network after security features have been enabled is a simple process. Using the Wireless Tools package for Linux, a simple command using the *iwconfig* tool will configure the client for access, including encryption, to the access point.

Most newer distributions of Linux include the Wireless Tools package on installation. The Wireless Tools package provides "man" pages for the tools it provides. Type **man iwconfig** for commands used with the iwconfig tool.

From a command prompt in Linux, escalate to root privileges. Type **iwconfig** to see a list of wireless devices that may be configured (see Figure 2.69).

Figure 2.69 Wireless Devices Using *iwconfig*

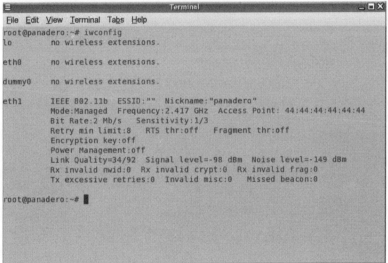

You can see that *eth1* is able to be configured and used as a wireless device. From the command prompt, you will specify the wireless card to use, the SSID, and the encryption key. The encryption key may be entered using the ASCII key or the hexadecimal key (see Figures 2.70 and 2.71).

Figure 2.70 Creating the Wireless Connection (Hex)

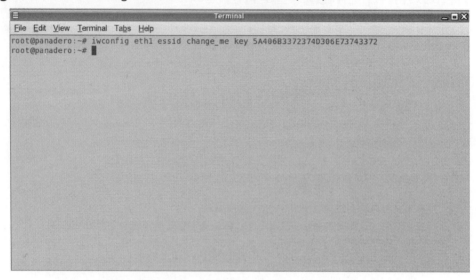

Figure 2.71 Creating the Wireless Connection (ASCII)

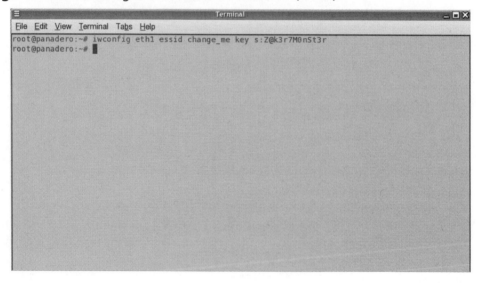

At this point, you have made the connection to the access point. You can either set a static IP address using the *ifconfig* command or via DHCP using the common *dhcpcd* or *dhclient* commands.

Configuring Linux Clients

The simplest way to configure Linux clients to connect to a WPA encrypted network is to use the *wpa_supplicant* open-source application. The *wpa_supplicant* is installed as a daemon that runs as a background process. It has a command-line interface as well as a graphical user interface. As of this writing, the latest stable release of the *wpa_supplicant* is version 0.4.9 and supports a wide variety of wireless PCMCIA cards. For this section, we use a Senao 802.11b PLUS EXT2 card, which uses the *host_ap* drivers.

The first step to installing the package is to download it from the Web. The official site of the package is located at http://hostap.epitest.fi/wpa_supplicant/. Download the file wpa_supplicant-0.4.9.tar.gz to your local host. Open a command prompt and change to the directory that contains the installation package. Untar the package by issuing the command *tar xvfz wpa_supplicant-0.4.9.tar.gz*. Change your directory to the newly created wpa_supplicant-0.4.9 directory (see Figure 2.72).

Figure 2.72 Extracting the *wpa_supplicant* Package

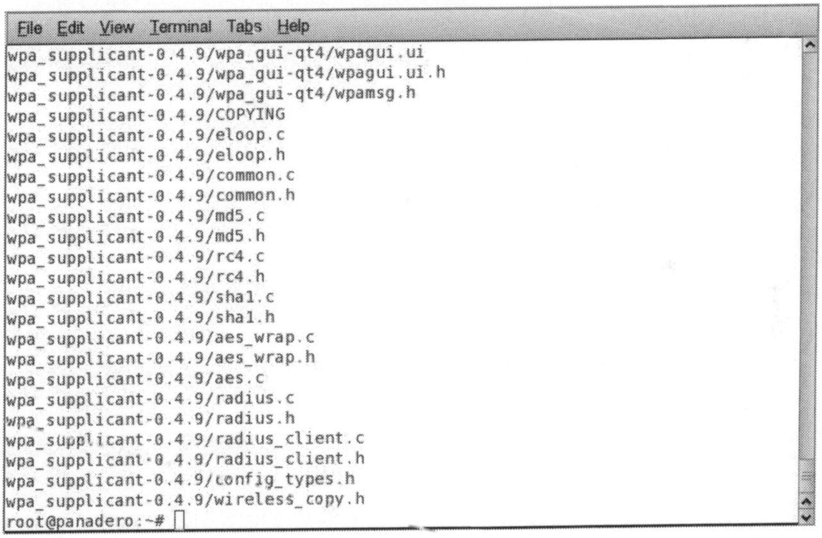

```
File  Edit  View  Terminal  Tabs  Help
wpa_supplicant-0.4.9/wpa_gui-qt4/wpagui.ui
wpa_supplicant-0.4.9/wpa_gui-qt4/wpagui.ui.h
wpa_supplicant-0.4.9/wpa_gui-qt4/wpamsg.h
wpa_supplicant-0.4.9/COPYING
wpa_supplicant-0.4.9/eloop.c
wpa_supplicant-0.4.9/eloop.h
wpa_supplicant-0.4.9/common.c
wpa_supplicant-0.4.9/common.h
wpa_supplicant-0.4.9/md5.c
wpa_supplicant-0.4.9/md5.h
wpa_supplicant-0.4.9/rc4.c
wpa_supplicant-0.4.9/rc4.h
wpa_supplicant-0.4.9/sha1.c
wpa_supplicant-0.4.9/sha1.h
wpa_supplicant-0.4.9/aes_wrap.c
wpa_supplicant-0.4.9/aes_wrap.h
wpa_supplicant-0.4.9/aes.c
wpa_supplicant-0.4.9/radius.c
wpa_supplicant-0.4.9/radius.h
wpa_supplicant-0.4.9/radius_client.c
wpa_supplicant-0.4.9/radius_client.h
wpa_supplicant-0.4.9/config_types.h
wpa_supplicant-0.4.9/wireless_copy.h
root@panadero:~# []
```

Once the files have been extracted, you will need to create a .config file to include the options you are interested in using. Included in the package is a file called *defconfig* that includes all the options to make the package. You can copy this file to the same directory, renaming it .config, with the following command:

```
cp defconfig .config
```

Once the file has been created, you may edit it to reflect any additional change or drivers you need. Once the .config file is ready, type **make** to compile the libraries (see Figure 2.73).

Figure 2.73 "Making" the *wpa_supplicant* Libraries

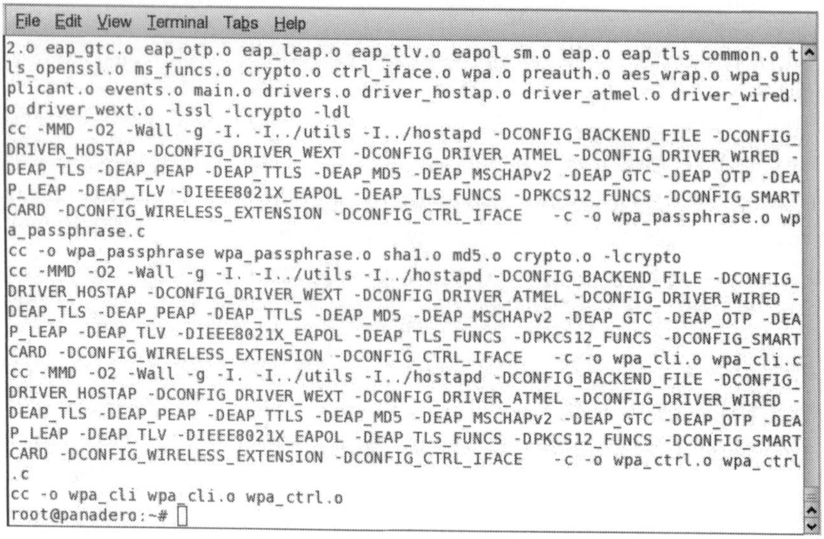

```
File  Edit  View  Terminal  Tabs  Help
2.o eap_gtc.o eap_otp.o eap_leap.o eap_tlv.o eapol_sm.o eap.o eap_tls_common.o t
ls_openssl.o ms_funcs.o crypto.o ctrl_iface.o wpa.o preauth.o aes_wrap.o wpa_sup
plicant.o events.o main.o drivers.o driver_hostap.o driver_atmel.o driver_wired.
o driver_wext.o -lssl -lcrypto -ldl
cc -MMD -O2 -Wall -g -I. -I../utils -I../hostapd -DCONFIG_BACKEND_FILE -DCONFIG_
DRIVER_HOSTAP -DCONFIG_DRIVER_WEXT -DCONFIG_DRIVER_ATMEL -DCONFIG_DRIVER_WIRED -
DEAP_TLS -DEAP_PEAP -DEAP_TTLS -DEAP_MD5 -DEAP_MSCHAPv2 -DEAP_GTC -DEAP_OTP -DEA
P_LEAP -DEAP_TLV -DIEEE8021X_EAPOL -DEAP_TLS_FUNCS -DPKCS12_FUNCS -DCONFIG_SMART
CARD -DCONFIG_WIRELESS_EXTENSION -DCONFIG_CTRL_IFACE   -c -o wpa_passphrase.o wp
a_passphrase.c
cc -o wpa_passphrase wpa_passphrase.o sha1.o md5.o crypto.o -lcrypto
cc -MMD -O2 -Wall -g -I. -I../utils -I../hostapd -DCONFIG_BACKEND_FILE -DCONFIG_
DRIVER_HOSTAP -DCONFIG_DRIVER_WEXT -DCONFIG_DRIVER_ATMEL -DCONFIG_DRIVER_WIRED -
DEAP_TLS -DEAP_PEAP -DEAP_TTLS -DEAP_MD5 -DEAP_MSCHAPv2 -DEAP_GTC -DEAP_OTP -DEA
P_LEAP -DEAP_TLV -DIEEE8021X_EAPOL -DEAP_TLS_FUNCS -DPKCS12_FUNCS -DCONFIG_SMART
CARD -DCONFIG_WIRELESS_EXTENSION -DCONFIG_CTRL_IFACE   -c -o wpa_cli.o wpa_cli.c
cc -MMD -O2 -Wall -g -I. -I../utils -I../hostapd -DCONFIG_BACKEND_FILE -DCONFIG_
DRIVER_HOSTAP -DCONFIG_DRIVER_WEXT -DCONFIG_DRIVER_ATMEL -DCONFIG_DRIVER_WIRED -
DEAP_TLS -DEAP_PEAP -DEAP_TTLS -DEAP_MD5 -DEAP_MSCHAPv2 -DEAP_GTC -DEAP_OTP -DEA
P_LEAP -DEAP_TLV -DIEEE8021X_EAPOL -DEAP_TLS_FUNCS -DPKCS12_FUNCS -DCONFIG_SMART
CARD -DCONFIG_WIRELESS_EXTENSION -DCONFIG_CTRL_IFACE   -c -o wpa_ctrl.o wpa_ctrl
.c
cc -o wpa_cli wpa_cli.o wpa_ctrl.o
root@panadero:~# []
```

Install the binaries by copying them to the /usr/local/bin directory with this command:

```
cp wpa_cli wpa_supplicant /usr/local/bin
```

A configuration file must be created in order for the *wpa_supplicant* to know which options to use. To connect to a WPA-PSK access point, you need to add several lines to the wpa_supplicant.conf file, to include the SSID and the PSK key.

To create the wpa_supplicant.conf file, at a command line, type **vi /etc/wpa_supplicant.conf** (or use your favorite text editor). You will be dumped into the vi text editor to add your access point information.

Press **i** to enter Insert Text mode. Type the following text, substituting your access point information for what's given:

```
network={
        ssid="change_me"
        key_mgmt=WPA-PSK
        psk="Z@k3r7M0nSt3r"
}
```

Press **Escape** to exit Insert Text mode and enter command mode. Type **:wq** to write the file and quit the vi editor, which will save the file to /etc/wpa_suppli-cant.conf. Figure 2.74 shows the wpa_supplicant.conf file as it looks in the vi editor, with the *wq* command to save and exit the editor.

Figure 2.74 Exiting and Saving the wpa_supplicant.conf File

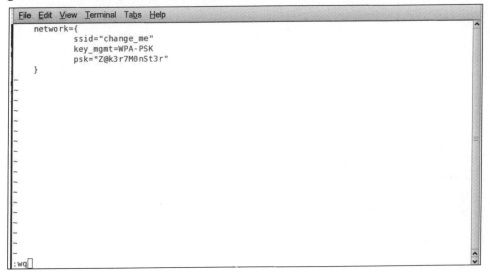

Once we have created the wpa_supplicant.conf file, it is time to make the WPA connection to the access point. This section uses the Senao NL-2511-CD PLUS EXT2 wireless PCMCIA card, which uses the *hostap* Linux drivers.

NOTE

For the Senao PLUS EXT2 card to support WPA, you need to make sure that the firmware has been updated to at least version 1.7.0. As of this publication, the most recent version is 1.8.4. Excellent instructions can be found on the Remote-Exploit.org Website. Comprehensive instructions can be found at the following address: http://remote-exploit.org/index.php/Prism_update.

Insert the card into the laptop's PCMCIA slot. From the command line, type **iwconfig** to ensure that the laptop can see the wireless card. You should see a listing of a *wifi0* interface and a *wlan0* interface (see Figure 2.75).

Figure 2.75 *iwconfig* Showing the *wifi0* and *wlan0* Interfaces

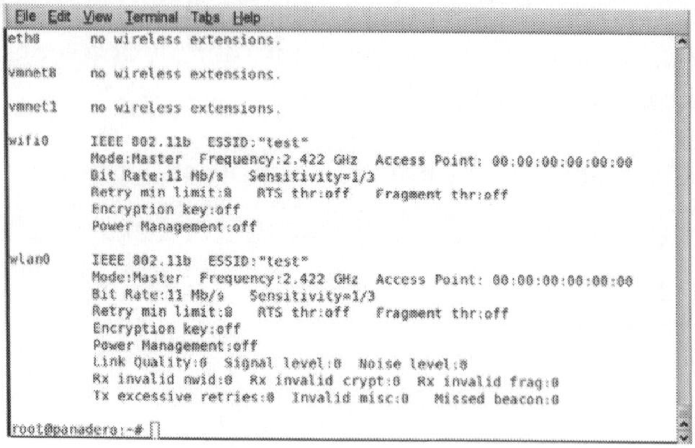

For the *wpa_supplicant* to work properly, you need to shut down all network interfaces available on the workstation by issuing the *ifconfig* command. At the command line, type **ifconfig** to see a list of interfaces, and type **ifconfig int_name down** to shut down the interface. Do this for at least these interfaces (see Figures 2.76 and 2.77):

- eth0
- wifi0
- wlan0

Figure 2.76 The List of Network Interfaces

Figure 2.77 Shutting Down the Interfaces

```
root@panadero:~# ifconfig eth0 down
root@panadero:~# ifconfig wifi0 down
root@panadero:~# ifconfig wlan0 down
root@panadero:~# ifconfig
lo        Link encap:Local Loopback
          inet addr:127.0.0.1  Mask:255.0.0.0
          UP LOOPBACK RUNNING  MTU:16436  Metric:1
          RX packets:226 errors:0 dropped:0 overruns:0 frame:0
          TX packets:226 errors:0 dropped:0 overruns:0 carrier:0
          collisions:0 txqueuelen:0
          RX bytes:11300 (11.0 KiB)  TX bytes:11300 (11.0 KiB)

vmnet1    Link encap:Ethernet  HWaddr 00:50:56:C0:00:01
          inet addr:192.168.226.1  Bcast:192.168.226.255  Mask:255.255.255.0
          UP BROADCAST RUNNING MULTICAST  MTU:1500  Metric:1
          RX packets:0 errors:0 dropped:0 overruns:0 frame:0
          TX packets:0 errors:0 dropped:0 overruns:0 carrier:0
          collisions:0 txqueuelen:1000
          RX bytes:0 (0.0 b)  TX bytes:0 (0.0 b)

vmnet8    Link encap:Ethernet  HWaddr 00:50:56:C0:00:08
          inet addr:192.168.227.1  Bcast:192.168.227.255  Mask:255.255.255.0
          UP BROADCAST RUNNING MULTICAST  MTU:1500  Metric:1
          RX packets:0 errors:0 dropped:0 overruns:0 frame:0
          TX packets:0 errors:0 dropped:0 overruns:0 carrier:0
          collisions:0 txqueuelen:1000
          RX bytes:0 (0.0 b)  TX bytes:0 (0.0 b)

root@panadero:~# []
```

Once the interfaces have been shut down, it's time to make the WPA connection to the access point. From the command line, you will specify the interface to use (*wlan0*) and the location of the .conf file (/etc/wpa_supplicant.conf), and the *–d* command provides extra debugging information if you experience problems. Here's the command (see Figure 2.78):

```
wpa_supplicant -i wlan0 -c /etc/wpa_supplicant.conf -d
```

When this command is executed and everything is operating correctly, you will have made the WPA connection to the access point (see Figure 2.79).

Figure 2.78 Creating the WPA Connection to the Access Point

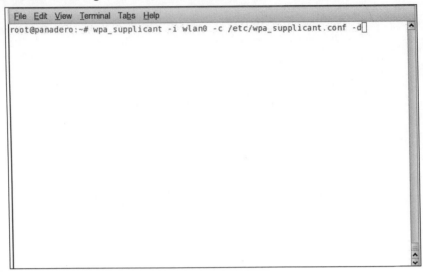

```
File  Edit  View  Terminal  Tabs  Help
root@panadero:~# wpa_supplicant -i wlan0 -c /etc/wpa_supplicant.conf -d
```

Figure 2.79 The WPA Connection Is Made

```
File  Edit  View  Terminal  Tabs  Help
  key_iv - hexdump(len=16): 94 35 78 82 4e a5 5c 63 af 11 8e de 3b 16 6e 59
  key_rsc - hexdump(len=8): 97 01 00 00 00 00 00 00
  key_id (reserved) - hexdump(len=8): 00 00 00 00 00 00 00 00
  key_mic - hexdump(len=16): 12 77 b8 5f 17 5c 1b 62 a3 94 31 5e f3 3c 28 89
State: GROUP_HANDSHAKE -> GROUP_HANDSHAKE
WPA: RX message 1 of Group Key Handshake from 00:13:10:69:aa:b6 (ver=1)
WPA: Group Key - hexdump(len=32): [REMOVED]
WPA: Installing GTK to the driver (keyidx=2 tx=0).
WPA: RSC - hexdump(len=6): 97 01 00 00 00 00
wpa_driver_hostap_set_key: alg=TKIP key_idx=2 set_tx=0 seq_len=6 key_len=32
WPA: Sending EAPOL-Key 2/2
WPA: Key negotiation completed with 00:13:10:69:aa:b6 [PTK=TKIP GTK=TKIP]
Cancelling authentication timeout
State: GROUP_HANDSHAKE -> COMPLETED
CTRL-EVENT-CONNECTED - Connection to 00:13:10:69:aa:b6 completed (auth)
EAPOL: External notification - portValid=1
EAPOL: External notification - EAP success=1
EAPOL: SUPP_PAE entering state AUTHENTICATING
EAPOL: SUPP_BE entering state SUCCESS
EAP: EAP entering state DISABLED
EAPOL: SUPP_PAE entering state AUTHENTICATED
EAPOL: SUPP_BE entering state IDLE
EAPOL: startWhen --> 0
```

At this point, you are running in a debug mode to ensure that everything is correct. Kill this connection with the **ctrl + c** command. Rerun the same command with the **−B** option instead of *−d* to run *wpa_supplicant* in daemon mode, which will show you no debug output and allows you to run additional commands from the command prompt (see Figure 2.80). You will need to set a static IP for the network you are connected to using the *ifconfig* command, or issue the *dhcpcd* or *dhclient* commands on the *wlan0* interface to get a DHCP address and make the connection.

Figure 2.80 Daemon Mode and Obtaining an IP Address

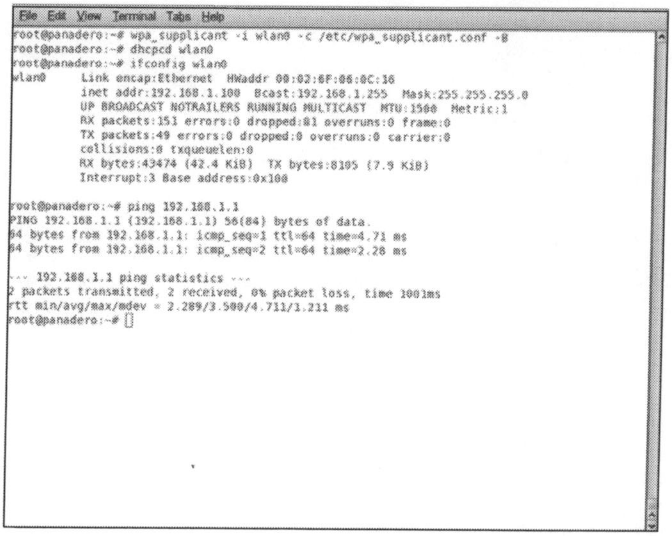

Notes from the Underground…

Enabling Security Features on the Xbox

Many Xbox owners like to take advantage of the Xbox Live feature. Xbox Live allows gamers to connect their Xboxes to the Internet and play selected games against online opponents. Since the Xbox is often connected to a TV that isn't necessarily in the same room with most of the household computer equipment, wireless networking is a natural choice for this connection.

Several available wireless bridges (such as the Linksys WET 11 Wireless Ethernet Bridge) will connect the Xbox to a home network. These devices must be configured to use the wireless network's security features.

First, log in to the WET 11. By default, the WET 11 is configured to use the IP address 192.168.1.251 (see Figure 2.81).

Continued

Figure 2.81 The Linksys WET 11 Initial Setup Screen

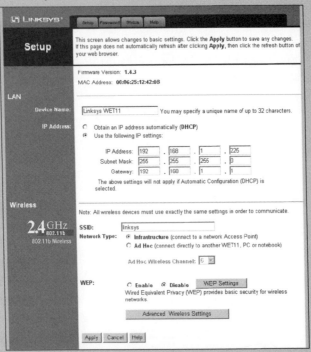

Enter the SSID for your wireless network in the **SSID** text box, and then select the **Enable** radio button next to **WEP** (see Figure 2.82).

Click the **WEP Settings** button to open the **Shared Keys** window (see Figure 2.83). Select **128 bit 26 hex digits** from the drop-down box, and then enter the WEP keys that your wireless network uses. The WEP keys can be entered in either of two ways:

- Generate the keys using the same passphrase used to generate the keys on your access point.
- Manually enter the WEP keys that your access point uses.

Next, click the **Apply** button on the **Shared Keys** window and the initial setup screen to save your settings. Finally, add the MAC address of your WET 11 to your allowed MAC address list on your access point.

Continued

Figure 2.82 Set the SSID and Enable WEP

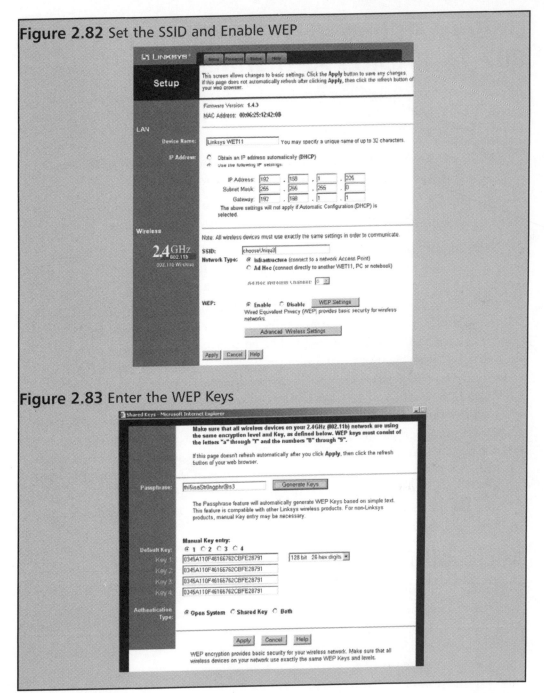

Figure 2.83 Enter the WEP Keys

Understanding and Configuring 802.1X RADIUS Authentication

To provide better security for wireless LANs and in particular to improve the security of WEP, a number of existing technologies used on wired networks were adapted for this purpose, including:

- **Remote Authentication and Dial-In User Service (RADIUS)** Provides for centralized authentication and accounting.

- **802.1X** Provides a method of port-based authentication to local area network (LAN) ports in a switched network environment.

These two services are used in combination with other security mechanisms, such as those provided by the Extensible Authentication Protocol (EAP), to further enhance the protection of wireless networks. Like MAC filtering, 802.1X is implemented at Layer 2 of the Open System Interconnection (OSI) model: It will prevent communication on the network using higher layers of the OSI model if authentication fails at the MAC layer. However, unlike MAC filtering, 802.1X is very secure, since it relies on mechanisms that are much harder to compromise than MAC address filters, which can be easily compromised through spoofed MAC addresses.

Although a number of vendors implement their own RADIUS servers, security mechanisms, and protocols for securing networks through 802.1X, such as Cisco's LEAP and Funk Software's EAP-TTLS, this section focuses on implementing 802.1X on a Microsoft network using Internet Authentication Services (IAS) and Microsoft's Certificate Services. Keep in mind, however, that wireless security standards are a moving target, and standards other than those discussed here, such as the PEAP, are being developed and might be available by the time this book is published or in the near future.

Microsoft RADIUS Servers

Microsoft's IAS provides a standards-based RADIUS server and can be installed as an optional component on Microsoft Windows 2000 and Net servers. Originally designed to provide a means to centralize the authentication, authorization, and accounting for dial-in users, RADIUS servers are now used to provide these services for other types of network access, including virtual private networks (VPNs), port-based authentication on switches, and, it's important to note, wireless network access. IAS can be deployed within Active Directory to use the Active Directory database to centrally manage the login process for users connecting over a variety of network types. Moreover, multiple RADIUS servers can be installed and configured so that secondary RADIUS servers will automatically be used in case the primary RADIUS server fails, thus providing fault tolerance for the RADIUS infrastructure. Although RADIUS is not required to support the 802.1X standard, it is a preferred method for providing the authentication and authorization of users and devices attempting to connect to devices that use 802.1X for access control.

The 802.1X Standard

The 802.1X standard was developed to provide a means of restricting port-based Ethernet network access to valid users and devices. When a computer attempts to connect to a port on a network device, such as switch, it must be successfully authenticated before it can communicate on the network using the port. In other words, communication on the network is impossible without an initial successful authentication.

802.1X Authentication Ports

Two types of ports are defined for 802.1X authentication: authenticator or supplicant. The *supplicant* is the port requesting network access. The *authenticator* is the port that allows or denies access for network access. However, the authenticator does not perform the actual authentication of the supplicant requesting access. The authentication of the supplicant is performed by a separate authentication service, located on a separate server or built into the device itself, on behalf of the authenticator. If the authenticating server successfully authenticates the supplicant, it will communicate the fact to the authenticator, which will subsequently allow access.

An 802.1X-compliant device has two logical ports associated with the physical port: an uncontrolled port and a controlled port. Because the supplicant must initially communicate with the authenticator to make an authentication request, an 802.1X-compliant device will make use of a logical *uncontrolled port* over which this request can be made. Using the uncontrolled port, the authenticator will forward the authentication request to the authentication service. If the request is successful, the authenticator will allow communication on the LAN via the logical *controlled port*.

The Extensible Authentication Protocol (EAP)

EAP is used to pass authentication requests between the supplicant and a RADIUS server via the authenticator. EAP provides a way to use different authentication types in addition to the standard authentication mechanisms provided by the Point-to-Point Protocol (PPP). Using EAP, stronger authentication types can be implemented within PPP, such as those that use public keys in conjunction with smart cards. In Windows, there is support for two EAP types:

- **EAP MD-5 CHAP** This allows for authentication based on a user-name/password combination. A number of disadvantages are associated with using EAP MD-5 CHAP. First, even though it uses one-way hashes in combination with a challenge/response mechanism, critical information is still sent in the clear, making it vulnerable to compromise. Second, it does not provide mutual authentication between the client and the server; the server merely authenticates the client. Third, it does not provide a mechanism for establishing a secure channel between the client and the server.

- **EAP-TLS** This is a security mechanism based on X.509 digital certificates that is more secure than EAP MD-5 CHAP. The certificates can be stored in the Registry or on devices such as smart cards. When EAP-TLS authentication is used, both the client and the server validate one another by exchanging X.509 certificates as part of the authentication process. Additionally, EAP-TLS provides a secure mechanism for the exchange of keys to establish an encrypted channel. Although the use of EAP-TLS is more difficult to configure in that it requires the implementation of a public key infrastructure (PKI)—not a trivial undertaking—EAP-TLS is recommended for wireless 802.1X authentication.

In a paper published in February 2002 by William A. Arbaugh and Arunesh Mishra, *An Initial Security Analysis of the IEEE 802.1x Standard,* the authors discuss how one-way authentication and other weaknesses made 802.1X vulnerable to man-in-the-middle and session-hijacking attacks. Therefore, although it might be possible to use EAP MD-5 CHAP for 802.1X wireless authentication on Windows XP (pre SP1), it is not recommended. EAP-TLS protects against the types of attacks described by this paper.

The 802.1X Authentication Process

For 802.1X authentication to work on a wireless network, the AP must be able to securely identify traffic from a particular wireless client. This identification is accomplished using authentication keys that are sent to the AP and the wireless client from the RADIUS server. When a wireless client (802.1X supplicant) comes within range of the AP (802.1X authenticator), the following simplified process occurs:

1. The AP point issues a challenge to the wireless client.
2. The wireless client responds with its identity.

3. The AP forwards the identity to the RADIUS server using the uncontrolled port.

4. The RADIUS server sends a request to the wireless station via the AP, specifying the authentication mechanism to be used (for example, EAP-TLS).

5. The wireless station responds to the RADIUS server with its credentials via the AP.

6. The RADIUS server sends an encrypted authentication key to the AP if the credentials are acceptable.

7. The AP generates a multicast/global authentication key encrypted with a per-station unicast session key and transmits it to the wireless station.

Figure 2.84 shows a simplified version of the 802.1X authentication process using EAP-TLS.

Figure 2.84 The 802.1X Authentication Process Using EAP-TLS

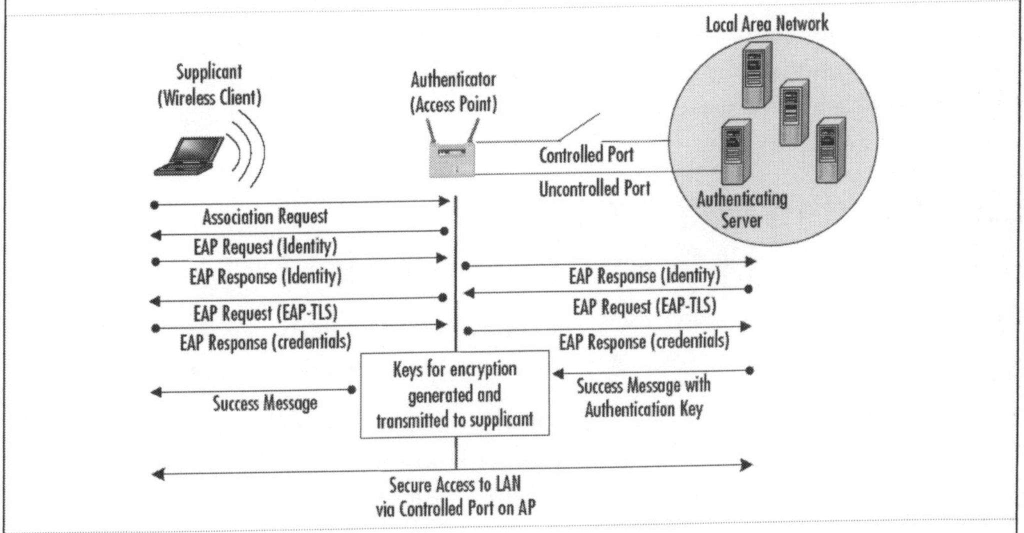

When the authentication process successfully completes, the wireless station is allowed access to the controlled port of the AP and communication on the network can occur. Note that much of the security negotiation in the preceding steps occurs on the 802.1X uncontrolled port, which is only used so that the AP can forward traffic associated with the security negotiation between the client and the RADIUS server. EAP-TLS is required for the process to take place. EAP-TLS, unlike EAP MD-5 CHAP, provides a mechanism to allow the secure transmission of the authentication keys from the RADIUS server to the client.

Advantages of EAP-TLS

There are a number of significant advantages to using EAP-TLS authentication in conjunction with 802.1X:

- The use of X.509 digital certificates for authentication and key exchange is very secure.

- EAP-TLS provides a means to generate and use dynamic one-time-per-user, session-based WEP keys on the wireless network.

- Neither the user nor the administrator knows the WEP keys that are in use.

For these reasons, using EAP-TLS for 802.1X authentication removes much of the vulnerability associated with using WEP and provides a high degree of assurance.

In the following section, we will look at how to configure 802.1X using EAP-TLS authentication on a Microsoft-based wireless network. If you are using other operating systems and software, the same general principles will apply. However, you might have additional configuration steps to perform, such as the installation of 802.1X supplicant software on the client. Windows XP provides this software within the operating system.

Configuring 802.1X Using EAP-TLS on a Microsoft Network

Before you can configure 802.1X authentication on a wireless network, you must satisfy a number of prerequisites. At a minimum, you need the following:

- **An AP that supports 802.1X authentication** You probably won't find these devices at your local computer hardware store. They are designed for enterprise-class wireless network infrastructures and are typically higher priced. Note that some devices will allow the use of IPSec between the AP and the wired network.

- **Client software and hardware that supports 802.1X and EAP-TLS authentication and the use of dynamic WEP keys** Fortunately, just about any wireless adapter that allows the use of the Windows XP wireless interface will work. However, older wireless network adapters that use their own client software might not work.

- **IAS installed on a Windows 2000 server** This provides a primary RADIUS server and, optionally, is installed on other servers to provide secondary RADIUS servers for fault tolerance.

- **Active Directory**

- **A PKI using a Microsoft stand-alone or Enterprise Certificate server to support the use of X.509 digital certificates for EAP-TLS** More certificate servers can be deployed in the PKI for additional security. An Enterprise Certificate server can ease the burden of certificate deployment to clients and the RADIUS server through auto-enrollment of client computers that are members of the Windows 2000 domain.

- **The most recent service packs and patches installed on the Windows 2000 servers and Windows XP wireless clients**

After you configure a PKI and install IAS on your Windows 2000 network, there are four general steps to configure 802.1X authentication on your wireless network:

1. Install X.509 digital certificates on the wireless client and IAS servers.

2. Configure IAS logging and policies for 802.1X authentication.

3. Configure the wireless AP for 802.1X authentication.

4. Configure the properties of the client wireless network interface for dynamic WEP key exchange.

Configuring Certificate Services and Installing Certificates on the IAS Server and Wireless Client

After you deploy Active Directory, the first step in implementing 802.1X is to deploy the PKI and install the appropriate X.509 certificates. You will have to install (at a minimum) a single certificate server, either a standalone or enterprise certificate server, to issue certificates. What distinguishes a standalone from an enterprise certificate server is whether it will depend on, and be integrated with, Active Directory. A standalone CA does not require Active Directory. This certificate server can be a *root* CA or a *subordinate* CA, which ultimately receives its authorization to issue certificates from a root CA higher in the hierarchy, either directly or indirectly through intermediate CAs, according to a *certification path*.

The root CA can be a public or commercially available CA that issues an authorization to a subordinate CA, or it can be one deployed on the Windows 2000 network. In enterprise networks that require a high degree of security, it is not recommended that you use the root CA to issue client certificates; for this purpose, you should use a subordinate CA authorized by the root CA. In very high-security environments, you should use intermediate CAs to authorize the CA that issues client certificates. Furthermore, you should secure the hardware and software of the root and intermediate CAs as much as possible, take them offline, and place them in a secure location. You would then bring the root and intermediate CAs online only when you need to perform tasks related to the management of your PKI.

In deploying your PKI, keep in mind that client workstations and the IAS servers need to be able to consult a *certificate revocation list* (CRL) to verify and validate certificates, especially certificates that have become compromised before their expiration date and have been added to a CRL. If a CRL is not available, authorization will fail. Consequently, a primary design consideration for your PKI is to ensure that the CRLs are highly available. Normally, the CRL is stored on the CA; however, additional distribution points for the CRL can be created to ensure a high degree of availability. The CA maintains a list of these locations and distributes the list in a field of the client certificate.

NOTE

It is beyond the scope of this chapter to discuss the implementation details of a PKI. For more information, please see the various documents available on the Microsoft Web site, in particular at www.microsoft.com/windows2000/technologies/security/default.asp, www.microsoft.com/windows2000/techinfo/howitworks/security/pki-intro.asp, and www.microsoft.com/windows2000/techinfo/planning/security/pki.asp.

Whether you decide to implement a stand-alone or an enterprise CA to issue certificates, you will need to issue three certificates: for both the computer and the user account on the wireless client as well as the RADIUS server. A certificate is required in all these places because mutual authentication has to take place. The computer certificate provides initial access of the computer to the network, and the user certificate provides wireless access after the user logs in. The RADIUS server will authenticate the client based on the wireless client's computer and user certificates, and the wireless client will authenticate the RADIUS server based on the server's certificate.

The certificates on the wireless client and the RADIUS server do not have to be issued by the same CA. However, both the client and the server have to trust each other's certificates. Within each certificate is information about the certificate path leading up to the root CA. If both the wireless client and the RADIUS server trust the root CA in each other's certificates, mutual authentication can successfully take place. If you are using a standalone CA that is not in the list of Trusted Root Certification Authorities, you will have to add it to the list. You can do this through a Group Policy Object, or you can do it manually. For information on how to add CAs to the Trusted Root Certification Authorities container, please see Windows 2000 and Windows XP help files. The container listing these trusted root certificates can be viewed in the Certificates snap-in of the MMC console, as shown in Figure 2.85.

Figure 2.85 Certificate Snap-In Showing Trusted Root Certification Authorities

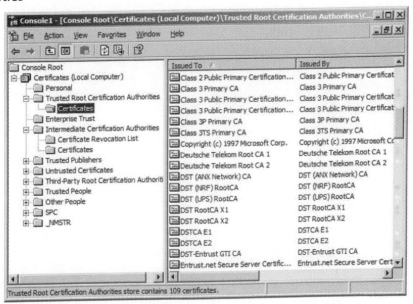

Using an enterprise CA will simplify many of the certificate-related tasks that you have to perform. An enterprise CA is automatically listed in the Trusted Root Certification Authorities container. Furthermore, you can use auto-enrollment to issue computer certificates to the wireless client and the IAS server without any intervention on the part of the user. Using an enterprise CA and configuring auto-enrollment of computer certificates should be considered a best practice.

If you put an enterprise CA into place, you will have to configure an Active Directory Group Policy to issue computer certificates automatically. You should use the **Default Domain Policy** for the domain in which your CA is located. To configure the **Group Policy** for auto-enrollment of computer certificates, do the following:

1. Access the **Properties** of the Group Policy object for the domain to which the enterprise CA belongs using **Active Directory Users and Computers**, and click **Edit**.

2. Navigate to **Computer Settings | Windows Settings | Security Settings | Public Key Policies | Automatic Certificate Request Settings**.

3. Right-click the **Automatic Certificate Request Settings**, click **New**, and then click **Automatic Certificate Request**, as shown in Figure 2.86.

Figure 2.86 Configuring a Domain Group Policy for Auto-Enrollment of Computer Certificates

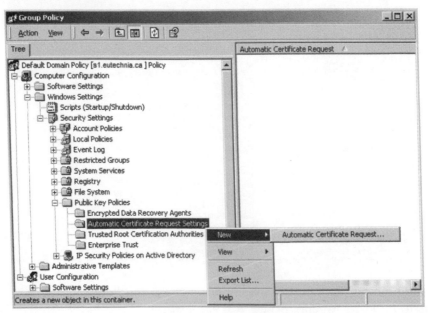

4. Click **Next** when the wizard appears. Click **Computer** in the **Certificate Templates**, as shown in Figure 2.87, and then click **Next**.

Figure 2.87 Choosing a Computer Certificate Template for Auto-Enrollment

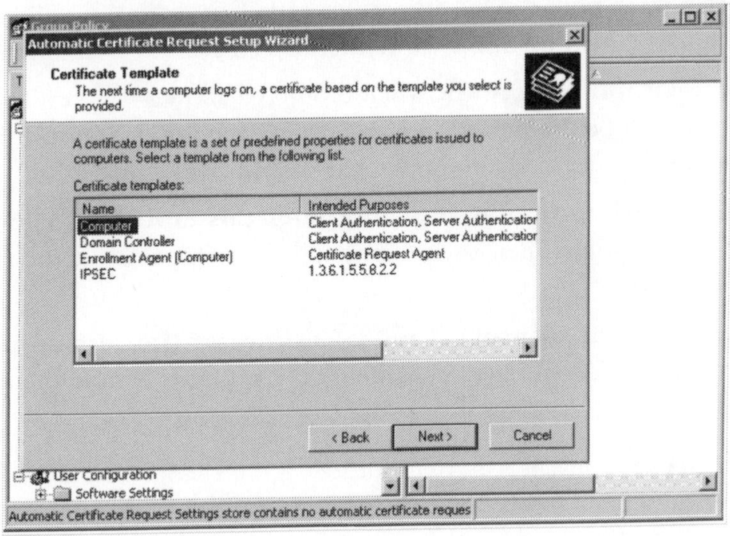

5. Click the enterprise CA, click **Next**, and then click **Finish**.

After you have configured a Group Policy for auto-enrollment of computer certificates, you can force a refresh of the Group Policy so that it will take effect immediately, rather than waiting for the next polling interval for Group Policy Changes, which could take as long as 90 minutes. To force Group Policy to take effect immediately on a Windows XP computer, type the command **gpupdate /target: computer**.

NOTE

On a Windows 2000 client, Group Policy update is forced using the *secedit/refreshpolicy* command.

Once you have forced a refresh of Group Policy, you can confirm whether the computer certificate is successfully installed. To confirm the installation of the computer certificate:

1. Type the command **mmc** and click **OK** from **Start | Run**.

2. Click **File** in the MMC console menu, and then click **Add/Remove Snap-in**.

3. Click **Add** in the **Add/Remove Snap-in** dialog box. Then select **Certificates** from the list of snap-ins and click **Add**. You will be prompted to choose which certificate store the snap-in will be used to manage.

4. Select **computer account** when prompted about what certificate the snap-in will be used to manage, and then click **Next**. You will then be prompted to select the computer the snap-in will manage.

5. Select **Local computer (the computer this console is running on)** and click **Finish**. Then click **Close** and click **OK** to close the remaining dialog boxes.

6. Navigate to the **Console Root | Certificates (Local Computer) | Personal | Certificates** container. The certificate should be installed there.

The next step is to install a user certificate on the client workstation and then map the certificate to a user account. There are a number of ways to install a user certificate: through Web enrollment, by requesting the certificate using the Certificates snap-in, by using a CAPICOM script (which can be executed as a login script to facilitate deployment), or by importing a certificate file.

The following steps demonstrate how to request the certificate using the Certificates snap-in:

1. Open an MMC console for **Certificates–Current User**. (To load this snap-in, follow the steps in the preceding procedure; however, at Step 5, select **My user account**.)

2. Navigate to **Certificates | Personal** and click the container with the alternate mouse button. Highlight **All Tasks** and then click **Request New Certificate**, as shown in Figure 2.88. The **Certificate Request Wizard** appears.

3. Click **Next** on the **Certificate Request Wizard** welcome page.

4. Select **User** and click **Next** on the **Certificate Types**, as shown in Figure 2.89. You can also select the **Advanced** check box. Doing so will allow you to select from a number of different cryptographic service providers (CSPs), to choose a key length, to mark the private key as exportable (the option might not be available for selection), and to enable strong private key protection. The latter option will cause you to be prompted for a password every time the private key is accessed.

Figure 2.88 Requesting a User Certificate

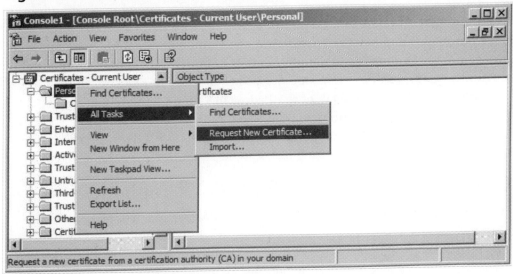

Figure 2.89 Choosing a Certificate Type

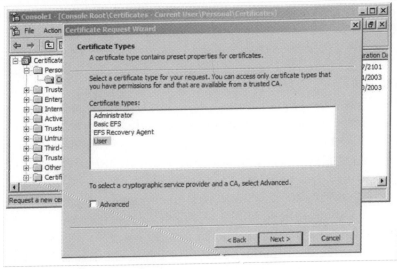

5. Type in a friendly name of your choosing and a description, and then click **Next**.

6. Review your settings and click **Finish**.

You now should have a user certificate stored on the computer used for wireless access. However, this user certificate will not be usable for 802.1X authentication unless it is mapped to a user account in Active Directory. By default, the certificate should be mapped to the user account. You can verify whether it has been mapped by viewing the **Properties** of the user account in **Active Directory Users and Computers**. The certificates that are mapped to the user account can be viewed in the **Published Certificates** tab of the **Properties** of the user account object.

After you configure certificate services and install computer and user certificates on the wireless client and a computer certificate on the RADIUS server, you must configure the RADIUS server for 802.1X authentication.

Configuring IAS Server for 802.1X Authentication

If you have configured RRAS for dial-in or VPN access, you will be comfortable with the IAS Server interface. It uses the same interfaces as RRAS for configuring dial-in conditions and policies. You can use IAS to centralize dial-in access policies for your entire network, rather than have dial-in access policies defined on each RRAS server. A primary advantage of doing this is easier administration and centralized logging of dial-in access.

Installing an IAS server also provides a standards-based RADIUS server that is required for 802.1X authentication. As with configuring RRAS, you will need to add and configure a Remote Access Policy to grant access. A Remote Access Policy grants or denies access to remote users and devices based on matching conditions and a profile. For access to be granted, the conditions you define have to match. For example, the dial-in user might have to belong to the appropriate group or connect during an allowable period. The profile in the Remote Access Policy defines such things as the authentication type and the encryption type used for the remote access. If the remote client is not capable of using the authentication methods and encryption strength defined in the profile, access is denied.

For 802.1X authentication, you will have to configure a Remote Access Policy that contains conditions specific to 802.1X wireless authentication and a Profile that requires the use of the Extensible Authentication Protocol (EAP) and strong encryption. After configuring the Remote Access Policy, you will have to configure the IAS server to act as a RADIUS server for the wireless AP, which is the RADIUS client.

Before installing and configuring the IAS server on your Windows 2000 or .NET/2003 network, you should consider whether you are installing it on a domain controller or member server (in the same or in a different domain). If you install it on a domain controller, the IAS server will be able to read the account properties in Active Directory. However, if you install IAS on a member server, you will have to perform an additional step to register the IAS server, which will give it access to Active Directory accounts.

There are a number of ways you can register the IAS server:

- The IAS snap-in
- The Active Directory Users and Computers admin tool
- The *netsh* command

NOTE

Perhaps the simplest way to register the IAS server is through the *netsh* command. To do this, log on to the IAS server, open a command prompt, and type the command **netsh ras add registeredserver**. If the IAS server is in a different domain, you will have to add arguments to this command. For more information on registering IAS servers, see Windows Help.

Once you have installed and, if necessary, registered the IAS server(s), you can configure the Remote Access Policy. Before configuring a Remote Access Policy, make sure that you apply the latest service pack and confirm that the IAS server has an X.509 computer certificate. In addition, you should create an Active Directory Global or Universal Group that contains your wireless users as members.

The Remote Access Policy will need to contain a condition for NAS-Port-Type that contains values for **Wireless-Other** and **Wireless-IEEE802.11** (these two values are used as logical *OR* for this condition) and a condition for **Windows-Groups=[*the group created for wireless users*]**. Both conditions have to match (logical *AND*) for access to be granted by the policy.

The Profile of the Remote Access Policy will need to be configured to use the Extensible Authentication Protocol and the Smart Card or Other Certificate EAP type. Encryption in the Profile should be configured to force the strongest level of encryption, if supported by the AP. Depending on the AP you are using, you might have to configure vendor-specific attributes (VSA) in the **Advanced** tab of the Profile. If you have to configure a VSA, you will need to contact the vendor of the AP to find out the value that should be used, if you can't find it in the documentation.

To configure the conditions for a Remote Access Policy on the IAS server:

1. Select **Internet Authentication Services** and open the IAS console from **Start | Programs | Administrative Tools**.

2. Right-click **Remote Access Policies**, and from the subsequent context menu, click **New Remote Access Policy**.

3. Enter a friendly name for the policy and click **Next**.

4. Click **Add** in the **Add Remote Access Policy Conditions** dialog box. Then select **NAS-Port-Type** in the **Select Attribute** dialog box and click **Add**, as shown in Figure 2.90.

Figure 2.90 Adding a NAS-Port-Type Condition to Remote Access Policy

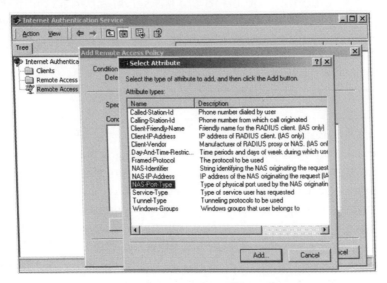

5. Select **Wireless–IEEE 802.11** and **Wireless–Other** from the left-hand window in the **NAS-Port-Type** dialog box, and click **Add>>** to move them to the Selected Types window, as shown in Figure 2.91. Click **OK**.

6. Add a condition for **Windows-Groups** that contains the group you created for wireless users after configuring the **NAS-Port-Type** conditions. Then click **Next**.

7. Click the radio button to **Grant remote access permission if user matches conditions** in the subsequent **Permissions** page for the new policy. The next step is to configure the **Profile** to support EAP-TLS and force the strongest level of encryption (128 bit).

Figure 2.91 Adding Wireless NAS-Port-Type Conditions

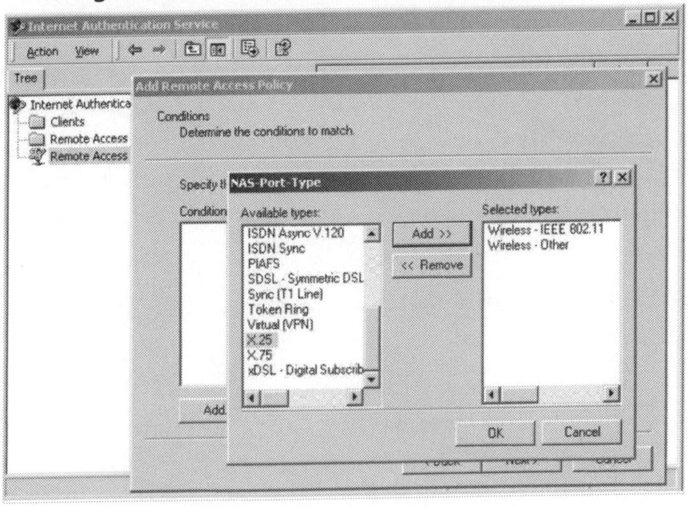

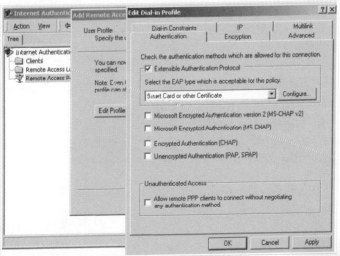

8. Click **Edit Profile** and click the **Authentication** tab.

9. Confirm that the check box for **Extensible Authentication Protocol** is selected and that **Smart Card or Other Certificate** is listed as the EAP type in the drop-down box. Clear all the other check boxes and click **Configure**.

10. Select the computer certificate you installed for use by the IAS server, and click **OK**. The resulting Authentication tab should look like the one in Figure 2.92.

Figure 2.92 Configuring the Dial-In Profile for 802.1X Authentication

11. Force the strongest level of encryption by clicking the **Encryption** tab and then clearing all the check boxes except the one for **Strongest**.

12. Save the policy by clicking **OK** and then **Finish**. Make sure that the policy you created is higher in the list than the default **Remote Access Policy**. You can delete the default policy if you like. Finally, you need to configure the IAS server for RADIUS authentication. To do this, you need to add a configuration for the RADIUS client—in this case, the AP—to the IAS server:

13. Right-click the **Clients** folder in the IAS console, and click **New Client** from the context menu.

14. Supply a friendly name for the configuration and click **Next**. The screen shown in Figure 2.93 appears.

Figure 2.93 Adding a RADIUS Client

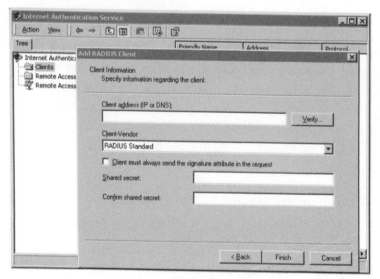

15. Configure the screen with the **Client address (IP or DNS)** of the wireless AP, and click the check box indicating that the **Client must always send the signature attribute in the request**. For the **Shared secret**, add an alphanumeric password that is at least 22 characters long for higher security.

16. Click **Finish**.

You can change the port numbers for RADIUS accounting and authentication by obtaining the properties of the Internet Authentication Service container in the IAS console. You can also use these property pages to log successful and unsuccessful authentication attempts and to register the server in Active Directory.

After installing certificates on the wireless client and IAS server and configuring the IAS server for 802.1X authentication, you will need to configure the AP and the wireless client. The following section shows the typical steps to complete the configuration of your wireless network for 802.1X authentication.

Configuring an Access Point for 802.1X Authentication

Generally, only enterprise-class APs support 802.1X authentication; this is not a feature found in devices intended for the SOHO market. Enterprise-class APs are not likely to be found in your local computer store. If you want an AP that supports 802.1X, you should consult the wireless vendors' Web sites for information on the features supported by the APs they manufacture. Vendors that manufacture 802.1X-capable devices include 3Com, Agere, Cisco, and others. The price for devices that support 802.1X authentication usually start at $500 (USD) and can cost considerably more, depending on the vendor and the other features supported by the AP. If you already own an enterprise-class AP, such as an ORiNOCO Access Point 500 or Access Point 1000, 802.1X authentication might not be supported in the original firmware but can be added through a firmware update.

Regardless of the device you purchase, an 802.1X-capable AP will be configured similarly. This section describes the typical configuration of 802.1X authentication on an ORiNOCO Access Point 500 with the most recent firmware update applied to it.

NOTE

For more information about the ORiNOCO device, see www.orinocowireless.com.

The configuration of the AP is straightforward and simple (see Figure 2.94). You will need to configure the following:

- **An encryption key length** This can be either 64 or 128 bits (or higher if your hardware and software support longer lengths).

- **An encryption key lifetime** When you implement 802.1X using EAP-TLS, WEP encryption keys are dynamically generated at intervals you specify. For higher-security environments, the encryption key lifetime should be set to 10 minutes or less.

- **An authorization lifetime** This is the interval at which the client and server will reauthenticate with one another. This interval should be longer than the interval for the encryption key lifetime but still relatively short in a high-security environment. A primary advantage here is that if a device is stolen, the certificates it uses can be immediately revoked. The next time it tries to authenticate, the CRL will be checked and authentication will fail.

- **An authorization password** This is the shared-secret password you configured for RADIUS client authentication on the IAS server. This password is used to establish communication between the AP and the RADIUS server. Thus, it needs to be protected by being long and complex. This password should be at least 22 characters long and use mixed case, numbers, letters, and other characters. You might want to consider using a random string generation program to create this password for you.

- **An IP address of a primary and, if configured for fault tolerance, a secondary RADIUS server** If the AP is in a DMZ and the RADIUS server is behind a firewall, this IP address can be the external IP address of the firewall.

- **A UDP port used for RADIUS authentication** The default port for RADIUS is port 1645. However, you can change this port on the IAS server and the AP for an additional degree of security.

Depending on your AP, you might have to go through additional configuration steps. For example, you might have to enable the use of dynamic WEP keys. On the AP 500, this configuration is automatically applied to the AP when you finish configuring the 802.1X settings. Consult your AP's documentation for specific information on configuring it for 802.1X authentication.

Figure 2.94 Configuring an ORiNOCO AP 500 for 802.1X Authentication

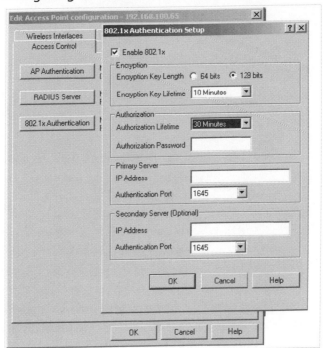

Configuring the Wireless Interface on Windows XP for 802.1X Authentication

If you have been following the preceding steps in the same order for configuring 802.1X authentication, the final step is to configure the properties of the wireless interface in Windows XP. You will have to ensure that the properties for EAP-TLS authentication and dynamic WEP are configured. To do this, perform the following steps:

1. Obtain the **Properties** of the wireless interface and click the **Authentication** tab.

2. Ensure that the check box for **Enable access control for IEEE 802.1X** is checked and that **Smart Card or other Certificate** is selected as the EAP type, as shown in Figure 2.95.

Figure 2.95 Authentication Properties for Wireless Client

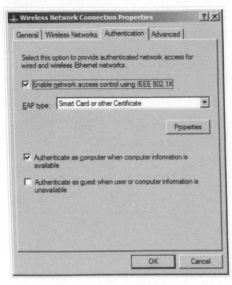

3. Click **Properties** to view the **Smart Card or other Certificate Properties** window. Ensure that the check box for **Validate server certificate** is checked, as shown in Figure 2.96.

Figure 2.96 Configure Smart Card or Other Certificate Properties

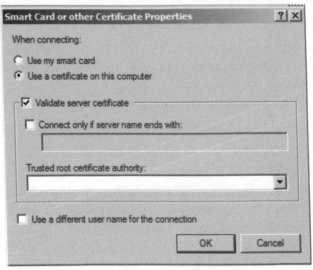

4. Select the root CA of the issuer of the server certificate in the **Trusted root certificate authority** drop-down box. If it is not already present, click **OK**. For additional security, you could select the check box for **Connect only if server name ends with** and type in the root DNS name—for example, tacteam.net.

5. Obtain the properties of the wireless interface and click the **Wireless Network** tab.

6. Confirm that the check box for **Use Windows for my wireless network settings** is selected in the **Preferred networks** dialog box. Highlight the SSID of the 802.1X-enabled AP, and click **Properties**.

7. Click the check box for **The key is provided for me automatically**, as shown in Figure 2.97, and then click **OK**.

Figure 2.97 Configuring Windows XP Wireless Properties for 802.1X Authentication

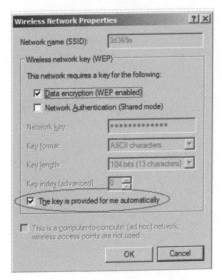

That's it. You're finished. The next time you attempt to authenticate and associate with the 802.1X-enabled AP, you might be presented with a prompt asking you to verify the identity of the IAS server certificate. By clicking **OK**, you will permit the authentication process to complete, thus allowing you secure access to the network.

Let's briefly review the steps to enable 802.1X authentication. We are assuming that you are using Active Directory, already have a PKI in place, can issue certificates from a Microsoft CA, and have installed and registered (if necessary) an IAS server. Your steps would be as follows:

1. Issue computer certificate to IAS server.

2. Issue computer certificate to wireless client.

3. Issue user certificate to wireless client user.

4. Create a Remote Access Policy on the IAS server for 802.1X authentication.

5. Configure RADIUS client settings on the IAS server.

6. Configure AP for 802.1X authentication.

7. Configure wireless client network interface for 802.1X authentication.

Although this might seem like a lot of work, the enhanced security provided by 802.1X might well justify the expense and effort of setting it up. Furthermore, much of the effort is up front. Since you don't have to worry about frequently rotating static WEP keys, you will realize significant savings in effort and time later.

802.1X authentication, in combination with EAP-TLS, is not the final word in wireless security. It mitigates many of the vulnerabilities associated with wireless networks, but other types of attacks might still be possible.

Summary

This chapter focused on the ins and outs of securing a wide range of wireless access points. We covered:

- A Unique SSID

- Disabling the SSID Broadcast

- Enabling Wired Equivalent Privacy (WEP)

- Enabling Wifi Protected Access (WPA)

- Filtering Clients by MAC Address

This chapter also discussed the necessary steps for configuring a workstation to connect to a secured access point, whether it's running a Windows, Linux, or Mac OS X operating system. The last section in the chapter touched on understanding and configuring 802.1X RADIUS Authentication.

Solutions Fast Track

Enabling Security Features on a Linksys WRT54G, a D-Link DI-624 AirPlus Xtreme G, an Apple Airport Extreme, and a Cisco 1100 Series Access Point

These have been consolidated because they are the recommendations for securing any AP/router and are not specific to a particular hardware:

☑ Assigning a unique SSID to your wireless network is the first security measure that you should take. Any attacker with a "default" configuration profile is able to associate with an access point that has a default SSID. Assigning a unique SSID in and of itself doesn't offer much protection, but it is one layer in your wireless defense.

☑ Many attackers use active wireless scanners to discover target wireless networks. Active scanners rely on the access point beacon to locate it. This beacon broadcasts the SSID to any device that requests it. Disabling SSID broadcast makes your access point "invisible" to active scanners. Because your access point can still be discovered by passive wireless scanners, this step should be used in conjunction with other security measures.

☑ Wired Equivalent Privacy (WEP) encryption, at a minimum, should be used on your home wireless network. Although there are tools available that make it possible to crack WEP, the fact that encryption is enabled on the access point may be the difference between an attack on your AP or your neighbor's. Adequate security for these networks is provided by 128-bit WEP.

☑ Enabling Wi-Fi Protected Access (WPA) on your home network is the most secure solution in use today. WPA uses enhanced encryption and dynamically changing keys that make the process of cracking your encryption key more difficult. Only a dictionary attack is possible at this time, so ensure that your passkey/passphrase is robust and not a common dictionary word.

☑ Filtering by Media Access Control (MAC) address allows only wireless cards that you specifically designate to access your wireless network. Again, it is possible to spoof MAC addresses; therefore, you shouldn't rely on MAC address filtering exclusively. It should be part of your overall security posture.

☑ Each of the four security steps presented in this chapter can be defeated. Fortunately, for most home users they do provide adequate security for a wireless network. By enacting a four-layer security posture on your wireless network, you have made it more difficult for an attacker to gain access to your network. Because the likelihood of a strong "return" on the attacker's time investment would be low, he is likely to move on to an easier target. Don't allow your wireless network to be a target of convenience.

Configuring Security Features on Wireless Clients

☑ Windows XP clients are configured using the Wireless Connection Properties and the Windows XP Wireless Client Manager. To associate with your access point once the security features have been enabled, your access point must be added as a Preferred Network. You need to enter the SSID and the WEP key during the configuration process. On the same token, you can also enable WPA during this process, including your passkey/passphrase for connection.

☑ Windows 2000 does not have a built-in wireless client manager like Windows XP. You need to enter the SSID and WEP key into a profile in the client manager software that shipped with your wireless card. Remember that Microsoft does not natively support WPA in Windows 2000. You must obtain client software from your network card vendor in order to use WPA with Windows 2000.

☑ Apple makes wireless connections seem trivial in their 10.x versions of their operating system. By simply adding the SSID and encryption key, in either WEP or WPA mode, you are able to gain access to the network in a small amount of time.

☑ Linux users now have the ability to install and use the Wireless Tools package for their distribution. This package includes the *iwconfig* binary that makes quick configuration of connecting to a WEP encrypted network. WPA can be easily implemented using the *wpa_supplicant* application, with supported wireless network cards and configuration files. There is plenty of information on the Internet for configuring wireless clients to use WEP and WPA in Linux.

Understanding and Configuring 802.1X RADIUS Authentication

☑ RADIUS provides for centralized authentication and accounting.

☑ 802.1X provides for a method of port-based authentication to LAN ports in a switched network environment.

☑ For 802.1X authentication to work on a wireless network, the AP must be able to securely identify traffic from a particular wireless client. This identification is accomplished using authentication keys that are sent to the AP and the wireless client from the RADIUS server.

Frequently Asked Questions

The following Frequently Asked Questions, answered by the authors of this book, are designed to both measure your understanding of the concepts presented in this chapter and to assist you with real-life implementation of these concepts. To have your questions about this chapter answered by the author, browse to **www.syngress.com/solutions** and click on the **"Ask the Author"** form.

Q: Why should I bother using Wired Equivalent Privacy (WEP) encryption if it can be cracked?

A: Due to the low amount of wireless network traffic that is usually generated on a home wireless network, it would take an attacker an extremely long time to capture enough packets to successfully crack the WEP key of your network. An attacker is unlikely to devote the required time and effort to cracking the WEP key on a home network when there are so many other home networks that have no security measures enabled.

Q: Should I use WEP or WPA in securing my wireless network?

A: That decision can be made by analyzing the different hosts on the network and whether they can handle WPA or not. Most older wireless network bridges (Linksys WET11, Microsoft Xbox wireless adapters, and the like) cannot handle WPA encryption, hence forcing you to use WEP encryption. Some older wireless network cards might not be able to handle WPA, and off-brand cards might not have the supported drivers and software to handle WPA.

Q: Aside from setting a default SSID, disabling SSID broadcast, enabling 128-bit WEP, and filtering by MAC address, are there other security measures I can take?

A: Of course there are. Some other easy security measures you can implement are to disable the Dynamic Host Configuration Protocol (DHCP) server on your router, use a nondefault IP address range, do not allow configuration changes to be made from a wireless client, and keep your firmware up-to-date.

Q: I have heard that WPA is vulnerable to dictionary attacks. What does this mean?

A: A dictionary attack tries to guess the preshared key, password, or passphrase in use by testing it against a list, or dictionary, of words and phrases. By using strong passphrases or, in the case of WPA, long preshared keys, you reduce your risk of being vulnerable to a dictionary attack. PV27

Chapter 3

Dangers of Wireless Devices in the Workplace

Solutions in this chapter:

- Intruders Accessing Legitimate Access Points
- Intruders Connecting to Rogue Access Points
- Intruders Connecting to WLAN Cards

☑ Summary

☑ Solutions Fast Track

☑ Frequently Asked Questions

Introduction

Convenience, ease of implementation, and cost are driving factors for the rapid introduction of wireless technologies into the workplace. The introduction and use of any new technologies must be carefully managed and risk must be accurately assessed in order to properly manage the security of any computer system or network. This chapter addresses the major threats wireless technologies introduce into the workplace and presents strategies to properly reduce and mitigate these threats.

Intruders Accessing Legitimate Access Points

At this stage in the game, the decision has been made to implement a wireless network at your organization. Hopefully the design and implementation of the wireless network was carefully thought out and security was a primary consideration, but we know that these factors are not always the reality. Unfortunately, in a rush to implement a new, cool technology, security is often an afterthought.

Since the radio frequency (RF) waves that carry a signal cannot be physically contained within the bounds of a specific office building or other geographic location, a wireless network essentially extends your organization's network as far as the wireless signal can travel. An intruder who takes advantage of this fact can be one of your neighbors in the office park, an employee at another organization on a different floor in your building, or a criminal sitting in a parked car in the parking lot.

The intruder's intent could be classified in one of two ways:

- Use of your network as an Internet "hotspot"—the *opportunist*
- The abuse of your network boundaries to steal data or use your Internet connection to attack others or conduct criminal behavior—the *criminal hacker*

The Opportunist

This type of intruder is typically looking for quick and easy access to the Internet to check e-mail, surf the Web, or connect to Internet Relay Chat (IRC) from a weird host mask to brag to his or her cyber buddies. You can usually prevent the opportunist from succeeding at this task by implementing the most rudimentary forms of security controls, such as:

- Disabling SSID broadcasts on the Wireless Access Points (WAPs)
- Enabling Wired Equivalent Privacy (WEP) or WiFi Protected Access (WPA)

If this intruder encounters encryption via WEP or WPA, he or she will likely move on to a "softer" target. Though this type of intruder might seem relatively harmless, keep in mind that the Internet sites the opportunist visits and the activities he or she conducts on the Web will be traced back to your organization, as though someone on the inside of your network conducted these activities—because for all intents and purposes, he or she *is* on the inside of your network.

How prevalent is this type of intruder? Well, the geek sport of "WarDriving" is fairly common, and large WarDriving events have been organized and successfully carried out, most with the purpose of simply collecting statistics on wireless networks. Websites are devoted to the topic of WarDriving, and some, such as www.wifimaps.com and www.wigle.net, even have searchable databases to locate wireless networks that other WarDrivers have identified. A distinction should be made between the WarDriver and the opportunist, in that the WarDriver merely collects and logs interesting statistics for analysis, whereas the opportunist will identify, then connect to and use, your network.

The Criminal Hacker

The criminal hacker, or *cracker*, has malicious intent, and aims to break into your network to steal your intellectual property or customer data or use your network for criminal conduct. The criminal hacker may be WarDriving—searching for targets of opportunity—or specifically targeting the theft of a particular piece of intellectual property from your organization. The cracker may also use your network to conduct credit card fraud, download child pornography, or any number of other Internet-based crimes. Why would an intruder conduct these types of activities from a network that could be attributed to him when there are large amounts of vulnerable wireless networks he can hide among?

NOTE

> In a rather bizarre criminal case in Canada, police officers stopped a man in his vehicle who was driving the wrong way down a one-way street. The man was found to be naked from the waist down, holding a laptop with a wireless card. To further complicate the situation, images of child pornography were displayed on his laptop screen. Although this is a bizarre case because the man was driving in his car, the reality of the situation is that this type of crime can often be committed from the comfort of the intruder's home or business.

In 2004 two young hackers were arrested for accessing and misusing the wireless network at a large U.S. home improvement store. They hopped onto a poorly configured wireless network, which was not properly segregated from the rest of the corporate network, and modified a program to record credit card numbers. They sat outside the store in the comfort of their car, antennae protruding from the windows, conducting their nefarious activities until the FBI nabbed them.

The first step in your security process should be preventing intruders from accessing the network; the second step is detecting and defeating intruders.

Preventing Intruders from Accessing the Network

The term *hard target* is one that is generally used in a military context. It refers to a target (for our purposes, think wireless network) that has been hardened as much as possible and deters would-be intruders from attacking it because it would be too much work, would be too difficult, or would be too costly in terms of time, money, or resources. If your only goal is to kill or maim two soldiers, would you attack an M-1 Abrams main battle tank or an unarmored Humvee?

Unless your organization is being specifically targeted by intruders, presenting a hard target will repel most wireless intruders.

BEST PRACTICES

To prevent intruders from accessing your network, you should make sure that you rely on the following security measures:

- **Disable SSID broadcasts.** On most modern WAPs, it is possible to disable the broadcasting of the SSID. This feature will prevent the detection of your wireless network from intruders using less sophisticated wireless tools such as Netstumbler. However, more advanced tools such as Kismet will still detect your network by analyzing client traffic. Chapter 2 covers the specifics of disabling SSID broadcasts on popular model WAPs.

- **Use an obscure SSID.** Choosing an obscure SSID that cannot be associated with your organization can aid in preventing a more targeted attack. For example, the fictitious company Dynamic Network Solutions is set up in an office park and has a sign on its building identifying the company. They company's IT staff have set up their wireless network to use the SSID *DYNAMIC*, which makes it trivial to correlate a wireless network to their organization.

- **Enable WEP or WPA.** Although these are not perfect implementations of security, some encryption is better than none at all, as long as your organization fully understands the limitations and flaws. WEP or WPA will more than likely prevent the opportunist from intruding on your network, but each should be considered one layer of the security, rather than a panacea.

- **Conduct MAC address filtering.** It's recommended that only recognized and approved wireless cards (clients) be permitted to associate with the wireless network. Wireless hardware vendors implement these controls in several different ways, but most modern access points support this feature. Additionally, the DHCP server should assign an IP address to approved wireless clients only. Again, don't rely on this one configuration item by itself to prevent attacks, because MAC address spoofing is a common and very possible attack method.

- **Control RF signal strength.** Some more advanced WAPs that are designed for enterprise use rather than home networks allow the control of the wireless signal strength. Even though it may take a fair bit of work to determine the signal boundaries, preventing the signal from reaching outside the bounds of your building or office can be a key configuration item that will significantly enhance the security of your wireless network.

- **Implement a wireless DMZ.** A properly implemented wireless DMZ will allow a wireless client the ability to connect to the network, but

Continued

only to specific devices such as a virtual private network (VPN) concentrator or a Web proxy server. This layered approach requires one more factor (for example, VPN credentials or a proxy account) from wireless users before from permitting network communications. Any intruder that is successful in cracking the WEP or WPA key would be severely restricted in his or her ability to communicate with other network hosts.

■ **Implement a wireless intrusion detection system (WIDS).** Several vendors now offer a wide array of wireless IDSes with varying feature sets. Ensure that the WIDS selected has the ability to detect rogue wireless network users along with the detection of rogue wireless access points.

NOTE

While the general consensus among security professionals seems to be that a WIDS is useful, in many circumstances they may be cost prohibitive. This is true with both commercial offerings and homegrown implementations. Depending on the size of the WIDS deployment, the costs of a commercial WIDS can quickly spiral out of control. The costs associated with a homegrown solution can appear to be cheaper, but one must consider the resources necessary to design, implement, and maintain the system. With either the commercial or the homegrown approach, the costs of monitoring and responding to this additional security tool must be factored into the equation. For this reason, many organizations choose to implement a hybrid form of WIDS. This can be a combination of WAP security logs, DHCP server logs, and other traditional IDS tools.

Case Study: Intruder's Introduction of a Wireless Sniffer/Cracker

A colleague of mine was given the task of breaking into a wireless network at a rather sensitive and secure facility as part of an authorized penetration test activity. Because the organization had its own private security force, the tester didn't have the option of sitting in a car for hours on end with various antennae pointing out the window. Instead, he designed and implemented what he calls a "wireless drop box."

The purpose of this drop box was to covertly sniff wireless network traffic in order to crack the WEP key. This box, which looked like any standard gray utility box, contained a single board PC, a wireless network card, a 5 gigabyte hard drive, a high-gain patch antenna, and a battery large enough to run the drop box for two days. The front of box contained a Plexiglas cutout that allowed the patch antenna to point directly at the target network. The Plexiglas was spray-painted gray to match the box, and the organization's logo was affixed to the front of the box, along with a telephone number in case of questions (see Figure 3.1).

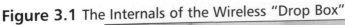

Figure 3.1 The Internals of the Wireless "Drop Box"

Visible in Figure 3.1 is the patch antenna attached to the lid with a Soekris single-board PC between the antenna and the large battery.

The box was quickly affixed to the utility pole using metal banding (see Figure 3.2) and was left to do its job of sniffing wireless network traffic. With the matching colors, the organization's logo (which has been removed from the picture) and a telephone number the box looked legitimate and went completely unnoticed. Two days later the box was removed from the utility pole and taken offsite for analysis and WEP key cracking. Within the two-day timeframe, the sniffer had collected enough interesting packets to successfully crack the encryption, and the project was a success. This feat was accomplished passively, with the tester never having to conduct an active attack such as packet injection that could have been detected by the target organization's WIDS.

Figure 3.2 The Drop Box Installed

Intruders Connecting to Rogue Wireless Access Points

Rogue WAPs present a very real threat to the security of your organization's network, whether you have an approved wireless network or not. Perhaps you've chosen not to allow wireless networking within your organization and think that due to this stance, you are safe from the technology's vulnerabilities. Or perhaps you've implemented a wireless network that follows the most stringent security recommendations possible. You've implemented all the basics and have gone one step further to set up a wireless DMZ that only allows IPSec traffic to your VPN concentrator. Unfortunately, all it takes is one uneducated employee plugging a WAP into his or her network jack to undermine all your organization's security efforts; in such an instance, your $100,000 firewall and your sophisticated IDS will no longer adequately protect your network from wireless attackers.

BEST PRACTICES

To make sure your network is free of rogue WAPs, you should be implementing the following security measures in your organization:

- **Implement clear organizational policy.** It's imperative that policy clearly dictate and define all acceptable user behavior, including behavior such as a user setting up an unauthorized wireless network.

- **Conduct user awareness training.** A continual user awareness training program is a key aspect of a complete IT security program. Users must be aware that their actions can affect the security of the overall organization. Would a user connect a WAP if she knew she was effectively circumventing the network's security? Risk accepted by one is imposed on all.

- **Control the procurement process.** Depending on how employees purchase equipment for the company, it might be possible to add a step where a member of the IT security department reviews the purchase for security concerns. If a procurement request or expense report is submitted for wireless equipment, this would raise a red flag. Of course, wireless equipment is so affordable these days that an employee may bring in equipment from home or purchase it himself.

- **Conduct periodic wireless assessments.** The only way to detect rogue access points is to implement a program to periodically conduct a wireless assessment. For an organization that doesn't have an authorized wireless network, this can be as simple as a wireless discovery using tools such as Netstumbler and Kismet. The assessment will be more complex where an authorized wireless network exists, but because of the risk factor associated with a rogue network, it's an activity that must be done.

- **Scan your network from the wired side.** Because it is possible to correlate MAC addresses to specific vendors, it could be possible and even simple to identify WAPs from the wired side. Let's say that your organization has standardized on a certain brand of Ethernet card, for example, 3Com 3509. Using the port scanner Nmap, we could identify non-3Com Ethernet cards and investigate. Of course, not everything on your network would have a 3Com card, but after some tweaking it should be relatively easy to spot wireless vendor's products such as Dlink or Linksys. A software program aptly named *wap-nmap* was developed to automate this type of wired-side scanning.

Chapter 4 of this book discusses rogue access point detection and mitigation in depth using a Cisco wireless network.

Case Study: Employees Using Accessible Wireless Networks to Circumvent Controls

So your organization has spent thousands of dollars on content-filtering software to restrict access to Internet sites and has spent countless hours managing and implementing outbound firewall rules to limit network communication with the outside world, yet the company next door in the office park has a poorly configured wireless network that allows your employees to surf the Internet and connect to any computer they chose—all while *still* connected to your network, thereby rendering your information security efforts pretty much useless. This realistic threat exists just about everywhere except in organizations that are geographically isolated or that exist inside an RF shielded facility.

Using an insecure wireless network, an employee can:

- **Circumvent content filters.** You may have implemented Web and e-mail content filters to prevent your employees from surfing certain Websites and to protect your organization from malicious e-mails. However, if one of your network users wants to browse restricted Websites, he can simply route through the wireless network.

- **Bypass egress filtering.** Your firewall may contain outbound rules that restrict communications with the outside world in total or in some restricted form. But if a user wants to SSH to his home system on a cable modem, he can simply route out through the wireless network rather than your corporate network that prevents outbound TCP port 22 traffic.

- **Bridge your network.** A malicious insider could completely expose your internal network by bridging the wireless and wired interfaces together.

Since it's usually not possible to prevent RF signals from entering your building, a different approach is necessary to ensure the security of your systems.

If wireless networking is not being used in your organization, the obvious choice would be to make sure that no system is capable of communicating via this medium. It's fairly easy to manage desktop systems, but laptops and mobile devices require more thought and planning. Perhaps the easiest method is to ensure that these systems have no built in wireless capabilities when they are procured. Controlling the procurement process can also be a useful tool in ensuring that wireless cards and devices are not purchased and implemented. For example, your organization could have a purchase request form that an employee is required to submit before purchasing equipment. This request could be screened by the IT department to ensure that the piece of equipment being procured is not capable of communication via 802.11 or other wireless mediums.

Knowing about other networks that are within range is the first step in protecting your organization from this threat. Conducting a wireless assessment of your own to identify these networks is a "must do" item. Your management might not want to invest the time and money in obtaining both wireless hardware and knowledge necessary to conduct a wireless assessment, especially if your organization doesn't implement wireless networking technologies. As a security-minded professional, you now have the task of educating your company's managers of the risks that these insecure networks pose to your organization and to propose solutions to mitigate these risks.

The Best Practices in the preceding section concerning rogue access points all help reduce the risk of this threat, but they do not completely eliminate it.

Shortcuts...

Quick and Easy Wireless Network Detection

Despite what you may have heard, you don't have to be a Linux guru to conduct a thorough WarDrive. The time it might take a Windows administrator to install and learn Linux, configure wireless networking, and install the latest wireless tools could be overwhelming. Luckily, other options are available. These range from using a bootable live CD such as BackTrack to purchasing commercial wireless security tools such as Airopeek. In addition, handheld devices, such as the YellowJacket, are available specifically for this type of activity.

Intruders Connecting to WLAN Cards

In many areas of information security, laptops greatly increase the threats to corporate security, and this is no exception in the wireless realm. With many vendors including built-in wireless chipsets as a standard option and with the widespread use of affordable wireless PCMCIA cards, the number of laptops vulnerable to such an attack is rapidly growing.

One tack an attacker might take is to attempt a peer-to-peer connection. In this scenario, an attacker probes or passively listens for wireless clients that are searching for their network. The attacker changes his network configuration to match the victim's laptop, and a connection is possible. The attacker then analyzes the network traffic to determine the IP address in use, and once again, configures his client to match the victims. Once a network connection is established, the attacker can then carry out any number of attacks to compromise the victim's laptop.

A second vector an attacker might use is to employ a tool such as Kismet to determine who (in other words, which laptops) are searching for their "home" network and then configure a fake WAP to match the request. The attacker then configures a DHCP server and assigns the victim computer an IP address, and network communications are possible, opening up the victim computer to any number of attacks.

Imagine an attacker camped out at a busy airport, providing wireless network access to your corporate users who very well might not even know their computer has connected to a wireless network. This could happen in the airport, or even at 35,000 feet! Without the protection of the corporate firewall, these laptops may be vulnerable to compromise.

Laptops are not the only mobile systems vulnerable to this type of attack. A growing number of handheld computers and mobile devices, including cellular telephones, are now shipping with built-in 802.11 capabilities. Those that don't include built-in wireless capabilities often have the capability of adding 802.11 to the device. These devices are vulnerable to the same types of issues as laptops except for the fact that most of these devices have less secure default wireless options, to keep things easy for the user. Though the attack might be complex, it is not inconceivable to imagine an attack in which an intruder connects via 802.11 to an insecure personal digital assistant (PDA) that is in turn connected to a desktop PC, which is connected to a corporate network. Could an attacker then essentially connect to the organization's network?

BEST PRACTICES

To prevent intruders from gaining access to your network via WLAN cards in laptops and other mobile devices, you need to implement the following security measures in your organization:

- **Implement clear organizational policy.** It's imperative that corporate policy clearly dictate and define all acceptable user behavior, including behavior such as a user adding wireless capabilities to laptops or other mobile devices.

Continued

- **Conduct user awareness training.** A continual user awareness training program is a key aspect of a complete IT security program. Users must be aware that their actions can affect the security of the overall organization. Hopefully, users would think twice about adding wireless capabilities to their systems, and when/if notification of a connection attempt occurred, they would know better than to form a connection. Educate users on the dangers of connecting to all untrusted networks, including wireless networks as well as other networks they may come across in their travels.

- **Utilize a host-based firewall.** Implement a host-based firewall solution on each laptop to prevent network communications if wireless communications are established by an attacker. Additionally, ensure that laptops are up to date with patches and are utilizing antivirus software.

- **Restrict privileges.** Mobile users should not have administrator privileges on their laptop computers. This level of privileges would allow them to override technical restrictions and install the software necessary to operate wireless cards.

- **Manage procurement.** Manage wireless hardware at the earliest stages possible. If wireless is prohibited via organizational policy, ensure that new mobile devices do not have built-in wireless cards.

- **Disable wireless networking.** Where possible, disable the built-in wireless card in the laptop's BIOS. As a backup control, disable the wireless card in the operating system. In Windows, disable the device using the Device Manager.

- **Enforce wireless network policies.** If using Windows Server 2003 and Active Directory, import the Wireless Network Policies extension to disable ad hoc access. Using this policy, you can restrict wireless networking to infrastructure mode only, thereby disabling peer-to-peer wireless networking. You can also essentially disable wireless networking by configuring only a specific type of configuration. For example, configuring Wireless Network Policies to use 802.1x authentication would require this type of configuration. A user attempting to connect to a wireless network not using 802.1x would be prevented from doing so.

NOTE

Laptops and mobile devices are vulnerable to a number of attack vectors that can contribute to the compromise of your organization's network and data. In an ideal InfoSec world, laptops would not exist. However, this is not reality for the growing mobile workforce. Therefore, it's important to treat laptops differently than computers that never leave the protective bounds of your organization's network. Be sure that laptops are kept up to date with software patches and are configured properly, and conduct vulnerability and antivirus scanning on laptops more frequently and once they return to the home network.

Shortcuts...

Install a Host-Based Firewall

One of the best methods to limit communications with your mobile devices is to simply install a host-based firewall product such as Norton Internet Security. In this case, even if an intruder is successful in forming a wireless connection he will still not be able to communicate with the laptop. With no network-level communications to the target computer the wireless connection becomes useless.

Summary

If we could identify a common theme throughout the best practices listed in this chapter, it would be that in order to have a complete and secure IT security program (which encompasses the wireless realm), two key components must be present:

- Sound organizational policy
- User awareness training

First we need an enforceable, achievable security policy that is clear in defining acceptable and unacceptable behavior in regard to computer usage. Second, a user education program will teach users about both the organizational policy and the latest threats regarding computer security. User education and training is a continual process that must be conducted on a periodic basis. Users should sign an agreement stating that they understand and accept responsibility for their actions and that they will not intentionally violate the policies.

Of course, no IT security program would be complete without technical controls to enforce the policy, so implementing the technical best practices listed in this and other chapters will help secure your organization's network.

Solutions Fast Track

Intruders Accessing Legitimate Access Points

- ☑ Disable SSID broadcasts
- ☑ Use an obscure SSID
- ☑ Enable encryption
- ☑ Filter MAC addresses
- ☑ Control RF signal strength
- ☑ Implement a wireless DMZ
- ☑ Implement wireless IDS

Intruders Connecting to Rogue Access Points

- ☑ Implement clear organizational policy
- ☑ Conduct user awareness training
- ☑ Control the procurement process
- ☑ Conduct periodic wireless assessments
- ☑ Scan your network from the wired side

Intruders Connecting to WLAN Cards

- ☑ Implement clear organizational policy
- ☑ Conduct user awareness training
- ☑ Utilize a host-based firewall
- ☑ Restrict administrator privileges
- ☑ Manage procurement
- ☑ Disable wireless networking
- ☑ Enforce wireless network policies

Frequently Asked Questions

The following Frequently Asked Questions, answered by the authors of this book, are designed to both measure your understanding of the concepts presented in this chapter and to assist you with real-life implementation of these concepts. To have your questions about this chapter answered by the author, browse to **www.syngress.com/solutions** and click on the **"Ask the Author"** form.

Q: If I enable WEP or WPA, won't this be enough to protect my wireless network?

A: No. Although it's a good start and should usually be implemented, wireless encryption is flawed and can be cracked using cracking tools commonly available on the Internet. No single action outlined in this chapter should be seen as a complete security solution. The best type of approach to security is a layered one—one that implements many different levels and types of protection tools.

Q: Implementing a wireless DMZ with a VPN is too expensive. Are cheaper solutions available?

A: Yes. If an enterprise VPN concentrator is out of reach and you still want to lock down your wireless network, you can restrict all wireless network traffic to a bastion host or two. Using a firewall, you can implement rules so that the only traffic permitted to pass is to a bastion host. Perhaps your bastion host is running only SSH or Remote Desktop.

Q: Why bother disabling SSID broadcasts if Kismet and other intelligent wireless hacking tools can still determine the SSID?

A: This step is one in a series of steps to protect your wireless network. Remember, it will stop potential intruders using less sophisticated tools such as Netstumbler.

Q: Controlling the procurement process in my organization is not a possible solution. Employees are free to purchase and expense what they like, with minimal controls.

A: This is probably the case in many organizations outside large enterprises. In this case, you will need to take a more active approach to find both rogue access points and rogue wireless cards.

Q: All my users have Administrator privileges on their PCs so they can install soft-ware and do routine tasks. How can I take this privelege away from them without causing too many problems?

A: Though each organization is different, in the vast majority of organizations I have audited, almost none of the users actually need Administrator-level privi-leges to go about their daily business. Taking away privileges is always a touchy subject but must be done for proper configuration management and control of systems.

Q: Will a host-based firewall really protect my mobile users?

A: Yes. If configured properly, a host-based firewall will prevent communications at the network layer, so it will stop an intruder from attempting to exploit a poorly configured or unpatched computer.

WLAN Rogue Access Point Detection and Mitigation

Solutions in this chapter:

- **The Problem of Rogue Access Points**
- **Preventing and Detecting Rogue APs**
- **IEEE 802.1x Port-based Security to Prevent Rogue APs**
- **Using Catalyst Switch Filters to Limit MAC Addresses per Port**

☑ Summary
☑ Solutions Fast Track
☑ Frequently Asked Questions

Introduction

This chapter discusses what may be the single greatest problem of wireless local area networks (WLANs): rogue access points and unauthorized people using otherwise legitimate access points. This chapter covers wireless-aware product features that address both of these problems, as well as how to set up and use them.

This chapter also we will take a closer look and discusses how to mitigate the threat of rogue access points that pose significant security threats to businesses and their networks.

Employees install wireless devices in their offices and cubicles for their own personal use because they are convenient and inexpensive. Installing access points is as easy as plugging into an Ethernet jack. Unauthorized wireless devices can expose protected corporate networks to attackers, allowing for a security breach. In this chapter, you will learn how personal access points can introduce such threats to your networks and how you can mitigate the threat of rogue access points by using both wireless- and wired-aware devices and their techniques.

You will study traditional techniques such as manual sniffing, physical detection, and wired detection to detect rogue access points, and will also use Cisco's new centralized solutions for detecting rogue access points. In a Cisco-aware infrastructure network, all wireless devices can work hand-in-hand to detect and report unauthorized access points to the central managing station. (Chapter 12 of this book details how to conduct a complete wireless penetration test using the Auditor Security Collection.)

The Problem with Rogue Access Points

A rogue access point is an unauthorized access point. Unauthorized access points can pose a significant threat by creating a back door into sensitive corporate networks. A back door allows access into a protected network by avoiding all front door access security measures. As discussed in previous chapters, wireless signals travel through the air and, in most cases, have no boundaries. They can travel through walls or windows, reaching long distances far outside of a corporate building parameter. Figure 4.1 shows a wireless signal from access points beaming through the air outside of a corporate building into the parking lot and nearby buildings across the street. These radio signal frequencies may represent both rogue and valid access points that carry sensitive confidential data from inside the corporation or from outside mobile workers. The difference between the radio frequencies from these two wireless access points is that the rogue unauthorized access point was installed by an employee with limited security protection, often leaving it at its default plug-and-play unsecured configuration, while the authorized access point was installed by a skilled engineer with full security support. Further, unlike authorized access points that are configured to protect radio signals confidentially with a robust authentication process, the rogue access point installed by the employee probably does not support such security options, as it does not have access to interact with third-party security servers to provide such services.

The bottom line is that rogue access points installed by employees pose a significant threat because they provide poor security measures while extending a corporate

Figure 4.1 Wireless Reachability

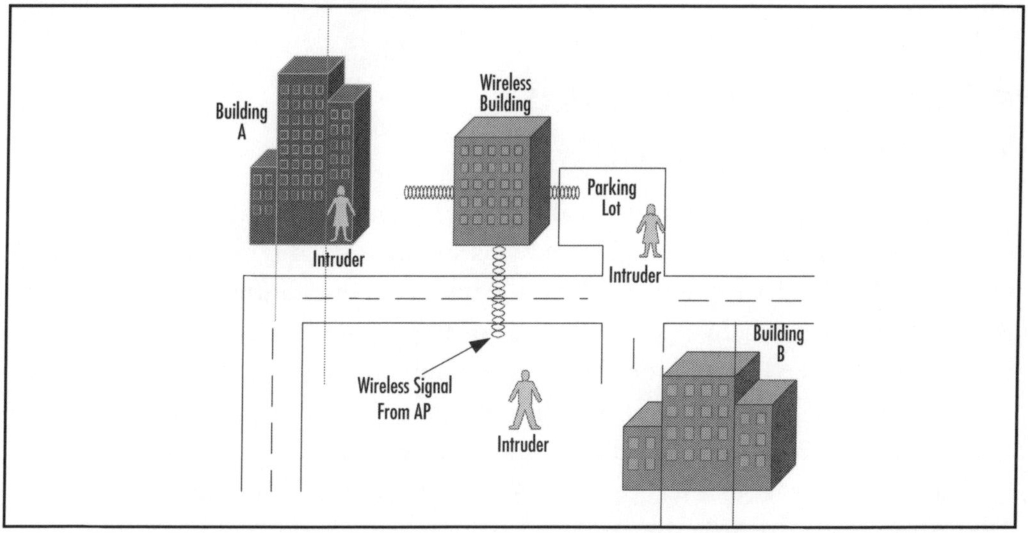

network's reachability to attackers from the outside.

Employees usually install unauthorized access points because of poor performance of current wireless infrastructure, because they may be located in a dead spot, or simply because their company does not provide wireless access. It is important to note that a rogue access point is most likely to be installed in an organization that does not support wireless networks for its employees.

NOTE

Audits to detect rogue wireless access points are required in all corporate network environments, even if they do not provide wireless access.

Unauthorized installed access points are unsecured. An average employee is not an expert on wireless security and does not realize the threat they pose with their newly installed rogue access point. Most rogue access points implement a plug-and-play feature allowing for minimal configuration by the user in the order of their use. Security settings are turned off by default, and default passwords are used that need to be reconfigured to prevent from intruders.

As covered in Chapter 2, the best security is implemented using 802.1x protocol features or virtual private networks (VPNs). Both of these security solutions require a third-party device that employees would not have access to; thus, rogue access points are not secure and can be easily attacked to gain access into the connected corporate network.

A Rogue Access Point is Your Weakest Security Link

A network is only as secure as its weakest security link. For example, consider that you have implemented a very stable and secure wireless and wired network. Your secure wireless local area network (LAN) includes per-user authentication using an 802.1x protocol, a dynamic Wired Equivalent Privacy (WEP) protocol key assignment with periodic key rotation for confidentiality, and logging for audit purposes.

Now consider that all of the time and money spent providing a secure wireless access can be diminished by a single rogue access point. Figure 4.2 represents a wireless DMZ in a secure wireless network topology. In order for valid User A to gain access onto the protected corporate network, they must go through the proper authentication process, pass the firewall and Intrusion Detection System (IDS), and use encryption. Unlike User A, User B does not need to go through any security measures in order to gain access to the corporate network. User B is simply taking advantage of a rogue access point that was most likely installed with a weak security policy and default settings.

This example represents a back door into a corporation that can be used by the employee who installed the rogue access point and by an intruder that may take advantage of the poorly secured rogue access point.

Figure 4.2 Bypassing Security with a Rogue Access Point

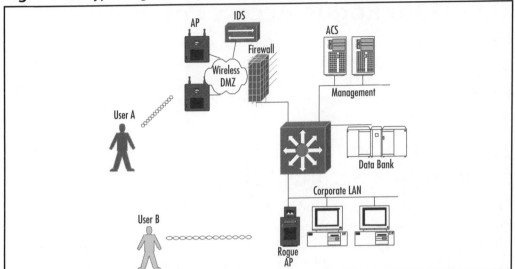

An Intruder's Rogue Access Point

An intruder can also install a rogue access point into a corporation. The difference between an intruder's access point and an employee's access point is that the intruder's is not connected to the wired network. How does this make it an unauthorized access point? It is still an unauthorized access point within the radio signal strength area that is used as the trap device to catch valid users. When a valid user tries to connect to an intruder's access point, the intruder's access point can trick the user into providing useful information such as the authentication type and credentials of the user, which can then be recorded and used later by the attacker to gain access to a valid access point.

One way to mitigate an intruder's rogue access point is to provide for dual authentication. In dual authentication, the user needs to authenticate the access point and the access point has to authenticate the user. Dual authentication is supported in the 802.1x protocol. Dual authentication allows the user to verify the validity of the access point before its use. The details of the 802.1x protocol are covered in Chapter 2.

Preventing and Detecting Rogue Access Points

Many techniques exist to prevent and detect rogue access points. Detecting rogue access points should be performed on every network audit to avoid possible back door exposure. As mentioned earlier, your security is only as strong as your weakest link. Do not let one rogue access point dismiss your entire security-configured infrastructure.

Preventing Rogue Access Points with a Security Policy

First and foremost, your security policy must include the use of wireless networks and prohibit the use of personal rogue access points. A security policy does not eliminate the threat of rogue access points, but it does set guidelines for current and future network installations and what steps to take if a rogue access point is detected. A security policy should mandate that all employees follow proper security measures for wireless networks and should also require written approval from the Information Technology (IT) and Security teams approving the installation of a personal access point. It is important that all employees know that freelance access points are prohibited, why they are prohibited, and what will happen if they break the rule. The risks are such that some companies will fire individuals for setting up their own access points.

For a security policy to be successful, it needs to be communicated to the users. If users are not aware of these security rules, they will not follow them. Continuous education and audits of the security policy are a must.

Provide a Secure, Available Wireless Network

Most rogue access points are installed by non-malicious employees who simply want wireless access in their work area. One way to prevent employees from installing such rogue access points is to provide wireless access to them. Installing stable wireless access throughout meeting rooms, the cafeteria, and the outdoor campus, allows you to control its access and security implementation. Doing so does not mean you can stop auditing and searching for rogue access points within your network, but it will decrease their detection count and improve overall security.

Sniffing Radio Frequency to Detect and Locate Rogue Access Points

Another technique for detecting rogue access points is to manually use a network sniffer to sniff the radio frequency within your organization's perimeter. A wireless sniffer allows you to capture all communication traveling through the air, which can then be used for later analysis such as Media Access Control (MAC) address comparison. Every wireless device has its own unique MAC address. If a new, unknown MAC address of an access point is detected in a wireless sniffer trace, it will be red flagged as a rogue access point and investigated further.

Designing & Planning…

Finding MAC Addresses

Every manufacturer programs a unique MAC address into their network card. Every network card has its own MAC address that it uses to communicate with. A MAC address is 48 bits long. The Institute of Electrical and Electronic Engineers (IEEE) controls the first 24 bits (3 octets) of the address. These first 3 octets are called the Organizationally Unique Identifier (OUI). OUIs are given to corporations that produce network devices such as network cards. These corporations must use the unique first 3 octets assigned to them in all of their network devices. The second 24 bits of the 48-bit long MAC address are controlled by the manufacturer. If the manufacturer runs out of unique addresses for the second half of the MAC address, it requests a new 3-octet address from the OUI.

If you detect a MAC address and want to look up its manufacturer, refer to the OUI database Web site at http://standards.ieee.org/regauth/oui/index.shtml

Knowing that every network device has a unique MAC address, you can find out a lot of useful specific information about each device. In Figure 4.3, MAC address 000CCE211918 has been detected. Entering 000CCE (the first half) into the OUI online database reveals that the device detected is a Cisco device.

Tools such as NetStumbler can be used as rogue access point detection sniffers. It displays a list of detected access points within the area of signal strength that can be compared to a friendly database of access points. NetStumbler can further be used to zero in on a physical rogue access point and its location by measuring the signal strength. Figure 4.3 shows a detected access point with MAC address

000CCE211918. After checking the list of friendly access points, we have determined that this detected MAC address does not match any of the authorized access points and thus is a possible rogue access point. To locate this rogue access point, we begin searching by walking around with a laptop and the NetStumbler utility following the signal strength. Notice that the signal strength increases as we close in on the physical location of the detected access point.

Figure 4.3 NetStumbler: Finding a Rogue Access Point with Signal Strength

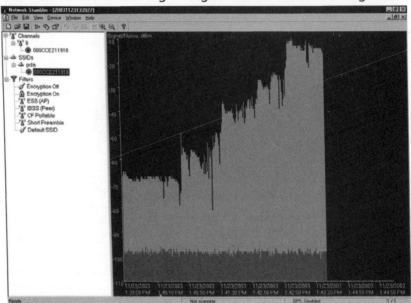

Tools such as Cisco's Aironet Client Utility (ACU) can also be used to follow the strength of a radio signal in order to find a detected rogue access point's physical location. The ACU is installed with Cisco's Aironet wireless adapter. Figure 4.4 shows the Link Status Meter tool in the ACU that displays the signal strength for MAC address 000CE211918, which was determined to be a rogue access point in the previous example. Another useful tracking tool within Cisco's ACU application is the Site Survey tool, as shown in Figure 4.5. Again, using the Site Survey tool, the closer you move to the physical location of a detected access point the higher the signal strength will be.

Figure 4.4 ACU: Link Status Meter

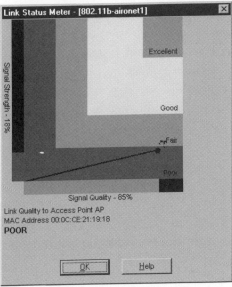

Figure 4.5 ACU: Site Survey

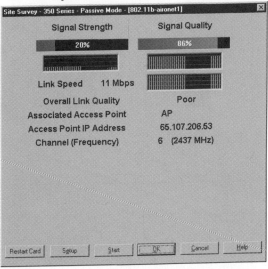

Cisco's Rogue Access Point Detection

Detecting rogue access points with a sniffer device can be a time-consuming and almost impossible task in large-scale wireless and wired environments. The administrator must walk throughout the entire area and manually compare friendly detected access points with possible rogue access points. This task must be repeated almost daily to assure security against rogue access points.

Cisco has developed a more robust solution to overcoming the manual work effort of sniffing for rogue access points. Instead of walking around with a laptop and antenna to detect possible rogue access points, Cisco's solution allows you to turn all of the wireless clients and access points into an army of sniffers that continually analyze and monitor the radio frequencies around them (see Figure 4.6). This allows you to perform 24 hours a day/7 days per week automatic detection of rogue access points throughout all locations where authorized wireless clients and access points are located. Rogue access points detected by wireless clients and access points are then sent to the central management station where the network administrator is alerted.

Figure 4.6 All Cisco-aware Devices Become Sniffers

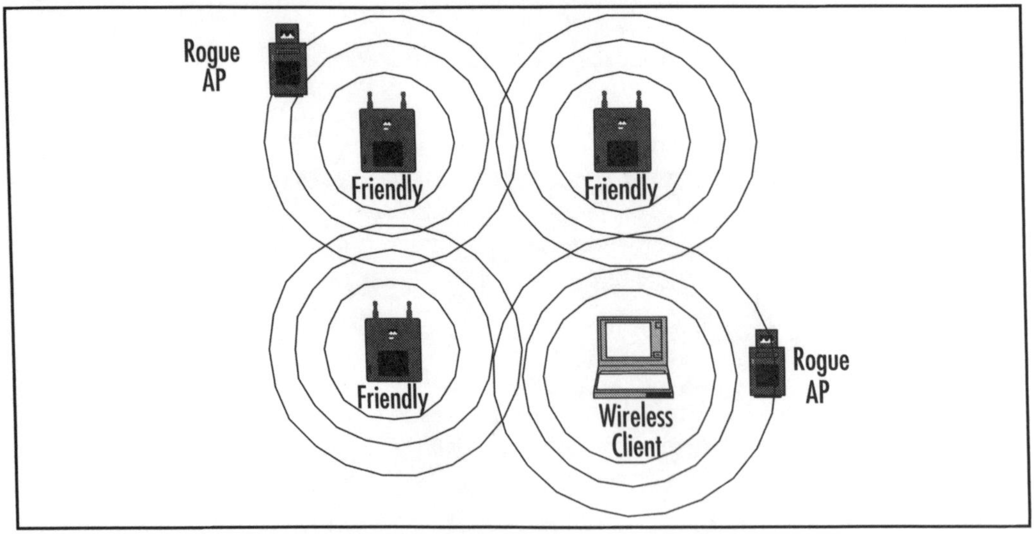

Central Management with WLSE to Detect Rogue Access Points

The Wireless LAN Solution Engine (WLSE) is a CiscoWorks application that provides central management for all Cisco-aware wireless devices. WLSE can be used to receive rogue access point-detected information from wireless clients and access points through Simple Network Management Protocol (SNMP). When a wireless client detects a possible rogue access point, it sends the information to a friendly

access point, which then sends it to WLSE engine via SNMP-trap
protocol to inform the management server of its findings (see Figure 4.7). WLSE
receives this information and compares it against a database of friendly access points.
If the WLSE cannot find the reported access point on its friendly list of valid access
points, it red flags it and alerts management that a possible rogue access point has
been detected.

A WLSE centralized solution is welcomed by administrators in large- and mid-
sized Cisco wireless-aware environments, as it provides scalability and central management and greatly improves the overall security against rogue access points, with
its automated process.

Figure 4.7 Rogue Access Point Detection by Client

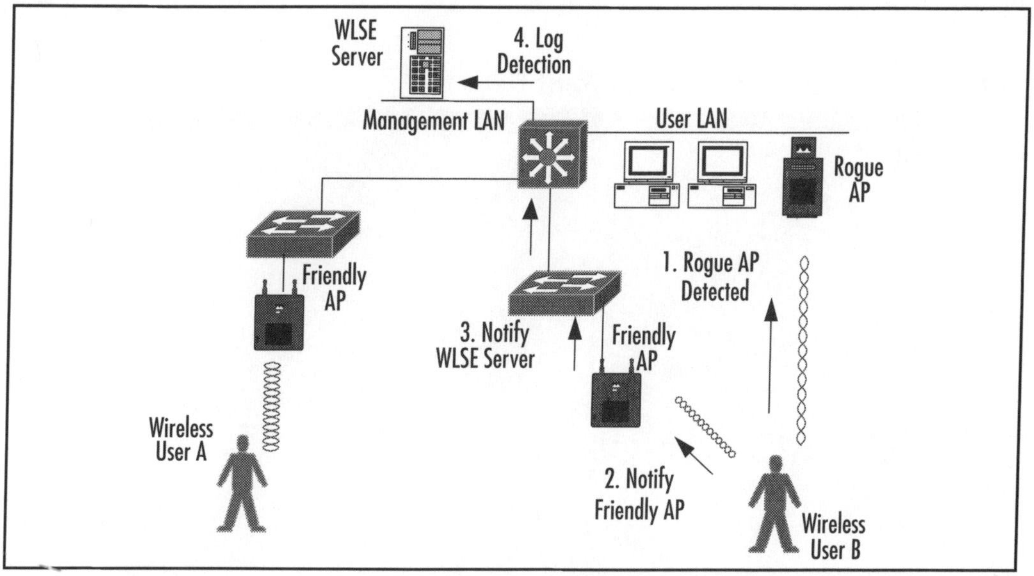

The WLSE can also use triangulation to calculate the physical location of rogue
access points, by using the signal strength of multiple wireless clients and access
points at the time of detection. This allows you to not only detect rogue access
points, but also to know its approximate physical location. WLSE is also capable of
providing the switch IP and port details into which the rogue access point is physically connected to, allowing you to quickly locate and disable the rogue access point
to eliminate its security threat.

Figure 4.8 shows a rogue access point detection alert from the WLSE that
reports that an unauthorized access point has been detected by four friendly access
points. Further information shows that the detected rogue access point is broad-

casting "ROGUE" SSID in its beacons. The Received Signal Strength Indicator (RSSI) next to each reporting access point represents the signal strength relationship between the rogue and the friendly access point, and is used to estimate the approximate physical location of the detected rogue access point.

One WLSE feature allows you to import and configure your floor blueprints, which can be used to provide a visual of the wireless clients and access points within the network wireless area. In Figure 4.9, a floor map is used along with RSSI information from friendly access points to visualize the location of a detected rogue access point. As you can see, the visual map shows four friendly access points reporting the detected rogue access points and their estimated physical location. Such automatic and detailed support from WLSE allows you to quickly find and terminate rogue access points.

Figure 4.8 WLSE Rogue Access Point Detected

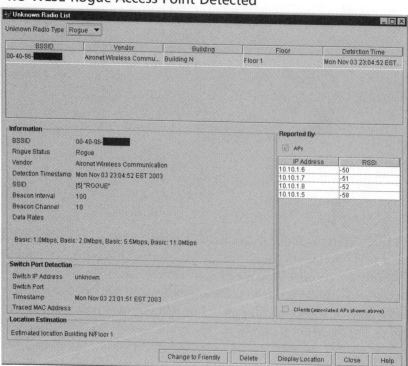

Figure 4.9 WLSE Rogue Access Point Location Map

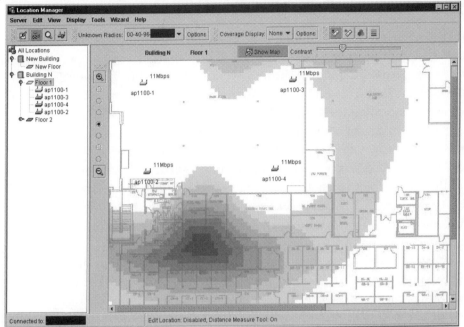

IEEE 802.1x Port-based Security to Prevent Rogue Access Points

This section reviews IEEE 802.1x protocol, its use in wireless and wired LANs, and how it can aid in mitigating the threat of rogue access points. For further details on the 802.1x protocol and its implementation in a wireless environment, refer to Chapter 2.

As discussed earlier, there are two different types of rogue access points: one that is installed by an employee with a physical connection to the corporate LAN or one that is installed by an intruder without any physical connection to the wired LAN. An intruder's rogue access point is used to trick valid users into establishing a connection in order to obtain confidential information. A valid user needs a method of validating an access point just as the access point needs a method that validates the user, to prevent connection to a rogue access point.

Prevent Users from Using Rogue Access Points with 802.1x

In a wireless environment, the 802.1x protocol provides mutual authentication that can be used to mitigate the threat of valid wireless users establishing a connection to rogue access points. Figure 4.10 shows a typical 802.1x Light Extendable Authentication Protocol (LEAP) dual authentication process, where the wireless client is authenticating the RADIUS Access Control Server (ACS) server at the same time that the server authenticates the client prior to establishing a successful connection. Both challenges are derived from the user's password that only the user and a valid RADIUS ACS server have, thus providing a successful challenge response.

If the access point in Figure 4.10 were a rogue access point, it would not have access to the RADIUS ACS server because it would have failed the user's authentication challenge and in turn the user would refuse to establish connection to the access point (see Figure 4.11).

Each authorized access point must be manually configured in the RADIUS ACS server in order to access the server for authentication purposes. Therefore, unauthorized devices such as the rogue access point in Figure 4.11 would not be allowed to query or use RADIUS ACS services because it was never added to the allowed list by the administrator.

Figure 4.10 802.1x Mutual Authentication

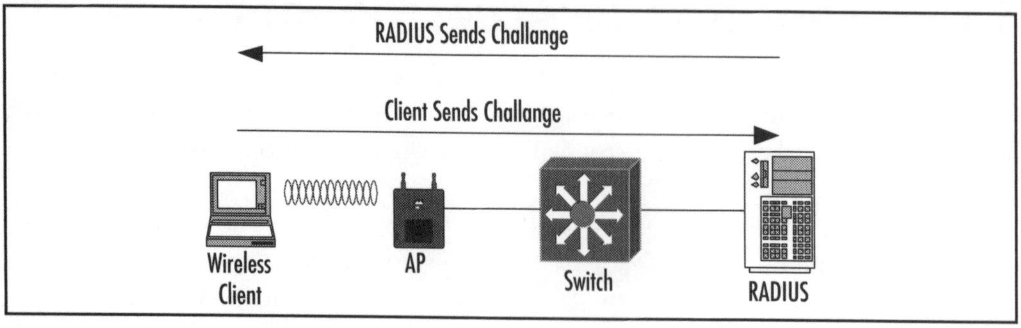

Figure 4.11 802.1x Failed Mutual Authentication

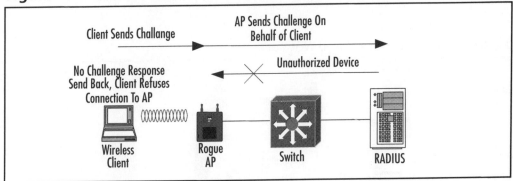

Mutual authentication is not supported in all 802.1x implementations or the Extensible Authentication Protocol (EAP). One of the supported methods of mutual authentication in EAP is Light Extensible Authentication Protocol (LEAP) and EAP-Transport Layer Security (EAP-TLS). In LEAP, authentication and challenges are derived from usernames and passwords. EAP-TLS is nearly identical to the LEAP process, but instead of using usernames and passwords it uses digital certificates. Refer back to Chapter 2 for a more in-depth review on both of these EAP types and their configurations.

Preventing Rogue Access Point from Connecting to Wired Network with 802.1x

Now that you know how to detect and track down rogue access points and avoid using them, you must learn how to prevent them from connecting to a wired LAN in the first place. The 802.1x protocol was originally designed to control access and restrict connection to physical wired ports. This newly developed protocol allows you to authenticate a device or user prior to using a physical port on a switch. Figure 4.12 shows three workstations that are able to communicate on the wired network, and a rogue access point that is not. As soon as one of the workstations is connected to the physical port, the switch sends an authentication challenge based on a username and password from the RADIUS server that the owner of the workstation must pass in order to successfully connect to the local LAN. When a rogue access point is connected to a physical port other than a workstation, it is unable to process a challenge request from the switch and thus will not be permitted to connect to the wired LAN. This is a great step towards security that allows you to authenticate a device or users before they are allowed to connect to a physical port. This mitigates the threat of unauthorized devices and users such as rogue access points from physically connecting into the LAN and possibly creating back doors into corporate networks.

Figure 4.12 802.1x in Wired Network

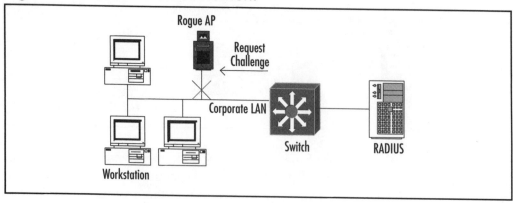

Understanding Devices and their Roles in Wired 802.1x Implementation

Each device in 802.1x plays a specific role. Figure 4.13 includes the following three main devices:

- Client (workstation)
- Switch
- Authentication Server

The client (workstation) requests access to the LAN by sending a request to the switch. The switch can also be configured such that it automatically requests an authentication challenge from a newly connected device without waiting for a request. The client must be compatible and support the 802.1x authentication process in order to process EAP requests and its challenges.

The switch controls the physical access to the LAN based on authentication messages from the authentication server and the client. The switch acts as a proxy between the authentication server and the client. Not all Cisco switches support 802.1x authentication. The switch allows the client to only send EAP traffic in order to authenticate. After successful authentication, the switch opens its port to allow all traffic from the client to pass through.

The authentication server performs the actual authentication of users. It holds the local or external user database and its restrictions. Each authenticating user must be configured in the authentication server in order to successfully authenticate. The authentication server must support RADIUS authentication protocol and EAP extensions. Cisco ACS version 2.6 and higher supports 802.1x and RADIUS authentication.

Configuring 802.1x Authentication on a Supported Switch

In this section you will configure 802.1x protocol on a supported Cisco Catalyst switch. Refer to Figure 4.13 for the topology. In this example, it is assumed that the client supports the 802.1x authentication process, and that the ACS – RADIUS server is configured with user database and authentication permissions.

NOTE

Make sure you have network connectivity between the switch and RADIUS server prior to configuring 802.1x support.

Figure 4.13 Implementing 802.1x Topology

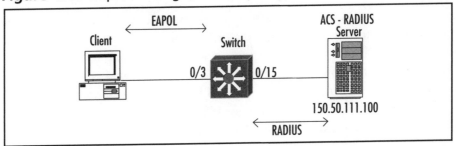

1. Configure a switch to RADIUS communication.

   ```
   Switch3550# configure terminal
   Switch3550(config)# radius-server host 150.50.111.100 key cisco
   ```

2. Configure 802.1x authentication.

   ```
   Switch3550(config)# aaa new-model
   Switch3550(config)# aaa authentication dot1x default group radius
   local
   ```

3. Configure the interface to request EAP authentication when the new
    device connects.

    ```
    Switch3550(config)# interface fastEthernet 0/3
    Switch3550(config-if) switchport mode access
    Switch3550(config-if)# dot1X port-control auto
    ```

4. Save all configurations.

    ```
    Switch3550(config-if)# end
    Switch3550# copy running-config startup-config
    ```

Now when a device connects into port 0/3 of the switch, the switch will
request authentication credentials from the device. By default, all traffic but the
authentication EAP protocol process will be blocked from the 0/3 port. After successful authentication the switch will allow all traffic to pass.

Enabling Multiple Host Authentication

The configuration above only allows one host to connect to port 0/3 at one time.
You can allow more than one device to authenticate and use the same port at one
time. By default, only one host MAC address is allowed to connect to an 802.1x-
configured port at one time, while other devices trying to use the same port are
dropped.

Using multiple host configurations, you can have more than one host connecting
to one port at the same time. In multi-host mode, it takes only one successful authentication to open up access to every other device connecting to the same port. If the
multi-host port becomes unauthorized due to an EAPOL-Logoff message or when reauthentication fails, it disables access for all hosts using the same port.

Multi-host port mode may be needed when clients are not connecting directly
to an 802.1x-compatible switch. Multi-mode host access can prove to be insecure as
it allows for only one EAP-compatible host to successfully pass the authentication
process, which could allow a rogue access point to slip by using the already authorized port with the previous user authentication.

If you need to use multi-host mode in 802.1x authentication, you should use it
in combination with a port-security feature to additionally restrict and permit hosts
by their MAC addresses to connect into the switch port. Using port-security features
in catalyst switches is covered later in this chapter.

1. To enable multi host support:

    ```
    Switch3550(config-if)# dot1x multiple-hosts
    ```

2. To disable multi-host support and go back to single-host only:

```
Switch3550(config-if)# no dot1x multiple-hosts
```

Viewing 802.1x Port Statistics

To display the configuration and port statistics of 801.1x-configured ports, use the **show dot1x** command in main privilege EXEC mode. Figure 4.14 shows the **show dot1x interface fastEthernet 0/3** command on the Catalyst 3550 switch configured in the previous examples. The port in Figure 4.14 is currently marked as "Unauthorized," which means that all traffic is blocked except 802.1x EAP protocol. When the client is plugged in and authenticates successfully, it will change to "Authorized" mode in which the switch will allow the client to communicate freely through the port.

Figure 4.14 show dot1x Command

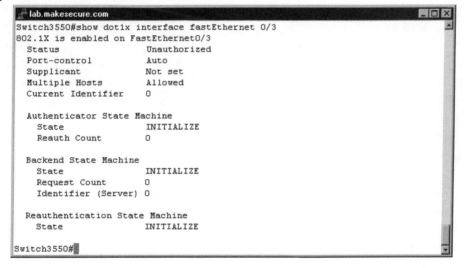

```
lab.makesecure.com
Switch3550#show dot1x interface fastEthernet 0/3
802.1X is enabled on FastEthernet0/3
  Status               Unauthorized
  Port-control         Auto
  Supplicant           Not set
  Multiple Hosts       Allowed
  Current Identifier   0

  Authenticator State Machine
    State              INITIALIZE
    Reauth Count       0

  Backend State Machine
    State              INITIALIZE
    Request Count      0
    Identifier (Server) 0

  Reauthentication State Machine
    State              INITIALIZE

Switch3550#
```

Fore more details on how to configure 802.1x support on Catalyst 3550 switches, refer to the documentation at www.cisco.com/univercd/cc/td/doc/product/lan/c3550/12119ea1/3550scg/sw8021x.htm.

802.1x is a dynamic protocol that can be used to accomplish mobility on wired and wireless networks. Ports can be dynamically configured and unconfigured on a per-user basis. Not only is this protocol used to restrict or permit devices based on its credentials, but it can also be used to configure per-user access lists or VLAN assignments based on individual user profiles that are stored on the authentication server.

Detecting a Rogue Access Point from the Wired Network

Although several rogue access point detection and prevention techniques were covered in previous sections, there are still many techniques that can be used on a network to detect rogue access points. The best solution for detecting wireless rogue access points is using Cisco's centralized management solutions such as the WLSE. There may be network environments where you do not have a WLSE engine or you may have a limited number of Cisco-aware wireless devices that do not cover your entire risk area. Manual sniffing and detection can only go so far, and must be physically performed in local areas.

Detecting rogue access points from a wired network is one of the alternative techniques used to detect unauthorized access points connected into corporate networks. Detection from a wired network works by scanning the user-wired LAN and identifying rogue devices that differ from a valid user's workstation signature. This signature is based on port numbers. For example, port 80 is used on Web servers to serve Hypertext Transfer Protocol (HTTP) content to users, and is also used on most wireless access points to provide administrative access. Other ports such as Telnet (23) and SSH (22) are also opened by default on most access points for user administration. How does this help us? Normal user workstations should not have these ports open, so when performing a large port scan of your user LAN, detecting ports such as 80 or 23 may indicate that the device running these ports may be a rogue device, not a user workstation.

There are many network scanners that can be used to scan large user LANs. One of the more popular scanners is called NMAP. NMAP is a free network scanner available at www.nmap.org website.

Detecting a Rogue Access Point with a Port Scanner

Figure 4.15 shows a typical user LAN with a large number of Windows workstations. The scanner is automatically run against these large user networks to detect any unique devices that do not match the typical workstation signature.

Figure 4.15 Port Scanning User LAN

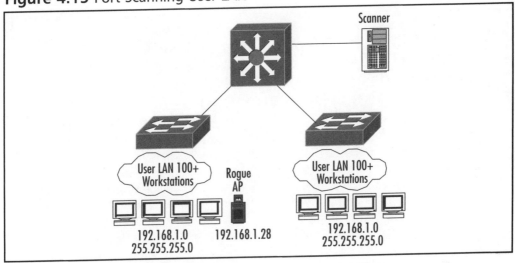

Figure 4.16 shows the actual scanner in action, scanning the 192.168.1.0 net-work. Notice that it found a device with IP 192.168.1.28 that has port 80, 22, and 23 open. It also detected that ports 22 and 23 are running on a Cisco device. By checking your list of Cisco network devices, you determine that 192.168.1.28 is not one of yours and thus you red flag it as a possible rogue device connected into your protected user LAN.

Figure 4.16 NMAP Scanner in Action

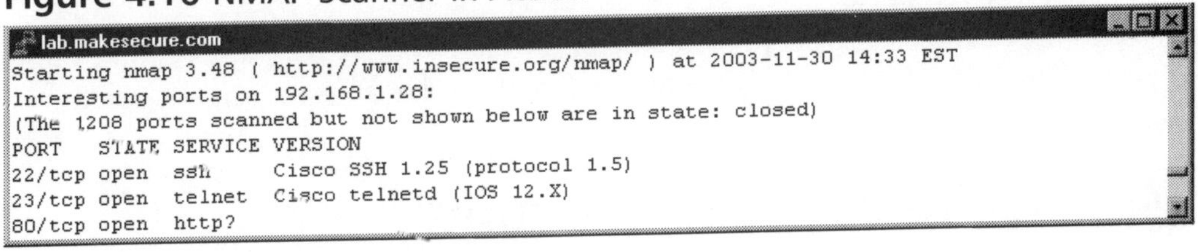

```
lab.makesecure.com                                              _ □ X
Starting nmap 3.48 ( http://www.insecure.org/nmap/ ) at 2003-11-30 14:33 EST
Interesting ports on 192.168.1.28:
(The 1208 ports scanned but not shown below are in state: closed)
PORT    STATE SERVICE VERSION
22/tcp open  ssh      Cisco SSH 1.25 (protocol 1.5)
23/tcp open  telnet   Cisco telnetd (IOS 12.X)
80/tcp open  http?
```

Once you detect a possible rogue access point on your network, you should track down its physical location by logging into the user switch and performing a reverse lookup on the detected IP to find its relative MAC address. Knowing the MAC address of the rogue device allows you to look through the MAC address table on the user switch and find out which port the device is connected to. When you know the actual port, you can trace down the physical cable to the device or disable the port.

Designing & Planning…

Extra Traffic and False Alarms

A network port scanner must connect to every device on the user LAN it is scanning, creating extra network traffic that can introduce unwanted congestion and slow performance on the overall network. You must make sure that the overall network performance is not affected when performing network scans.

Port scanners also require a connection to each device's port number. Such a connection can trigger security alarms such as personal workstation firewalls or security devices such as an IDS. Make sure network scans are coordinated between the groups that need to be aware of them order to avoid confusion and unwanted problem tickets.

Using Catalyst Switch Filters to Limit MAC Addresses per Port

Another technique for preventing rogue access points is successfully connecting to a wired network using switch port security. Switch port security uses security features on the catalyst switch to restrict connections to a port interface based on a configured list of allowed devices.

This list of allowed devices is represented by hardware MAC addresses. Each port must be configured with its own list of MAC addresses to prevent unauthorized devices from connecting to the port.

MAC Addresses in Port Security

There are three different types of MAC addresses that can be configured in the port security feature on a catalyst switch. These are:

- Static MAC
- Dynamic MAC
- Sticky MAC

Static MAC

Static MAC addresses must be manually configured on each device MAC address on switch ports that are allowed to connect. Configuring a static MAC address on an IOS Catalyst switch is accomplished using the **switchport port-security mac-address <MAC>** command. By default, you are only allowed to configure one static MAC address. If you have multiple hosts using the same port, you must increase the number of allowed devices with the **switchport port-security maximum <NUM>** command. If you try to configure more than one static MAC without first increasing the number of allowed MAC addresses on a port, you will receive an error message. Static MAC addresses are saved in a configuration file so that when the switch reboots, it does not lose its MAC port security configuration.

Dynamic MAC

Dynamic MAC addresses are learned dynamically from connected devices. If a switch port is configured to allow a maximum of three devices, it learns the first three MAC addresses dynamically and stores them in the memory table. Dynamic MAC addresses are not saved in a configuration. When the switch reboots, all dynamically learned MAC addresses are reset. Dynamic configuration is generally not used to defeat rogue access points.

Sticky MAC

Sticky MAC addresses use a combination of static and dynamic methods to configure its list. MAC addresses are learned dynamically, but they can also be saved in a configuration file as static. This becomes useful when you have a LAN of 200 plus users. You can dynamically learn all of 200 workstation MAC addresses and then turn them into a static MAC list. Sticky port security mode is accomplished using the **switchport port-security mac-address sticky** command in the IOS catalyst switch.

Security Violation

A port security violation occurs when an unknown device that is not on a MAC address list tries to access the switch port. Cisco Catalyst supports three different configured actions you can take when violation occurs. Each switch port can use one of the following three settings:

- Protect mode
- Restrict mode
- Shutdown mode

Protect Mode

When a violation occurs in Protect mode, the device that is trying to gain connectivity to the port on the switch is blocked and not allowed to connect and all pockets coming from the unauthorized device are dropped. When using Protect mode, no alert message is sent out to notify the administrator of the incident.

Restrict Mode

Restrict mode is similar to Protect mode in that all packets from the unauthorized detected device are dropped. The difference between Restrict mode and Protect mode is that Restrict mode logs the incident. It can generate an SNMP trap-to-management station alerting the administrator of a violation. It can also send a syslog message and increase the violation counter on the switch port setting.

Shutdown Mode

In Shutdown mode, the switch port shuts down when it detects an unauthorized device trying to connect to it. The switch sends out an SNMP-trap-to-management station or a syslog message and increases the port violation counter as it would in Restrict mode. When a port is shut down it must be manually re-enabled.

Configuring Port Security in an IOS Catalyst Switch

Figure 4.17 shows our network topology. We have user A's and B's workstations connecting to a corporate LAN. We want to make sure that only those two workstations are allowed to connect to the LAN and no other device.

Figure 4.17 Port Security Topology

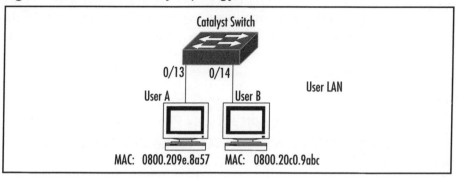

1. Configure port 0/13 on a catalyst switch with **static** MAC assignment and **restrict** violation mode:

```
Switch# configure terminal
Switch(config)# interface fastethernet 0/13
Switch(config-if)# switchport mode accesses
Switch(config-if)# switchport port-security maximum 1
Switch(config-if)# switchport port-security mac-address
0800.209e.8a57
Switch(config-if)# switchport port-security violation restrict
Switch(config-if)# switchport port-security
Switch(config-if)# end
Switch#
```

2. Configure port 0/14 on a catalyst switch with **sticky** MAC assignment and **shutdown** violation mode:

```
Switch# configure terminal
Switch(config)# interface fastethernet 0/14
Switch(config-if)# switchport mode accesses
Switch(config-if)# switchport port-security maximum 1
Switch(config-if)# switchport port-security mac-address sticky
Switch(config-if)# switchport port-security violation shutdown
Switch(config-if)# switchport port-security
Switch(config-if)# end
Switch# copy running-config startup-config
```

3. After configuring port security, verify your settings with show commands such as **show run**, **show port-security**, **show port-security address**, and **show port security interface**. Figure 4.18 shows switch **show port-security** and **show port-security address** commands. Notice that the port 0/14 MAC address was learned dynamically by using a sticky MAC list. Both ports show no violations thus far.

Figure 4.18 show port-security Commands

```
lab.makesecure.com

Switch#show port-security
Secure Port    MaxSecureAddr    CurrentAddr    SecurityViolation    Security Action
               (Count)          (Count)        (Count)
----------------------------------------------------------------------------------
    Fa0/13         1                1               0                Restrict
    Fa0/14         1                1               0                Shutdown
----------------------------------------------------------------------------------
Total Addresses in System : 2
Max Addresses limit in System : 128

Switch#
Switch#show port-security address
          Secure Mac Address Table
--------------------------------------------------------------------------
Vlan    Mac Address      Type              Ports    Remaining Age
                                                    (mins)
----    -----------      ----              -----    -------------
  1     0800.209e.8a57   SecureConfigured  Fa0/13   -
  1     0800.20c0.9abc   SecureSticky      Fa0/14   -
--------------------------------------------------------------------------
Total Addresses in System : 2
Max Addresses limit in System : 128

Switch#
```

Now look at what will happen if user A using switch port 0/13 tries to unplug their workstation and use the cable to connect to a rogue access point or a personal hub instead (see Figure 4.19).

Figure 4.19 Connecting Rogue Access Points

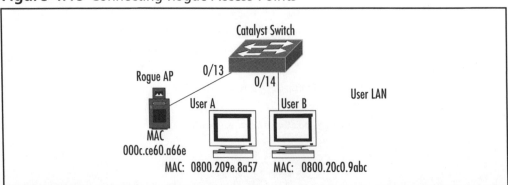

Port 0/13 goes down and up as the user unplugs and plugs the cable from their workstation to the rogue access point. As soon as the switch detects a new device that is not allowed on the configured MAC address list, it will act based on its port's security configuration. In the port 0/13 configuration the switch restricts and drops all packets coming from the unauthorized rogue access point, and sounds an alarm by sending an SNMP alert to the management station. See Figure 4.20 for the actual violation error messages taken from the switch when connecting a rogue access point to port 0/13.

Figure 4.20 Violation Error Messages

```
lab.makesecure.com
Switch#
Switch#
Switch#
00:17:53: %LINEPROTO-5-UPDOWN: Line protocol on Interface FastEthernet0/13, changed state to do
wn
00:18:55: %LINK-3-UPDOWN: Interface FastEthernet0/13, changed state to up
00:18:56: %LINEPROTO-5-UPDOWN: Line protocol on Interface FastEthernet0/13, changed state to up
00:19:16: %PORT_SECURITY-2-PSECURE_VIOLATION: Security violation occurred, caused by MAC addres
s 000c.ce60.a66e on port Fa0/13.
00:19:26: %PORT_SECURITY-2-PSECURE_VIOLATION: Security violation occurred, caused by MAC addres
s 000c.ce60.a66e on port Fa0/13.
00:19:36: %PORT_SECURITY-2-PSECURE_VIOLATION: Security violation occurred, caused by MAC addres
s 000c.ce60.a66e on port Fa0/13.
00:19:46: %PORT_SECURITY-2-PSECURE_VIOLATION: Security violation occurred, caused by MAC addres
s 000c.ce60.a66e on port Fa0/13.
Switch#
```

Figure 4.21 shows the **show port-security interface fastEthernet 0/13** and **show port security** commands. Notice that the "Security Violation" number count from the previous example has increased from 0 to 60. This means that the rogue access point is trying to gain access but the switch keeps denying it.

Figure 4.21 Security Violation Counter Increase

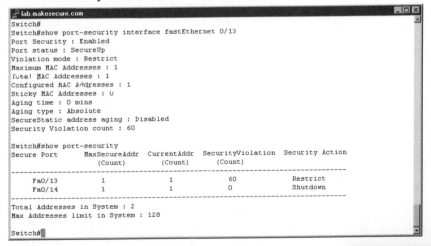

```
lab.makesecure.com
Switch#
Switch#show port-security interface fastEthernet 0/13
Port Security : Enabled
Port status : SecureUp
Violation mode : Restrict
Maximum MAC Addresses : 1
Total MAC Addresses : 1
Configured MAC Addresses : 1
Sticky MAC Addresses : 0
Aging time : 0 mins
Aging type : Absolute
SecureStatic address aging : Disabled
Security Violation count : 60

Switch#show port-security
Secure Port    MaxSecureAddr    CurrentAddr    SecurityViolation    Security Action
                  (Count)          (Count)          (Count)
---------------------------------------------------------------------------
   Fa0/13           1                1               60              Restrict
   Fa0/14           1                1                0              Shutdown
---------------------------------------------------------------------------
Total Addresses in System : 2
Max Addresses limit in System : 128

Switch#
```

Summary

Throughout this chapter you learned different techniques for detecting and preventing rogue access points. Rogue access points are unauthorized access points that are installed by employees without approval from the IT and security departments, or by an intruder trying to trick valid users in order to gain sensitive information.

It is important to mitigate the threat of rogue access points as they have the ability to demolish an entire security architecture. A single rogue access point installed by an employee can create a back door into a corporate network, ignoring and bypassing all border security architecture such as firewalls and IDS engines.

Several manual techniques exist for detecting rogue access points, such as wireless sniffers and wired network scanners. A wireless sniffer has the ability to detect wireless radio signals and access points within a reachable signal area that can then be compared against a list of known authorized access points. If the new MAC address of an access point that does not match any of authorized access points is detected within the radio signal area, it must be tracked down as a possible rogue access point. Several wireless application tools exist that can measure the signal strength of an access point, which can then be used to locate an access point's physical location.

Wired network scanners can be used to scan user LANs to detect possible rogue devices. A rogue access point-scanned signature differs from a user's workstation. Signatures such as a workstation's open port numbers differ from access point port numbers. Ports such as 80 (HTTP) on a detected device or 23 (Telnet) can reveal that a device is something other than a user workstation.

Among these manual rogue access point detection techniques, Cisco offers a robust centralized detection solution. A management station WLSE can be used to control all Cisco-aware devices such as access points and wireless clients to perform automatic and periodic scans of radio signals. They then report any findings back to the central management engine that are then matched against a database of authorized access points.

Preventing rogue access points from connecting into a protected wired LAN is as important as its detection. Throughout this chapter you learned several techniques that can be used to prevent an unauthorized access point from connecting to a wired network. A good prevention technique should eliminate detection. Techniques such as 802.1x port-based authentication and MAC port security features can be used on Cisco switches to control which device can and cannot establish physical connection to the network.

It is important to note that rogue access points are most likely found in organizations where no wireless network is supported. Every network and security administrator needs to worry about rogue devices that can dismiss the entire implemented security measures.

Solutions Fast Track

The Problem with Rogue Access Points

- ☑ A rogue access point is an unauthorized access point installed by an employee without permission from the IT or Security departments.

- ☑ One rogue access point can dismiss an entire security architecture.

- ☑ Employees install rogue access points for their own benefit without realizing that they have created a back door to the corporate LAN.

Preventing and Detecting Rogue Access Points

- ☑ The first step in protecting against rogue access points is having a security policy. A security policy should outline the rules against unauthorized wireless devices and employees must be educated about the policy.

- ☑ A wireless sniffer can aid in the detection of wireless access points throughout an area that can then be compared against a list of authorized access points.

- ☑ Cisco offers a centralized solution with a WLSE engine where all Cisco-aware wireless devices work together to detect possible rogue access points and report them to the central management station.

- ☑ Rogue access points can be detected from the wired network by using a network port scanner. Unlike a user's workstation, rogue access points usually have port 80 (HTTP) and 23 (Telnet) open for administration purposes.

- ☑ A port scanner can trigger false alarms and extra traffic on already congested traffic by scanning every device. Coordinated scanning should be performed to avoid confusion.

IEEE 802.1x Port–based Security to Prevent Rogue Access Points

☑ The 802.1x protocol allows mutual authentication where the access point authenticates the user and the user authenticates the access point, to ensure that the user is connecting to a valid, not a rogue, access point.

☑ In 802.1x protocol, users are prompted for authentication credentials as soon as they plug their workstation into the switch port. Devices such as rogue access points that do not support such authentication will not be allowed to connect to the wired port.

☑ A third-party authentication server that supports RADIUS protocol is required to store all user credentials and perform the actual authentication. The access point or the catalyst switch is used as a proxy server between the authenticating client and the RADIUS server.

Using Catalyst Switch Filters to Limit MAC Addresses per Port

☑ Port security in catalyst switches allows you to restrict devices that can physically connect to the port by their MAC addresses.

☑ The three types of MAC addresses in port security feature are static, dynamic, and sticky.

☑ When an unauthorized device connects to a secured port, a violation occurs. The three configurable reactions to a violation are protect, restrict, and shutdown modes.

☑ In shutdown violation mode the port is shut down and requires the administrator to manually bring it back up.

Frequently Asked Questions

The following Frequently Asked Questions, answered by the authors of this book, are designed to both measure your understanding of the concepts presented in this chapter and to assist you with real-life implementation of these concepts. To have your questions about this chapter answered by the author, browse to **www.syngress.com/solutions** and click on the **"Ask the Author"** form.

Q: Can 802.1x protocol be implemented in wired network devices just as it can be in a wireless network?

A: Yes. 802.1x was originally designed for a wired network. Cisco supports 802.1x protocol in many of their hardware devices and is expected to expand its collection in the future. There are, however, a few differences in how 802.1x is implemented that you should review prior to its implementation.

Q: I have over 1,000 users that I need to move over to a Cisco ACS RADIUS server. What is the best way to do this?

A: Cisco ACS supports multiple external databases. If your user database is one of them, you can link it up to provide user authentication. Refer to Cisco's ACS product details for a list of supported external databases.

Q: Is a once-a-week detection scan sufficient to mitigate the threat of rogue access points?

A: Rogue access point detection and awareness should be performed constantly to protect your networks from intruders.

Q: What is the best way to protect against rogue access points?

A: The best protection should be a combination of multiple techniques. One of the best techniques is to use Cisco's centralized WLSE engine solution along with its wireless devices to perform continues rogue access point detection scans.

Chapter 5

Wireless LAN VLANs

Solutions in this chapter:

- **Understanding VLANs**
- **VLANs in a Wireless Environment**
- **Wireless VLAN Deployment**
- **Configuring Wireless VLANs in IOS**
- **Broadcast Domain Segmentation**
- **Primary (Guest) and Secondary SSIDs**
- **Using RADIUS for VLAN Access Control**

- ☑ Summary
- ☑ Solutions Fast Track
- ☑ Frequently Asked Questions

Introduction

Virtual local area networks (VLANs) represent the logical separation of physical LANs. A VLAN allows you to split up your physical network devices such as Cisco's switches and access points into different virtual local area networks (LANs) in which each VLAN takes on its own unique characteristics. Up until now only one group policy has been supported by Cisco's wireless access. Due to this one access point/one policy limitation, any wireless client group not compatible with the main policy settings have to use their own separate compatible access points. With the introduction of VLANs into a wireless network, you can define multiple compatible group policies such as voice and data groups that allow you to use one access point for all of your unique wireless client groups. VLANs can also be characterized and used to represent a group of devices on different physical LAN segments that can then communicate with each other as if they are on the same physical LAN.

WLANs have been widely used in the wired LAN industry since the 1990s. The VLANs proven support of scalability and cost savings are required by network administrators and architects in LAN network deployments.

Cisco has taken VLAN technology from wired LANs and its standards and incorporated it into its wireless devices to offer some of the advantages of a WLAN such as scalability, security, and per-VLAN policy, thus making wireless networks more scalable and appealing to corporations and more cost effective.

This chapter reviews the basic workings of the VLANs used on wired networks. It covers specific protocols and functions that make up a VLAN and its technology. You will learn how VLANs help overall network design scalability, security, availability, and cost savings.

Wireless VLAN deployment and configuration differs a bit from a wired LAN. This chapter takes a closer look at these differences and similarities. You will learn how to deploy and configure VLANs in wireless networks using Command Line Interface (CLI) in IOS and using a Web browser.

This chapter covers broadcast domain segmentation and its advantages for overall performance on the LAN and WLAN networks. A broadcast domain benefits from performance and also from certain security aspects.

You will review techniques on how users are assigned into allowed VLANs by using Service Set Identifier (SSID) and Remote Authentication Dial-In User Service (RADIUS). You will learn the differences between the two and how to configure them.

VLAN support is found mostly in multi-group support designs. It allows for differentiation in policy such as security or Quality of Service (QOS) among multiple devices such as wireless users using laptops, IP phones, or personal digital assistants

(PDAs). The introduction of VLANs in wireless technology makes wireless networks more intercompatible with wired networks and more appealing to corporations.

Understanding VLANs

This section reviews VLANs and some of their standard protocols such as VLAN Trunking Protocol (VTP) and trunk ports.

VLANs are incredibly flexible due to their logical rather than physical implementation. Logical implementation can be used to split one physical switch device into multiple Layer 2 segments representing different domain groups (see Figure 5.1). Logical separation is done by configuring VLANs. These different Layer 2 segments may then further span across multiple switches allowing one or more segments to coexist in different geographical areas (see Figure 5.2).

> **NOTE**
>
> Layer 2 segmentation is synonymous with multiple VLANs. By creating VLANs on a switch, you logically separate them into multiple Layer 2 domains.

A VLAN is a logical separation of a LAN. One separated segment is restricted to interact between the other segments unless a Layer 3–aware device such as a router is used to route and restrict traffic between them. In Figure 5.1, all three computers are connected to the same switch: two are configured in VLAN A and one is configured in VLAN B. Communication between VLANs A and B is restricted, even though they are connected to the same switch.

Figure 5.1 VLAN Logical Segmentation

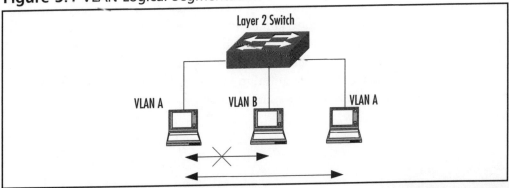

Figure 5.2 VLANs across Multiple Switches

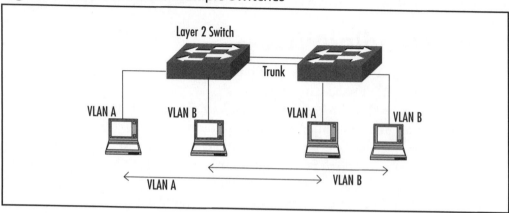

Before VLANs were introduced, two separate switches were required in order to create two separate LAN segments. If you needed to separate segments into more than two segments, it came with a heavy price tag. By creating VLAN segments, you also separate each logical LAN segment into its own broadcast domain. (Broadcast domains and their benefits are covered later in this chapter in the "Broadcast Domain Segmentation" section.) Other benefits of logical separation are per-VLAN compatibility requirements such as a security policy. For example, you may need to separate employees into multiple groups throughout multiple LAN locations and restrict specific access between them. Different VLANs can have different QOS policies configured. A voice-critical application VLAN may have a higher priority QOS set over a user VLAN group. A slight delay in sensitive Voice over IP (VoIP) traffic may be unacceptable and should be prioritized over user traffic such as Internet browsing, which can bare some delay during congestion.

VLANs are usually associated with different IP subnetworks. For example, all devices in the same VLAN usually belong to the same subnet, which differs from other VLAN subnets. You need a router or Layer 3-aware device to route between these VLAN subnetworks. In Figure 5.3, the PC configured in VLAN A needs to pass through the router (Layer 3) in order to reach VLAN B. New devices used today to route between VLANs include Layer 3-aware switches. Layer 3-aware switches have the capability to route and thus eliminate the need for an external router. In Layer 3-aware switches, the router is built inside the switch itself.

The International Standard Organization (ISO) has created a layered model called the Open System Interconnect (OSI) model. The purpose of the OSI model is to describe and define each layer in the network system. The seven OSI layers are: (7) Application, (6) Presentation, (5) Session, (4) Transport, (3) Network, (2) Data Link, and (1) Physical.

A network switch works on Layer 2 (Data Link) of the OSI model. It uses frame data such as MAC addresses to direct traffic. A network router works on Layer 3 (Network) of the OSI model. It uses packet data such as IP addresses to direct traffic.

Cisco has combined both switch (Layer 2) and router (Layer 3) technology so that one device can both switch and route. These are called Layer 3-aware switches. Layer 3-aware switches include routing capabilities and can be used to route between VLANs.

Figure 5.3 Using a Router To to Route Between between VLANs

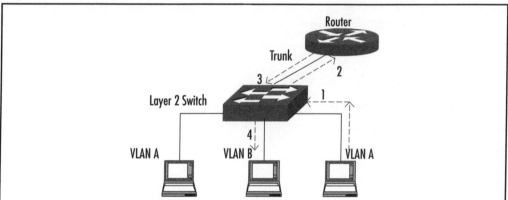

By default, every port on a Cisco catalyst switch is assigned to VLAN 1. VLAN 1 is called the native VLAN. In most devices VLAN 1 poses configuration limitations such as that it cannot be deleted. For this reason, when configuring VLANs try to avoid adding devices to VLAN 1, as it can cause potential network issues and security leaks. VLANs may be numbered from 1 through 4096 on a Cisco catalyst switch.

Configuring & Implementing…

VLAN Numbers

Although VLANs 1 through 4096 are supported in Cisco devices, they arefurther grouped and restricted.

- VLAN 1 The default VLAN for every port on a switch.

- VLANs 2 through 1001 These VLANs are used for Ethernet and can be deleted or created at the will of the network administrator.

- VLANs 1002 through 1005 These VLAN IDs are used for Fiber Distributed Data Interface (FDDI) and Token Ring in Cisco devices. They cannot be deleted and are restricted.

- VLANs 1006 through 4096 The VLANs in this range are called the extended VLANs. Additional requirements must be enabled in order to use extended VLANs.

VTP in a Wired Network

VTP is a Layer 2 management protocol that allows the administrator to create, delete, or modify VLANs on a server switch, that are then populated throughout the client network of switches in the same VTP domain. VTP allows for better central management compatibility and avoids problems such as configuring inconsistencies of duplicate VLAN IDs, names, and types. In Figure 5.4, the administrator modifies VLAN A or B on the master switch after which new configurations will be propagated to its clients. Both the master and clients need to share identical VTP settings such as the domain name, version, and authenticating password to operate correctly.

Figure 5.4 VTP in Virtual LAN

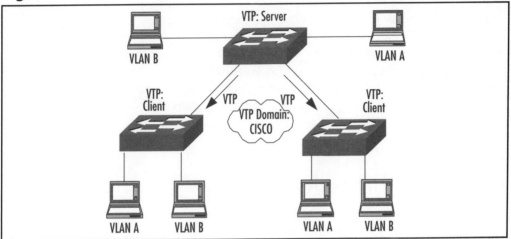

VTP Modes

After configuring a VTP domain name, you must choose from the following three different VTP modes:

- **Server Mode** Server mode is the default configuration mode on a switch. When in server mode, the administrator has the ability to create, delete, or modify VLANs. Server mode devices advertise their VLAN configurations to other connected network devices.

- **Client Mode** When in Client mode, the administrator cannot create, delete, or modify VLAN settings.

- **Transparent Mode** A device in Transparent mode can create, delete, and modify VLAN settings. It does not, however, participate in a VTP domain nor does it advertise or sync up VLAN configuration from server-configured devices. A transparent configured device will, however, accept VTP advertisement from servers and pass it along.

WARNING

Use caution when adding new switches to your environment. Always check the VLAN revision number of the switch. Devices in the same VTP domain and configuration always use configurations from the highest revision number server switch. If you add a new default server config- ured switch with a high revision number, you risk having your other switches automatically copy the default switch VLAN configuration.

Dealing with Trunk Ports

Trunk ports are ports that carry VTP messages. Trunk ports are also used to send multiple VLANs over a single link across two devices. In Figure 5.4, all switches are interconnected with a single Ethernet cable configured as a trunk port to carry both VLAN A and VLAN B information. Without trunk ports, you would need one Ethernet cable per VLAN to interconnect between switches. A trunk port is consid- ered a point-to-point type link between switches, routers, and wireless access points. Without trunk ports, subnetworks would not be able to be partitioned across mul- tiple switches or other devices. Trunks extend LAN domain where users connected to VLAN A can communicate to users on different switches that are also connected to VLAN A.

Two following Cisco-supported trunking encapsulation protocols are available for configuration:

- Inter-Switch Link (ISL)
- IEEE 802.1q

ISL is a Cisco proprietary trunking encapsulation protocol that supports span- ning tree on a per-VLAN basis (PVST). PVST can block or forward based on the specific VLAN. It treats each VLAN separately. By treating each VLAN on a per- VLAN basis, the administrator can load-balance traffic by forwarding different VLANs over different trunks. ISL supports Ethernet, FDDI, and Token Ring frames with up to 1000 VLAN configurations. One implication of ISL is that it includes 30 bytes of encapsulated header on top of the network packet and also includes an additional Frame Check Sequence (FCS) that is added at the end of the ISL-encap- sulated frame. When an ISL header is added to an already large Ethernet packet (1518 bytes), the packet will go over its Ethernet standard allowed limit size and will ask to be fragmented into smaller packets, thus possibly lowering the overall performance and requiring fragmentation support. The FCS is used to ensure that the frame has not been damaged during transmission and that it is added at the end of every frame. With the additional FCS around the ISL encapsulation, the switch must check the frame twice, which in a switch environment should not have any noticeable impact, but may be more difficult on routers if used for trunking.

The IEEE 802.1q is the industry standard trunk encapsulating protocol. If you are interconnecting vendor devices other than Cisco, you must use 802.1q to have the compatible standard. IEEE's 802.1q is also capable of supporting PVST in the same Cisco environment as ISL. However, unlike ISL, it only supports Ethernet frames. The header overhead in 802.1q is much smaller than in the ISL protocol, as it only includes 4 extra bytes. Further, the 802.1q does not include the extra FCS at the end of the frame. 802.1q simply recalculates the existing FCS of the frame after it adds its 4 extra bytes.

VLANs in a Wireless Environment

VLANs in a wireless environment have the same purpose as in a wired network. They are used to separate devices into groups of specific services. Whether trying to accomplish better security, performance, or scalability, a VLAN can help.

In a wired network you statically configure VLAN settings on the switch port that the user is connected to. In a wireless network there are no cables or ports that users must connect to. Therefore, you need a mechanism to separate and identify wireless users or devices belonging in different VLANs. This fundamental of identifying to which VLAN user a device needs to be mapped to in a wireless environment is accomplished using SSIDs. A different and more secure way of identifying a user or device VLAN assignment on a wireless network is with the use of a RADIUS server (covered later in this chapter in "Using RADIUS for VLAN Access Control" section).

SSIDs are used to map to unique VLAN IDs to help the access point recognize and connect users to its proper VLAN assignment. SSIDs are not used for security purposes, rather their purpose is to separate users into groups so that access points can recognize and match an individual device into the properly configured VLAN. After separation is recognized and the user is mapped to the proper VLAN, the device must pass the VLAN-configured security policy (that may mandate for EAP, WEP, or MAC authentication) before the user is allowed to fully use its mapped VLAN on the wired side.

Cisco access points have the ability to support up to 16 different SSIDs; therefore, you can configure up to 16 different VLANs. Each VLAN can have its own policy, allowing the network administrator to use one access point to configure up to 16 different groups of devices, each with its own unique compatible service policy. In the past, without VLAN support you would need a new separate access point for each different group with unique policy settings.

Per-VLAN Settings

As mentioned previously, an access point has the ability to support up to 16 VLANs that can be configured with different policies and restrictions. These individual policies may include:

- **Authentication Type** Open, Shared, and EAP.
- **MAC Authentication** In, Open, Shared, or EAP.
- **Maximum Associations** The maximum number of clients allowed to be connected at one time.
- **Encryption Key** Every VLAN must have a unique WEP key that is used for broadcast and multicast traffic. Broadcast domain segmentation is discussed later in this chapter.
- **Enhanced Message Integrity Check (MIC)** Used for WEP packet verification.
- **Temporary Key Integrity Protocol (TKIP)** A WEP per-packet keying mechanism.
- **Filters** Allows different filters per VLAN.
- **QOS** Allows for a different Class of Service (COS) priority per VLAN.

By allowing for such dynamic and specific settings on a per-VLAN basis, you can support multiple unique groups of users and devices on one access point, as shown in Figure 5.5. As shown in the drawing, a wireless PDA, a wireless IP phone, and different groups of users can all use the same access point but have unique configuration policies and restrictions to fit their compatible requirements. Each wireless client device must have a proper SSID configured in order for the access point to recognize its VLAN assignment. If you have two VLANs configured with identical security authentication policies but a unique restrictive access policy exists for each VLAN, you must ensure that the client device from one VLAN does not change its assigned SSID in order to be mapped by the access point into the other VLAN. To mitigate the threat of unauthorized VLAN hopping between wireless client devices, use a RADIUS server to monitor and assign a VLAN SSID other than relying on the clients to have the proper configured SSIDs. RADIUS servers and their use in

VLAN assignments are covered later in this chapter in "Using RADIUS for VLAN Access Control" section.

Figure 5.5 Multiple Devices Connected to One Access Point

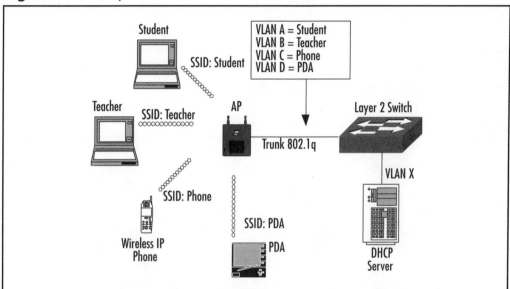

VTP in a Wireless Network

The dynamic VTP protocol used to manage dynamic VLAN assignment settings is not supported in wireless access point devices, but is supported in wired–switch network devices. Wireless devices work as stub devices and require manual static VLAN configuration from the wired and wireless network. Currently, the server/client VTP relationship is not supported in wireless devices.

Trunk Ports

Trunk ports are used to carry multiple VLANs over a single communication line. This allows the access point to support and map multiple devices from a wireless environment into a wired LAN using just one interface. In Figure 5.5, the trunk is configured between the access point and the switch using 802.1q encapsulation. It allows for multiple wireless VLAN devices to map into wired VLAN domains.

Trunk Ports between Bridges

Trunk ports can be configured in wireless bridges to extend wired or wireless LANs across two or more different areas. A trunk is configured on both the radio and Ethernet interfaces for transferring multiple VLANs. Figure 5.6 shows a trunk implementation between two wireless bridges, supporting multiple VLAN communications over 802.1q trunk encapsulation protocol. This allows wireless networks to have the same advantages to extend and scale as wired networks. Additionally, wireless bridges support Spanning Tree Protocol (STP) as regular wired bridges, allowing for scalable loop-free wireless architecture.

Figure 5.6 A Trunk Port across a Wireless Bridge

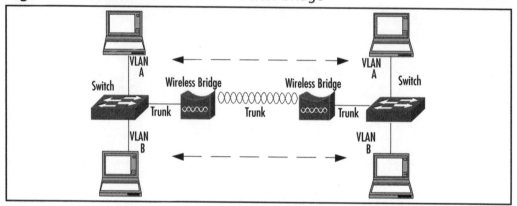

Keep in mind that, just as in a wired network, if VLAN A wants to talk to VLAN B on its local or extended LAN, you will need a Layer 3-aware device such as a router to route between VLANs.

Wireless VLAN Deployment

Wireless VLANs are supported in Cisco 1200,1100,350,340 access points and Cisco 350,1400 wireless bridges. The supporting versions required are VxWorks Firmware release 12.00T or later and Cisco IOS release 12.2.4JA or later.

Native VLAN

Native VLAN is the default configured VLAN. An access point must match the configuration of the native VLAN of the opposite connected device over the trunk port. If you have multiple access points or bridges in the same wireless LAN Extended Sub System (ESS) that need to communicate with each other, they must match their native VLAN configuration. All wired Cisco switches and bridges use VLAN 1 as their default-configured VLAN, therefore you will need to configure the same value on your access points to allow for interaction compatibility with connecting switches.

Any administration traffic such as Telnet, SSH, or RADIUS, directed from the access point or to the access point IP address, will be tagged using the native VLAN. An IP filter list is recommended to restrict and allow only authorized network administrators.

Routing between VLANs

Figure 5.1 shows that two VLANs cannot talk to one another unless they use a Layer 3-aware device such as a router. Can you see a possible problem in Figure 5.5? Wireless users that are required to use services such as dynamic IP assignment will not be able to reach their dynamic host control protocol (DHCP) server located on VLAN X because they are configured in different VLANs. To overcome this problem, you must use a Layer 3-aware switch or external router that allows for inter-VLAN routing.

If you are using services such as RADIUS in your design to allow for EAP authentication, the access point is the one initiating authentication connection to the RADIUS server on behalf of the users. With that in mind, you must configure your RADIUS server to be in your native VLAN, or allow native VLAN-tagged traffic from the access point to reach RADIUS's VLAN using routing.

Per-VLAN Filters

As discussed earlier, one of the many benefits of creating multiple VLANs for different classes of users is that you have the ability to apply different security filters. Filters restrict or allow communication to enter from a wireless LAN into a wired network.

Each VLAN works as if it has its own interface. The main interface is split into several subinterfaces. Each subinterface can have specific settings configured to it.

Figure 5.7 Per-VLAN Filters

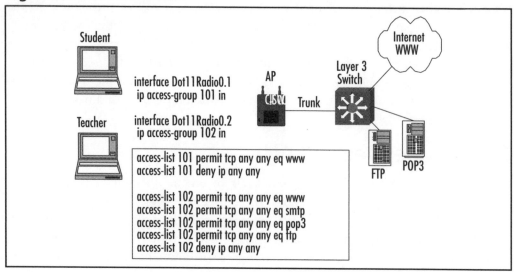

As shown in Figure 5.7, radio interface "0" has been split into "0.1" and "0.2" sub-interfaces in which unique access groups 101 and 102 have been applied. The dot "." in the interface represents a sub-interface. Sub-interfaces are used to accomplish multiple VLAN configurations with unique policies such as filters. According to the drawing, the Student group is bound to the interface with access list 101, which is only permitting HTTP access to be sent to the wired network from the Student wireless VLAN. The Teacher group with filter list 102 is allowed to access the World Wide Web (WWW), mail, and the File Transfer Protocol (FTP) on the wired network.

Per-VLAN QOS

QOS policies can be applied on a per-VLAN basis. For example, you may want to give a higher priority to the wireless IP phone's traffic VLAN than to the student VLAN. VoIP may not work properly during congestion, therefore it is important to prioritize it. Or you may want to prioritize teachers' communication over students or guests when an access point becomes congested. You can specify different QOS policies on a per-VLAN basis where different groups are mapped.

Per-VLAN Authentication and Encryption

Each VLAN can have its own authentication and encryption policy. You can support a guest network for your students without an authentication or WEP encryption policy, while at the same time use Cisco EAP authentication with WEP+TKIP policy for teachers. Also, your PDA devices may not support the same authentication policy as the teachers, and will require a compatible policy of its own. Just like filters and QOS, these settings are configured on per sub-interface VLAN basis.

If you need to support two different groups that share identical authentication types but require different restrictions on the wired network, you need a way to prevent the wireless user from simply changing its SSID in order to be mapped into the restricted VLAN after passing authentication. How to mitigate such a threat is discussed later in this chapter.

Configuring Wireless VLANs Using the IOS: A Case Study

A local university has asked you to implement wireless technology for its faculty, students, and maintenance workers. After conducting a site survey and developing security policy requirements for the university, you have come up with a solution. Since students, faculty, and maintenance workers require different security policies and restrictions, your design will include three different VLANs in every access point. Refer to Figure 5.8 for part of the network topology map used in this scenario.

Faculty and students require strict per-user authentication in order to map into their specified VLANs. The faculty needs to access the Internet to surf the Web and access the student grades system to update records. Students will only be allowed to surf the Web. The maintenance workers will take advantage of the new wireless design to allow communication and report back to the maintenance server using wireless PDA devices. Refer to Table 5.1 for a listing of the requirements.

Figure 5.8 School Topology

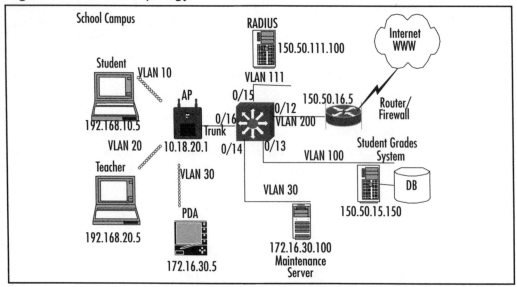

Table 5.1 Table of Requirements

	Teacher	Student	Maintenance
SSID	Teacher	Student	PDA
VLAN ID	10	20	30
Authentication	LEAP	LEAP	MAC/WEP
Encryption	Dynamic 128-bit WEP	Dynamic128-bit WEP	Static 40-bit WEP
Filter List	Yes #101	Yes #102	Yes #103

The following steps are required to configure the access point to support the network topology from Figure 5.8.

1. Configure SSIDs for all three groups and their authentication types. The first two authentication types for VLANs 10 and 20 are configured using the EAP method. VLAN 30 is authenticated using an open static WEP and MAC address list. (Refer to Chapter 7 for details on authentication types.)

```
AP# configure terminal
AP(config)# interface DotRadio 0
AP(config-if)# ssid teacher
```

```
AP(config-if-ssid)# vlan 10
AP(config-if-ssid)# authentication open eap eap_methods
AP(config-if-ssid)# authentication network-eap eap_methods
AP(config-if-ssid)# exit

AP(config-if) ssid student
AP(config-if-ssid)# vlan 20
AP(config-if-ssid)# authentication open eap eap_methods
AP(config-if-ssid)# authentication network-eap eap_methods
AP(config-if-ssid)# exit

AP(config-if) ssid pda
AP(config-if-ssid)# vlan 30
AP(config-if-ssid)# authentication open mac-address 798
```

2. Configure the native VLAN interface. You can configure the native VLAN only on the Ethernet interface to avoid administration access directly to the access point's IP address from wireless clients. We configure native VLAN on both the radio and Ethernet interfaces. The VLAN number is followed by the key word **native**.

```
AP(config)# interface DotRadio0.1
AP(config-if)# encapsulation dot1Q 1 native
AP(config-if)# bridge-group 1
AP(config-if)# exit
AP(config)# interface FastEthernet0.1
AP(config-if)# encapsulation dot1Q 1 native
AP(config-if)# bridge-group 1
```

3. Configure VLANs 10, 20, and 30 by creating sub-interfaces and enabling encapsulation on radio and Ethernet interfaces.

```
AP(config)# interface DotRadio0.10
AP(config-if)# encapsulation dot1Q 10
AP(config-if)# bridge-group 10
AP(config-if)# exit
AP(config)# interface FastEthernet0.10
AP(config-if)# encapsulation dot1Q 10
AP(config-if)# bridge-group 10

AP(config)# interface DotRadio0.20
```

```
AP(config-if)# encapsulation dot1Q 20
AP(config-if)# bridge-group 20
AP(config-if)# exit
AP(config)# interface FastEthernet0.20
AP(config-if)# encapsulation dot1Q 20
AP(config-if)# bridge-group 20

AP(config)# interface DotRadio0.30
AP(config-if)# encapsulation dot1Q 30
AP(config-if)# bridge-group 30
AP(config-if)# exit
AP(config)# interface FastEthernet0.30
AP(config-if)# encapsulation dot1Q 30
AP(config-if)# bridge-group 30
```

4. Configure WEP keys. Two 128-bit WEP keys will be used for VLANs 10 and 20. These two keys will be used for broadcast and multicast traffic only, as unicast WEP keys are dynamically derived on a per-user basis in the 802.1x EAP authentication process. There will be one static 40-bit WEP key to support the maintenance worker's wireless PDA compatibility. This key will be used for unicast encryption between PDAs and access points. For security purposes, the broadcast key is rotated in VLANs 10 and 20 using the **broadcast-key** command. Broadcast key rotation is currently only supported in LEAP authentication.

```
AP(config)# interface DotRadio 0

AP(config-if)# encryption vlan 10 key 1 size 128bit <key-here> transmit-key
AP(config-if)# encryption vlan 10 mode ciphers wep128
AP(config-if)# broadcast-key vlan 10 change <# of seconds>

AP(config-if)# encryption vlan 20 key 1 size 128bit <key-here> transmit-key
AP(config-if)# encryption vlan 20 mode ciphers wep128
AP(config-if)# broadcast-key vlan 10 change <# of seconds>
```

```
AP(config-if)# encryption vlan 30 key 1 size 40bit  <key-here>  transmit-key
AP(config-if)# encryption vlan 30 mode ciphers wep40
```

5. Configure filter lists to restrict the types of communication accepted from wireless groups into the wired network. Part of the campus requirement is to restrict student access to surf the Internet only and prevent them from accessing the student grades database. A unique filter list can be applied on each VLAN radio sub-interface. Filter lists and its configuration have been covered. (Refer to Chapter 7 for how to configure and apply filter lists to restrict or permit traffic.)

6. Apply identical configurations to the secondary radio interface. If you are using access points such as the 1200 series that support up to two installed radios such as 802.11b, 802.11g, or 802.11a, you must repeat all of the configurations for interface "DotRadio 1" as you configured for interface "DotRadio 0." This includes SSIDs and the creation of sub-interfaces, WEP keys, and IP filters.

NOTE

In a Web-based access point administrator graphical user interface (GUI) you can use the "Apply-all" button in the interface configuration menu to apply your settings to both of the installed radios at once. The 1200 series access point supports up to two installed radios including 802.11a, 802.11b, and 802.11g. Each radio can have unique or identical settings.

There is one big security concern and risk in the current school campus design called *VLAN hopping*. To mitigate VLAN hopping you must use a RADIUS server to authenticate VLANs. This concept is covered later in this chapter and must be considered in the design to prevent students from accessing their confidential records.

In Figure 5.8, a Catalyst 3550 Layer 3-aware switch with IP routing was enabled. Part of the switch configuration is displayed below for reference purposes. Notice that the trunk port configured under the FastEthernet 0/16 interface only allows VLANs required on the wireless side. Also, access filters can be configured that can be applied on the switch VLAN interfaces to restrict traffic communication between VLANs.

As shown in Figure 5.8, topology map Interface 0/12 is configured to be part of VLAN 200.

```
interface FastEthernet0/12
description Port to Internet Router
switchport access vlan 200
switchport mode access
no ip address
```

Interface 0/13 is part of VLAN 100 and is used as a student records server.

```
interface FastEthernet0/13
description Student Records Server
switchport access vlan 100
switchport mode access
no ip address

interface FastEthernet0/14
description Maintenance Server
switchport access vlan 30
switchport mode access
no ip address

interface FastEthernet0/15
description Radius Server
switchport access vlan 111
switchport mode access
no ip address
```

Interface 0/16 is used to establish a trunk port to carry multiple VLANs between the access point and the switch connection. The trunk is encapsulated with 802.1Q protocol to support access point compatibility. Further, VLANs that are allowed to pass the trunk with the **allowed vlan** command have been restricted. This will ensure that only required VLANs from the switch are allowed to cross to the wireless side.

```
interface FastEthernet0/16
description Trunk Port to AP
switchport trunk encapsulation dot1q
switchport trunk allowed vlan 1,10,20,30
switchport mode trunk
no ip address
```

Logical VLAN interfaces are assigned with IP addresses that are used for Layer 3 routing between the different VLANs. They are also used as default gateways for devices on each VLAN.

```
interface Vlan1
ip address 10.18.20.3 255.255.255.0
interface Vlan10
ip address 192.168.10.1 255.255.255.0
interface Vlan20
ip address 192.168.20.1 255.255.255.0
interface Vlan30
ip address 172.16.30.1 255.255.255.0
interface Vlan100
ip address 150.50.15.1 255.255.255.0
interface Vlan111
ip address 150.50.111.11 255.255.255.0
interface Vlan200
ip address 150.50.16.1 255.255.255.0
```

The default gateway is configured with the **ip route 0.0.0.0 0.0.0.0** command to match and route all traffic not directed to any specific VLAN on the switch, such as Internet browsing towards the Internet router.

```
ip classless
ip route 0.0.0.0 0.0.0.0 150.50.16.5
```

Broadcast Domain Segmentation

Broadcast domain segmentation prevents broadcast and multicast traffic from one group from entering other segmented groups. One of the advantages of separating LANs with VLANs includes the creation of separate broadcast domains. A broadcast domain assures performance and scalability and prevents users from different logical domains from exchanging broadcast or multicast traffic.

Traffic Types

There are many different traffic types. To understand broadcast domain segmentation and its benefits, a review of the three fundamental traffic types—unicast, broadcast and multicast— is required.

Unicast

Unicast traffic is when traffic is directly directed to one individual. An example of this one-to-one relationship can be found at www.cisco.com. Only the client and the Web site are involved in receiving and sending traffic.

Broadcast

In a broadcast network, the client sends only one packet that is directed to everyone. This is a one-to-all relationship. As shown in Figure 5.9, one server sends a broadcast message and everyone on the LAN receives it. A broadcast can be stopped by logically separating the LAN with VLANs, or by a Layer 3 device. Every client receiving broadcast messages must process them, thus lowering the overall performance of a LAN.

Broadcast frames contain the broadcast MAC address (ff:ff:ff:ff:ff:ff). When the switch sees this address it forwards it out of every LAN port. Servers make use of broadcast traffic to announce information services they provide. The broadcast domain is the group of logical network devices where broadcast messages are flooded.

Figure 5.9 Broadcast Traffic

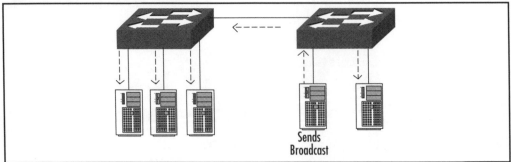

Sends
Broadcast

Multicast

Multicast traffic is similar to broadcast traffic. Its intentional relationship is one-to-many. Unlike broadcast traffic, multicast traffic is sent to a set of users in a group. It is still forwarded like broadcast traffic; however, unlike in a broadcast environment where each device must process the broadcast, multicast devices that are not listening in to the specific multicast group being advertised will disregard the multicast traffic. How can multicast benefit your network? Unlike in unicast traffic where the server is required to send a copy of the same packet to every server it needs to communicate with, in multicast it only needs to send one multicast packet that will reach all of the users listening in on a specific multicast group.

Broadcast Domain in Wireless

Now that you understand the different types of traffic and benefits of broadcast domain segmentation in wired networks, we will take a closer look at broadcast segmentation in wireless networks. In a wired network, VLANs are used to separate broadcast domains.

As discussed earlier, every packet traveling through the air can be seen by its neighbors as long as they are within signal reach. Thus, for this reason, every wireless client regardless of VLAN assignment will receive broadcast and multicast traffic. This is the difference between a wired and wireless network and their treatment of broadcasts in VLANs. You cannot prevent broadcast messages from reaching other VLAN segments on the wireless side because no physical separation (such as an Ethernet cable) exists.

Not being able to prevent broadcast messages from reaching multiple wireless users from different VLANs requires a workaround solution. Cisco wireless access point devices allow you to configure a different WEP key for the broadcast traffic for each unique VLAN. This WEP key differs from the unicast traffic key and is communicated to the wireless clients. When the access point sends out a broadcast message on its wireless side, other wireless users will still receive those broadcast messages, but because they do not share the same broadcast WEP VLAN key, devices not belonging to the same VLAN will discard them.

A broadcast WEP key can be dynamically derived or statically configured and is synced up between the users and the access point. A broadcast key shares some of the same ability as a WEP unicast key, including the ability to rotate when used with LEAP protocol within a configured timeout. Figure 5.10 shows a broadcast sent from the access point to the teachers VLAN. Anyone not on this broadcast VLAN will still receive the packet but will discard the broadcast traffic because they do not share a common broadcast WEP key. If this was a wired network, the students would never receive the broadcast from the teacher, as it is in different VLAN.

Figure 5.10 Wireless Broadcast

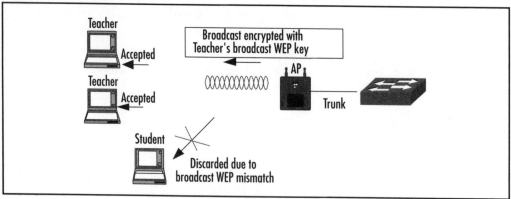

Primary (Guest) and Secondary SSIDs

The SSID is a unique case-sensitive 32-alphanumeric character used in VLAN mappings. Up to 16 SSIDs can be configured. Hence, the limit of 16 VLANs is due to the limit of the SSID, as each VLAN must contain a unique SSID.

Each SSID can be configured with different policy characteristics. All SSIDs are active, allowing clients to use and pick from all 16 SSIDs at once. Some of the characteristics that can be configured based on a unique SSID include the authentication type, VLAN, guest mode, and RADIUS accounting among others. SSIDs are not used for any type of security purpose. SSIDs travel in cleartext through radio frequency (RF), which anyone can capture. Its use is purely to separate and recognize multiple group policy requests.

Guest SSID

Guest SSID allows wireless users without any configured SSID to associate with the access point. Guest SSID is also used to broadcast unsolicited beacons from the access point to advertise its presence to the wireless community. The default configured SSID is **tsunami** on Cisco wireless devices and is enabled as a guest SSID. Broadcasting beacons should be disabled if you do not plan to use the access point for guest network access.

Only the primary SSID in multiple VLAN configurations can be included in broadcast beacons. Clients will still be allowed to request all different SSIDs from the access point, and the access point will respond with the proper SSID. However, in environments such as guest access networks where clients do not know the SSID, only one SSID can be used as the primary that is advertised in broadcast beacons. Figure 5.11 shows how to enable SSID as guest mode in a Web administration interface.

Figure 5.11 Enabling Guest Mode SSID

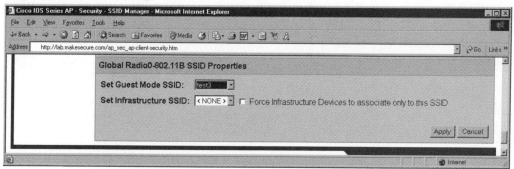

Using RADIUS for VLAN Access Control

A RADIUS server can be used to control VLAN and SSID assignments. In previous examples, all SSIDs were configured on the access point. These SSIDs are used to map wireless devices into certain policy groups, whether it for security or QOS requirements.

Refer back to Figure 5.8 for the school campus implementation. Students and teachers share an identical authentication type. Both of these groups will require to authentication using LEAP protocol in order to be mapped to the proper VLAN base on the SSID. Further, each VLAN in this scenario has a unique access filter that allows teachers greater access on the wired network.

What will happen if a student decides to configure his adapter with the teacher's SSID? It will still be mapped to the VLAN with the LEAP authentication policy, which the student passes, after which the student will be mapped into the teacher's VLAN using the teacher's SSID. This is called VLAN hopping. VLAN hopping happens when an identical authentication type is used in multiple VLAN groups, where two or more groups can pass the identical authentication process.

To prevent VLAN hopping, a third-party service such as a RADIUS server is required to perform SSID or VLAN check assignments based on a user's record. It can be accomplished in two methods:

- RADIUS-based SSID
- RADIUS-based VLAN

In a RADIUS SSID-based verification, after a user successfully authenticates, the RADIUS sends a list of SSIDs that the user is allowed to use. If the SSID that user is using matches the list, the user is mapped into its proper VLAN. If it does not match, the user is not mapped into the VLAN and is disconnected. In Figure 5.12, student John Doe tries to access the network with teacher SSID. Student John Doe is rejected because it does not match the allowed SSID list profile on the RADIUS server.

In RADIUS VLAN-based verification, after the user successfully authenticates, RADIUS assigns the user to a VLAN based on its profile settings. For this method, no SSID is required to be sent by the user. RADIUS statically maps the user to its allowed VLAN. VLAN information is sent back instead of the allowed SSID list.

Figure 5.12 Radius VLAN Control

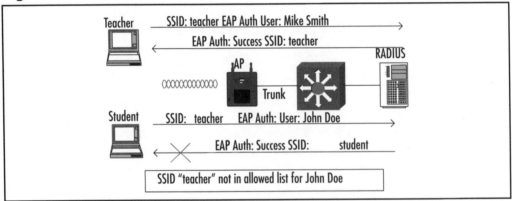

RADIUS verification can only be used when using protocols such as EAP for authentication. You need a per-user authentication method where VLAN restrictions can be verified. If you rely on static WEP key authentication only between multiple VLAN settings, each device or user can hop VLAN by changing the clients SSID.

Configuring RADIUS Control

The RADIUS user attributes used for VLAN-based assignments are:

1. IETF 64: set this to "VLAN"

2. IETF 65: set this to "802" as the tunnel mode type

3. IETF 81: set this to the VLAN ID number you want the user to assume

For a RADIUS SSID control list configure the Cisco's 009/001 **cisco-av-pair**. This Vendor Specific Attribute (VSA) allows you to enter a list of SSIDs that the user is allowed to use in order to authenticate.

To enable and configure a list of allowed SSIDs in a Cisco ACS RADIUS server, go into User Settings and scroll down to "Cisco IOS/PIX RADIUS Attributes." Figure 5.13 shows the enabled attribute with the **ssid=student** value. This will prevent this particular student account from choosing any other SSIDs other than **student** and thus mitigate the VLAN hopping threat. You can add multiple allowed SSIDs per user.

Figure 5.13 Configuring an SSID List in ACS

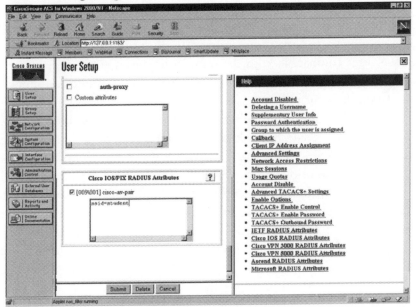

Summary

Wireless VLANs and its technology bring wireless technology closer to acceptance with wired networks. Its integration ability with wired networks allows for scalable wireless solutions. This chapter covered the basic fundamentals of wired and wireless VLANs.

The creation of a VLAN allows you to logically separate network devices into multiple domains. These domains are unique because they work independently from other VLANs, which allow you to configure each of them with a unique characteristics policy. Some of the characteristics you can configure for per-VLAN in wireless network are an authentication method, security filters, and an encryption method. You can configure up to 16 different VLANs with unique characteristics. Each VLAN is represented by a unique SSID. In the past, without VLAN technology, there was only support for one static policy. This prohibited different devices or groups of users not compatible with the static policy from connecting. Administrators needed to purchase extra equipment if they wanted to support multiple groups with different policies.

Access points or bridges with multiple configured VLANs require a connection to a trunk port to the wired side. A trunk port is an interface port configured to transfer more than one VLAN. Since there are multiple VLAN mappings from the wireless users, the access point or bridge needs a way to communicate with the wired network on all of the VLANs. A trunk port uses the 802.1Q encapsulation standard to communicate VLAN information between access points and switches. The access point must also include a native VLAN. The native VLAN tag is used for all traffic coming directly from the access point or to the access point IP address such as SSH, Telnet, or RADIUS administration.

When designing VLANs it is important to remember that you need a Layer 3-aware device such as a router to route between VLANs. For example, you may have a DHCP server that all wireless users need to connect to on the wired network regardless of the VLAN settings.

Each VLAN has its own broadcast domain. A broadcast sent from VLAN A cannot reach users on VLAN B on a wired network. Although this concept is applied to wired networks, it works differently in wireless communication. You cannot prevent a broadcast sent out through the air from reaching a group of users configured on a different VLAN. In wireless networks, you need to configure a unique WEP key for each VLAN to protect your broadcast and multicast traffic. When a broadcast is send out, it is encrypted with the VLAN broadcast WEP key, so that only users belonging to that broadcast domain will recognize its content.

A RADIUS server is used to support and assign users to the proper VLAN. It is required when using an identical authentication policy in more than one VLAN. A RADIUS server prevents users from changing their SSID and hopping to an unauthorized VLAN. RADIUS works only when per-user authentication is used, such as in EAP. It verifies the user's SSID credentials that are used to map VLAN.

Solutions Fast Track

Understanding VLANs

- ☑ A VLAN is used to define the logical separation of a LAN network into multiple broadcast domains.

- ☑ Two configured VLANs cannot interact with each other unless they are routed with a Layer 3–aware device such as router.

- ☑ A trunk port is a configured interface port that allows for multiple VLAN communications. A trunk port is used between the access point and the switch to transfer multiple VLANs using the 802.1q encapsulation standard.

VLANs in a Wireless Environment

- ☑ SSID is used to bind a wireless user to the proper VLAN.

- ☑ Each VLAN can have unique characteristics such as the authentication method, IP filters, and the encryption method. This allows one access point or bridge to support multiple groups of users and devices.

- ☑ A native VLAN is used to tag traffic originating and directed to the IP address of the access point or bridge, such as SSH and HTTP administration.

Wireless VLAN Deployment

- ☑ Currently you can configure up to 16 VLANs. You can only configure up to 16 SSIDs on Cisco's wireless devices.

- ☑ VLANs are supported in VxWorks 12.00T release and IOS 12.2.4-JA release and later.

☑ Av 802.1q trunk port must be configured between two bridges supporting multiple VLAN communications.

Configuring Wireless VLANs in IOS

☑ Multiple SSID configurations using the **ssid** command are configured under interface configuration mode.

☑ Radio and Ethernet interfaces are split into logical sub-interfaces to represent each VLAN configuration.

☑ You should always copy the running configuration and startup configuration to save your configuration in case the device reboots.

Broadcast Domain Segmentation

☑ A broadcast domain segmentation prevents broadcast-directed traffic from one VLAN reaching other VLANs that are considered to be in a separate broadcast domain.

☑ Unlike in wired broadcast segmentation, in 802.11 all broadcasts are seen and processed by every wireless user, even if they are in a different VLAN.

☑ To overcome the differences between 802.11 and a wired network, a broadcast WEP key configuration is required per VLAN. This still does not prevent broadcasts from reaching every wireless user, but it allows only specific VLAN users who know the broadcast key to read its content.

Primary (Guest) and Secondary SSIDs

☑ A guest mode SSID allows users without any SSID to associate to the access point.

☑ The access point sends out a guest SSID in its broadcast beacon to announce its presence.

☑ Only the primary (Guest) SSID can be used in beacons.

Using RADIUS for VLAN Access Control

☑ RADIUS can be used to verify user VLAN mapping and prevent VLAN hopping using unauthorized SSIDs.

☑   RADIUS can either send a list of SSIDs to the user that they are allowed to use, or statically assign a user to a specific VLAN without the need for an SSID.

☑   You can only use RADIUS in a per-user authentication environment such as EAP.

Frequently Asked Questions

The following Frequently Asked Questions, answered by the authors of this book, are designed to both measure your understanding of the concepts presented in this chapter and to assist you with real-life implementation of these concepts. To have your questions about this chapter answered by the author, browse to **www.syngress.com/solutions** and click on the **"Ask the Author"** form.

Q: Why is there a limit on the number of VLANs in wireless networks?

A: Because each VLAN must be represented by a unique SSID and Cisco's wireless devices only support 16 SSIDs.

Q: Why use VLANs if I only have one group of users that share identical policies?

A: VLANs are an optional configuration, and even though you may not require one now, it allows for a future growing scalable environment without the extra expense.

Q: How can I block traffic between wireless users in the same VLAN connecting to the same access point?

A: You can configure Public Secure Packet Forwarding (PSPF) on a per-VLAN basis. PSPF prevents wireless clients in the same VLAN from communicating with each other through the access point.

Q: In multiple VLAN EAP authentication, do I need to make sure that all wireless VLANs can reach the RADIUS server through a Layer 3-aware device?

A: No. The RADIUS authentication that you provide for authentication is between you and the access point. The access point then initiates the RADIUS request to the RADIUS server on behalf of the client, using its native VLAN tag over the trunk port. The only requirement is that your native VLAN can reach RADIUS server.

Chapter 6

Designing a Wireless Network

Solutions in this chapter:

- **Exploring the Design Process**
- **Identifying the Design Methodology**
- **Understanding Wireless Network Attributes from a Design Perspective**

☑ **Summary**

☑ **Solutions Fast Track**

☑ **Frequently Asked Questions**

Introduction

Up to this point in the book, we've explained the technologies behind wireless networking, as well as some of the essential components used to support a wireless network. Now it's time to begin applying what you have learned thus far to network design. This chapter outlines the framework necessary to design a wireless network. We will also discuss the *process* associated with bringing a network design to fruition.

Initially, we will evaluate the design process with a high-level overview, which will discuss the preliminary investigation and design, followed by implementation considerations and documentation. The goal is to provide the big picture first, and then delve into the details of each step in the process. There are numerous steps—diligently planning the design according to these steps will result in fewer complications during the implementation process. This planning is invaluable because often, a network infrastructure already exists, and changing or enhancing the existing network usually impacts the functionality during the migration period. As you may know, there is nothing worse than the stress of bringing a network to a halt to integrate new services—and especially in the case of introducing wireless capabilities, you may encounter unforeseen complications due to a lack of information, incomplete planning, or faulty hardware or software. The intention of this chapter is to provide you with design considerations to help avoid potential network disasters.

The final portion of this chapter will discuss some design considerations and applications specific to a wireless network. These include signal budgeting, importance of operating system efficiency, signal-to-noise ratios, and security.

Exploring the Design Process

For years, countless network design and consulting engineers have struggled to streamline the design and implementation process. Millions of dollars are spent defining and developing the steps in the design process in order to make more effective and efficient use of time. Many companies, such as Accenture (www.accenture.com), for example, are hired specifically for the purpose of providing processes.

For the network recipient or end user, the cost of designing the end product or the network can sometimes outweigh the benefit of its use. As a result, it is vital that wireless network designers and implementers pay close attention to the details associated with designing a wireless network in order to avoid costly mistakes and forego undue processes. This section will introduce you to the six phases that a sound design methodology will encompass—conducting a preliminary investigation regarding the changes necessary, performing an analysis of the existing network environment, creating a design, finalizing it, implementing that design, and creating the necessary documentation that will act as a crucial tool as you troubleshoot.

Conducting the Preliminary Investigation

Like a surgeon preparing to perform a major operation, so must the network design engineer take all available precautionary measures to ensure the lifeline of the network. Going into the design process, we must not overlook the network that is already in place. In many cases, the design process will require working with an existing legacy network with preexisting idiosyncrasies or conditions. Moreover, the network most likely will be a traditional 10/100BaseT wired network. For these reasons, the first step, conducting a preliminary investigation of the existing system as well as future needs, is vital to the health and longevity of your network.

In this phase of the design process, the primary objective is to learn as much about the network as necessary in order to understand and uncover the problem or opportunity that exists. What is the impetus for change? Almost inevitably this will require walking through the existing site and asking questions of those within the given environment. Interviewees may range from network support personnel to top-level business executives. However, information gathering may also take the form of confidential questionnaires submitted to the users of the network themselves.

It is in this phase of the process that you'll want to gather floor-plan blueprints, understand anticipated personnel moves, and note scheduled structural remodeling efforts. In essence, you are investigating anything that will help you to identify the *who*, *what*, *when*, *where*, and *why* that has compelled the network recipient to seek a change from the current network and associated application processes.

In this phase, keep in mind that with a wireless network, you're dealing with three-dimensional network design impacts, not just two-dimensional impacts that commonly are associated with wireline networks. So you'll want to pay close attention to the *environment* that you're dealing with.

Performing Analysis of the Existing Environment

Although you've performed the preliminary investigation, oftentimes it is impossible to understand the intricacies of the network in the initial site visit. Analyzing the existing requirement, the second phase of the process, is a critical phase to understanding the inner workings of the network environment.

The major tasks in this phase are to understand and document all network and system dependencies that exist within the given environment in order to formulate your approach to the problem or opportunity. It's in this phase of the process that you'll begin to outline your planned strategy to counter the problem or exploit the opportunity and assess the feasibility of your approach. Are there critical interdependencies between network elements, security and management systems, or billing and accounting systems? Where are they located physically and how are they interconnected logically?

Although wireless systems primarily deal with the physical and data-link layers (Layers 1 and 2 of the OSI model), remember that, unlike a traditional wired network, access to your wireless network takes place "over the air" between the client PC and the wireless access point (AP). The point of entry for a wireless network segment is critical in order to maintain the integrity of the overall network. As a result, you'll want to ensure that users gain access at the appropriate place in your network.

Creating a Preliminary Design

Once you've investigated the network and identified the problem or opportunity that exists, and then established the general approach in the previous phase, it now becomes necessary to create a preliminary design of your network and network processes. All of the information gathering that you have done so far will prove vital to your design.

In this phase of the process, you are actually transferring your approach to paper. Your preliminary design document should restate the problem or opportunity, report any new findings uncovered in the analysis phase, and define your approach to the situation. Beyond this, it is useful to create a network topology map, which identifies the location of the proposed or existing equipment, as well as the user groups to be supported from the network. A good network topology will give the reader a thorough understanding of all physical element locations and their connection types and line speeds, along with physical room or landscape references. A data flow diagram (DFD) can also help explain new process flows and amendments made to the existing network or system processes.

It is not uncommon to disclose associated costs of your proposal at this stage. However, it would be wise to communicate that these are estimated costs only and are subject to change. When you've completed your design, count on explaining your approach before the appropriate decision-makers, for it is at this point that a deeper level of commitment to the design is required from both you and your client.

It is important to note that, with a wireless network environment, terminal or PC mobility should be factored into your design as well as your network costs. Unlike a wired network, users may require network access from multiple locations, or continuous presence on the network between locations. Therefore, additional hardware or software, including PC docking stations, peripherals, or applications software may be required.

Finalizing the Detailed Design

Having completed the preliminary design and received customer feedback and acceptance to proceed, your solution is close to being implemented. However, one last phase in the design process, the detailed design phase, must be performed prior to implementing your design.

In the detailed design phase, all changes referenced in the preliminary design review are taken into account and incorporated into the detailed design accordingly. The objective in this phase is to finalize your approach and capture all supporting software and requisite equipment on the final Bill of Materials (BOM). It is in this phase that you'll want to ensure that any functional changes made in the preliminary design review do not affect the overall approach to your design. Do the requested number of additional network users overload my planned network capacity? Do the supporting network elements need to be upgraded to support the additional number of users? Is the requested feature or functionality supported through the existing design?

Although wireless networking technology is rapidly being embraced in many different user environments, commercial off-the-shelf (COTS) software is on the heels of wireless deployment and is still in development for broad applications. As a result, you may find limitations, particularly in the consumer environment, as to what can readily be supported from an applications perspective.

Executing the Implementation

Up to this point, it may have felt like an uphill battle; however, once that you've received sign-off approval on your detailed design and associated costs, you are now ready to begin the next phase of the design process—implementing your design. This is where the vitality of your design quickly becomes evident and the value of all your preplanning is realized.

As you might have already suspected, this phase involves installing, configuring, and testing all supporting hardware and software that you have called for in your network design. Although this may be an exhilarating time, where concept enters the realm of reality, it is vital that you manage this transition in an effective and efficient manner. Do not assume that the implementation is always handled by the network design engineer. In fact, in many large-scale implementations, this is rarely the case.

The key in this phase of the process is minimizing impact on the existing network and its users, while maximizing effective installation efforts required by the new network design. However, if your design calls for large-scale implementation efforts or integration with an existing real-time network or critical system process, I would highly recommend that you utilize skilled professionals trained in executing this phase of the project. In doing so, you'll ensure network survivability and reduce the potential for loss in the event of network or systems failure.

There are many good books written specifically on the subject of project management and implementation processes that outline several different approaches to this key phase and may prove useful to you at this point. At a minimum, from a wireless network perspective, you'll want to build and test your wireless infrastructure as an independent and isolated network, whenever possible, prior to integrating this segment with your existing network. This will aid you in isolating problems inherent to your design and will correct the outstanding issue(s) so that you may complete this phase of the process. Similarly, all nodes within the wireless network should be tested independently and added to the wireless network in building-block fashion, so that service characteristics of the wireless network can be monitored and maintained.

Capturing the Documentation

Although the last phase of this process, capturing the documentation, has been reserved for last mention, it is by no means a process to be conducted solely in the final stages of the overall design process. Rather, it is an iterative process that actually is initiated at the onset of the design process. From the preliminary investigation phase to the implementation phase, the network design engineer has captured important details of the existing network and its behavior, along with a hardened view of a new network design and the anomalies that were associated with its deployment.

In this process phase, capturing the documentation, the primary focus is to preserve the vitality and functionality of the network by assembling all relevant network and system information for future reference. Much of the information you've gathered along the way will find its way into either a user's manual, an instructional and training guide, or troubleshooting reference material. Although previous documentation and deliverables may require some modification, much can be gleaned from the history of the network design and implementation process. Moreover, revisiting previous documentation or painstakingly attempting to replicate the problem itself may result in many significant findings.

For these reasons, it is crucial to your success to ensure that the documentation procedures are rigorously adhered to throughout the design and implementation process. Beyond network topology maps and process flow diagrams, strongly consider using wire logs and channel plans wherever possible. Wire logs provide a simple description of the network elements, along with the associated cable types, and entry and exit ports on either a patch panel or junction box. Channel plans outline radio frequency (RF) channel occupancy between wireless access points. Trouble logs are also invaluable tools for addressing network issues during troubleshooting exercises. In all cases, the information that you have captured along the way will serve to strengthen your operational support and system administration teams, as well as serve as an accurate reference guide for future network enhancements.

Identifying the Design Methodology

There are many ways to create a network design, and each method must be modified for the type of network being created. At the beginning of this chapter, we outlined the necessary phases for a sound design methodology (preliminary investigation, analysis, preliminary design, detailed design, implementation, and documentation). Nevertheless, network types can vary from service provider to enterprise, to security, and so on. As wireless networking becomes more commonplace, new design methodologies tuned specifically for the wireless environment will be created.

In this chapter, we give you an overview of the piece of the engagement methodology that provides Lucent consultants a framework for applying their technical expertise during the various stages of the network lifecycle. Referred to as the Network Engagement Methodology (NEM), it is a tool developed by the consultants of Lucent ESS and provides best practices, procedures, and tools from their most successful projects. What you will see in this chapter is the basis for what makes up the final network design (the other phases of NEM include business development, initiation and definition, planning, execution and control, and finally, closeout. This section provides information on the execution and control phase, specifically tuned for *a service provider network*. The execution and control phase has been broken down into five stages: *plan*, *architect*, *design*, *implement*, and *operate*. The next several sections provide a high-level description of what makes up the plan, architect, and design stages of NEM.

Creating the Network Plan

Every good network design begins with a well thought out plan. The *network plan* is the first step in creating a network design. It is where information regarding desired services, number of users, types of applications, and so forth is gathered. This phase is the brainstorming phase during which the initial ideas are put together. The planning stage can be one of the longest segments of a network design, because it is dependent on several factors that can be very time consuming. However, if each planning step is thoroughly completed, the architecture and design stages move along much more quickly.

Gathering the Requirements

The first and most important step in creating a network plan is to gather the requirements. The requirements will be the basis for formulating the architecture and design. If a requirement is not identified at the beginning of the project, the entire design can miss the intended goal of the network. The requirements include:

- **Business Requirements** A few examples of possible business requirements are budget, time frame for completion, the impact of a network outage, and the desired maintenance window to minimize the negative effects of an outage.

- **Regulatory Issues** Certain types of wireless networks (such as MMDS) require licenses from the FCC. If the wireless network is going to operate outside of the public RF bands, the regulatory issues need to be identified.

- **Service Offerings** These are the primary justification for the design of a new network or migration of an existing network. Simply, these are services or functionality the network will provide to the end users.

- **Service Levels** Committed information rate (CIR) is an example of a service level agreement (SLA). This involves the customer's expectation of what the service provider guarantees to provide.

- **Customer Base** This establishes who the anticipated end users are, and what their anticipated applications and traffic patterns are.

- **Operations, Management, Provisioning, and Administration Requirements** These identify how the new network will impact the individuals performing these job functions, and whether there will be a need to train these individuals.

- **Technical Requirements** These can vary from a preferred equipment vendor to management system requirements.

- **Additional Information** Any additional information that can affect the outcome of the design.

Once all of the requirements have been collected, it is recommended that a meeting be set up with the client to ensure that no key information is missing. This is important because it not only keeps the client involved, but also allows both the client and network architect to establish and understand the expectations of the other. Once you get client buy-off on the goals and requirements of the network, you can proceed with baselining the existing network.

Baselining the Existing Network

The reason you need to baseline the existing network is to provide an accurate picture of the current network environment. This information will be used later on to identify how the new design will incorporate/interface with the existing network. When conducting the baseline, be sure to include the following considerations:

- Business processes

- Network architecture

- IP addressing

- Network equipment

- Utilization

- Bandwidth

- Growth

- Performance

- Traffic patterns

- Applications

- Site identification/Surveys

- Cost analysis

With proper identification of these items, you will gain a good understanding of both the existing network and get an idea of any potential issues or design constraints. In the case of utilization—that is, *overutilization*—unless kept under a watchful eye, it can contribute to a less-than-optimized network. Therefore, by evaluating the health of the existing network, you can either eliminate or compensate for potential risks of the new network. In addition to monitoring network conditions, it is also a good idea to perform site surveys in this step, to identify any possible problems that are not identified in either the requirements collection or the baseline monitoring.

Analyzing the Competitive Practices

When you compare the client's business and technology plan to the competitors' in the same industry, you can learn what has and hasn't worked and why. Once you have evaluated and understand the industry practices, you can identify what not to do as well. This is a potential opportunity for a network architect to influence the functionality, in terms of services and choice of technology, that will facilitate the desired network. The primary reason the architect is involved is because of his or her knowledge of the technology—not only how it works, but also how it is evolving.

Beginning the Operations Planning

The operations systems support daily activities of telecommunications infrastructures. The purpose of this step is to identify all of the elements required for the operations system. Depending on the needs of the client, any or all of the following processes need to be identified:

- Pre-order
- Order management
- Provisioning
- Billing
- Maintenance
- Repair
- Customer care

If your client is not planning on offering any services with the new design, then this step can be skipped. Once the operations planning step is complete, you can move on to the *gap analysis*.

Performing a Gap Analysis

The *gap analysis* will be a comparison of the existing network to the future requirements. The information obtained through the gathering of requirements and baselining of the current network provide the data needed to develop a gap analysis.

The gap analysis is a method of developing a plan to improve the existing network, and integrate the new requirements. The documented result should include the following items:

- Baseline
- Future requirements
- Gap analysis
- Alternative technology options
- Plan of action

Once the client reviews and accepts the requirements' definition document and gap analysis, the time frame required to complete the project becomes more evident. At this point, the client should have a good understanding of what the current network entails and what it will take to evolve into their future network. Once this step is complete, the next task is to create a *technology plan*.

Creating a Technology Plan

This step involves identifying the technology that will enable the business goals to be accomplished. There can be several different technology plans—a primary plan and any number of alternatives. The alternative plans can be in anticipation of constraints not uncovered yet, such as budget. Being able to provide alternatives allows the client some options; it provides them with a choice regarding the direction of their network and the particular features that are of top priority. Oftentimes, until a plan is devised and on paper, the "big picture" (the process from ideas to a functioning network) can be somewhat difficult to realize fully.

The *technology plan* should identify what types of equipment, transport, protocols, and so on will be used in the network. Make sure that the plan has both a short-term focus (usually up to a year), and a long-term outlook (typically a 3 to 5 year plan). Creating a good technology plan requires that you understand the existing technology, migration paths, and future technology plans. There are several steps you can take when creating a technology plan. Some of the more important steps include:

- Business assessment
- Future requirements analysis
- Current network assessment
- Identifying technology trends and options
- Mapping technology to client needs

The technology plan will not contain specific details about how the new network will operate—it will identify the technologies that will enable the network.

Creating an Integration Plan

Whenever a new service, application, network component, or network is added to an existing network, an integration plan needs to be created. The *integration plan* will specify what systems will be integrated, where, and how. The plan should also include details as to what level of testing will be done prior to the integration. Most importantly, the integration plan must include the steps required to complete the integration. This is where the information from the gap analysis is utilized. As you may recall, the gap analysis provides information on what the network is lacking, and the integration plan provides the information on how the gaps will be resolved.

Beginning the Collocation Planning

If the network needs to locate some of its equipment off the premises of the client, collocation agreements will need to be made. Specifically with wireless networks, if you plan on connecting buildings together and you lease the buildings, you will need to collocate the equipment on the rooftops. Depending on the amount of collocation required, this step can be skipped or it can be a significantly large portion of the plan phase.

Performing a Risk Analysis

It is important to identify any risks that the client could be facing or offering its perspective customers. Once the risks have been identified, you will need to document and present them to the client. The way to identify risks is by relating them to the return they will provide (such as cost savings, increased customer satisfaction, increased revenue, and so on). An easy way to present the various risks is in a matrix form, where you place risk on the horizontal axis and return on the vertical axis. Assign the zero value of the matrix (lower left corner) a low setting for both risk and return, and assign the max value (upper right corner) a high setting. This provides a visual representation of the potential risks. Once the matrix is created, each service can be put in the matrix based on where they fit. An example of this would be providing e-mail service, which would be put in the lower left corner of the matrix (low risk, low return).

This is important because you are empowering the client to make certain decisions based on industry and technological information. For example, if the client is planning on offering a service and is unaware that the service is high risk with low return, the client will need to offset or eliminate the risk. Perhaps the client could offer a service package pairing the high risk, low return with a low risk, high return service. After all, the goal is to help make your client successful. Once the client accepts the risk analysis, the *action plan* can be created.

Creating an Action Plan

Once all of the previous planning steps have been completed, an action plan needs to be created. The *action plan* identifies the recommended "next steps." The recommended next steps can either identify what needs to be done to prepare for the architecture phase (such as a project plan), or what action needs to be taken to clarify/correct any problems encountered during the planning phase. For example, with a situation as indicated in the risk analysis section previously, the action plan may need to provide a solution to a particular risk. Basically, the action plan functions to address any open issues from the information gathering stages. This step is to ensure all of the required information has been obtained in order to provide the best solution for the client. As soon as the action plan is created and approved, the planning deliverables can be prepared.

Preparing the Planning Deliverables

The last step in the plan phase is to gather all information and documentation created throughout the plan and put them into a deliverable document. This is somewhat of a sanity checkpoint, in terms of making the client fully aware of the plans you have devised and what to expect for the remainder of the project. Some of the items to include in the document are:

- Requirements document
- Current environment analysis
- Industry practices analysis
- Operations plan
- Gap analysis
- Technology plan
- Collocation plan
- Risk analysis
- Action plan

Once the planning deliverable document is complete and has been presented to the client, the next phase of the network design can begin.

Developing the Network Architecture

The *network architecture* is also referred to as a *high-level* design. It is a phase where all of the planning information is used to begin a conceptual design of the new network. It does not include specific details to the design, nor does it provide enough information to begin implementation. (This will be explained in greater detail in the following sections.) The architecture phase is responsible for marrying the results of the planning phase with the client's expectations and requirements for the network.

Reviewing and Validating the Planning Phase

The first step in developing a network architecture is to review and validate the results of the planning phase. Once you have thoroughly gone through the results of the planning phase, and you understand and agree to them, you are finished with this step and can move on to creating a high level topology. The reason that this step is included here is that many times teams on large projects will be assembled but the architecture team can consist of people that were *not* in the plan team. This step is to get everyone familiar with what was completed prior to his or her participation.

Creating a High-Level Topology

A *high-level topology* describes the logical architecture of a network. The logical architecture should describe the functions required to implement a network and the relationship between the functions. The logical architecture can be used to describe how different components of the network will interoperate, such as how a network verifies the authentication of users. The high-level topology will not include such granularity as specific hardware, for example; rather, it illustrates the desired functionality of the network. Some of the components to include in the high-level topology are:

- Logical network diagrams
- Functional network diagrams
- Radio frequency topology
- Call/Data flows
- Functional connectivity to resources
- Wireless network topology

Creating a Collocation Architecture

Once the *collocation plan* has been complete, a more detailed architecture needs to be created. The architecture should include information that will be used as part of the requirements package that you give to vendors for bids on locations. Information to include in the requirements includes:

- Power requirements in Watts
- Amperage requirements
- Voltage (both AC and DC) values
- BTU dissipated by the equipment
- Equipment and cabinet quantity and dimensions
- Equipment weight
- Equipment drawings (front, side, top, and back views)
- Environmental requirements

The intention of this type of architecture is to provide information to assist in issuing either a request for information (RFI) or a request for proposal (RFP) to a vendor(s). It is in the best interest of the client to include enough information about the network requirements to evoke an adequate response from the vendor, but not give away information that potentially could be used for competitive intelligence.

Defining the High-Level Services

The services that the client plans on offering their customers will usually help determine what the necessary equipment requirements will be. These services should match up with the services identified in the risk portion of the plan phase. Once the services have been identified, they need to be documented and compared against the risk matrix to determine what services will be offered. The client typically will already have identified the types of services they are interested in providing, but this is an opportunity to double-check the client's intentions. Any services that will not be offered need to be removed from the architecture. Once you have presented the documented services and get the client's service offering list, you can move on to creating a high-level physical design.

Creating a High-Level Physical Design

The *high-level physical design* is the most important step in the architecture phase and is usually the most complicated and time consuming. A lot of work, thought, and intelligence go into this step. It defines the physical location and types of equipment needed throughout the network to accomplish its intended operation. It does not identify specific brands or models of equipment, but rather functional components such as routers, switches, access points, etc. The high-level physical design takes the RF topology, for example, completed in the high-level topology step, and converts that to physical equipment locations. Due to the many unknowns with RF engineering, several modifications and redesigns may be necessary before this step is complete. Upon acceptance of the high-level physical design, the operations services needs to be defined.

Defining the Operations Services

The purpose of defining the *operations services* is to identify the functionality required within each operations discipline. Some of the more common operations disciplines include:

- Pre-order
- Order management
- Provisioning
- Billing
- Maintenance
- Repair
- Customer care

Once the functionality for each discipline has been defined, documented, and accepted, you are ready to create a high-level operations model.

Creating a High-Level Operating Model

If a network can't be properly maintained once built, then its success and even its life can be in jeopardy. The purpose of creating a *high-level operating model* is to describe how the network will be managed. Certainly a consideration here is how the new network management system will interoperate with the existing management system. Some of the steps that need to be considered when creating a high-level operating model include:

- Leveraging technical abilities to optimize delivery of management information

- Providing an easily managed network that is high quality and easy to troubleshoot

- Identifying all expectations and responsibilities

The high-level operating model will be used later to create a detailed operating model. Once the high-level operating model has been developed and accepted by the client, you can proceed with evaluating the products for the network.

Evaluating the Products

In some cases, the step of evaluating the products can be a very lengthy process. Depending on the functionality required, level of technology maturity, and vendor availability/competition, this can take several months to complete. When evaluating products, it is important to identify the needs of the client and make sure that the products meet all technical requirements. This is where the responses from the RFI/RFP will be evaluated. However, if the project is not of a large scale, it may be the responsibility of the design engineer to research the products available on the market. Once the list of products has been identified, an evaluation needs to be performed to determine which vendor will best fit the client. There are several factors that affect the decision process including:

- Requirement satisfaction

- Cost

- Vendor relationship

- Vendor stability

- Support options

- Interoperability with other devices

- Product availability

- Manufacturing lag time

The result of this step should leave you with each product identified to the model level for the entire network. Once the products have been identified, an action plan can be created.

Creating an Action Plan

The *action plan* will identify what is necessary to move on to the design phase. The action plan's function is to bridge any gaps between the architectural phase and the actual design of the network. Some of the items for which an action plan can be given are:

- Create a project plan for the design phase
- Rectify any problems or issues identified during the architecture phase
- Establish equipment and/or circuit delivery dates

This is another checkpoint in which the network architect/design engineer will verify the progression and development direction of the network with the client. Once the action plan is complete and approved by the client, the network architecture deliverables can be created.

Creating the Network Architecture Deliverable

During this step, all of the documents and information created and collected during the architecture phase will be gathered and put into a single location. There are several different options for the location of the deliverable, such as:

- Master document
- CD-ROM
- Web page

Any and all of the methods listed can be used for creating the architecture deliverable. One thing to include in this step is the deliverables from the plan phase as well. This lets the client reference any of the material up this point. Also, as new documents and deliverables are developed, they should be added. Once the architecture deliverable has been completed and it has been presented to the client, the detailed design phase can begin.

Formalizing the Detailed Design Phase

The *detailed design phase* of the NEM is the last step before implementation begins on the network. This phase builds on the architecture phase and fills in the details of each of the high-level documents. This is the shortest and easiest phase of the design (assuming the plan and architecture phase was completed thoroughly and with accurate information). Basically, the detail design is a compilation of the entire planning process. This is absolutely where the rewards of the prior arduous tasks are fully realized.

Reviewing and Validating the Network Architecture

The first step of a detailed design phase is to review and validate the network architecture. The network architecture is the basis for the design, and there must be a sanity check to ensure that the architecture is on track. This involves making sure all of the functionality is included. As you did at the beginning of the architecture phase, you may be validating work done by other people. Once the network architecture has been validated, you begin the detailed design by creating a detailed topology.

Creating the Detailed Topology

The *detailed topology* builds on the high-level topology, adding information specific to the network topology, such as:

- Devices and device connectivity
- Data/voice traffic flows and service levels
- Traffic volume
- Traffic engineering
- Number of subscribers
- IP addressing
- Routing topology
- Types of technology
- Location of devices
- Data-link types
- Bandwidth requirements
- Protocols
- Wireless topology

The detailed topology is a functional design, not a physical design. The detailed topology is where client dreams become a reality. By this point the client should be fully aware of what they would like the network to offer, and your job is to make it happen. In addition to the documented results, you should have detailed drawings of the various topologies listed earlier. Once the detailed topology is complete, a detailed collocation design can be created.

Creating a Detailed Service Collocation Design

As with the detailed topology, the detailed service collocation design builds on the collocation architecture. This step will provide the details necessary to install equipment in collocation facilities. Include the following information with the design:

- Network Equipment Building Standards (NEBS) compliance
- Facilities
- Cabling

Once the detailed service collocation design is complete and accepted by the client, it can be presented to the collocation vendor for approval. Once the vendor approves the design, the implementation phase for collocation services can begin.

Creating the Detailed Services

This step will define and document the specific services that the client will offer to its customers. The services offered are a continuation of the services list identified in the high-level services design step. When creating the design, be sure to include information such as timeline for offering. This information will most likely be of interest to the client's marketing department. You can easily understand that in a service provider environment, the customers and the resulting revenue justify the network. Some of the information to provide with each service includes:

- Service definition
  - Service name
  - Description
  - Features and benefits
  - SLAs
- Service management
- Functionality
- Configuration parameters
- Access options
- Third-party equipment requirements
- Service provisioning
- Network engineering

- Customer engineering
- Service options

Not only do you need to provide information regarding when these services will be available, but you should include how they will be offered and how they will interface with the network. Once the detailed services have been created, they can be put to the implementation process.

Creating a Detailed Physical Design

The detailed physical design builds on the high-level physical design. It specifies most of the physical details for the network including:

- Equipment model
- Cabling details
- Rack details
- Environment requirements
- Physical location of devices
- Detailed RF design

The detailed physical design builds on information identified in the following documents:

- High-level physical design
- Detailed topology
- Detailed service collocation
- Product evaluation
- Site survey details

The detailed physical design is a compilation of these items as well as finalized equipment configuration details including IP addressing, naming, RF details, and physical configuration. When you finish this step you should have a detailed physical drawing of the network as well as descriptions of each of the devices.

Creating a Detailed Operations Design

The *detailed operations design* builds on the high-level operations design. The purpose of this step is to specify the detailed design of the support systems that will be implemented to support the network. Some of the results of this step include determining vendor products, identifying technical and support requirements, and determining costs. Major steps in this phase include:

- Develop systems management design

- Develop services design

- Develop functional architecture

- Develop operations physical architecture analysis and design

- Develop data architecture

- Develop OSS network architecture

- Develop computer platform and physical facilities design

The detailed operations design is complete when it is documented and reviewed. After it is complete, the detailed operating model can be designed. Due to the fact that the operations network can be very small (or nonexistent), or that it could be an entirely separate network with its own dedicated staff, the specific details for this step in the design process has been summarized. In large network projects, the operations design can be a completely separate project, consisting of the full NLM process.

Creating a Detailed Operating Model Design

This step is intended to describe the operating model that will optimize the management of the network. The detailed design builds on the high-level operating model. When creating the detailed design you should answer as many of the following questions as possible:

- Which organizations will support what products and services, and how?

- Who is responsible for specific tasks?

- How will the organization be staffed?

- How do the different organizations interact?

- How long will a support person work with an issue before escalating it?

- How will an escalation take place?

- Which procedures will be automated?

- What tools are available to which organization?
- What security changes are required?

Depending on the size of the network, the management network may be integrated in the main network, or it could be its own network. Additionally, the management network might run on the single network administrator's PC (for a very small network), or it could be run in a large Network Operations Center (NOC) staffed 24 hours a day, or anywhere in between. Because of the variations in size and requirements to network management, only a brief description is provided on what needs to be done. On larger networks, often the management design is an entirely separate design project deserving its own NLM attention.

Creating a Training Plan

Depending on the size of the new network and the existing skill set of the staff, the *training plan* can vary greatly. Interviewing existing staff, creating a skills matrix, and comparing the skills matrix to the skills needed to operate the network can help determine training needs. If the client wants to perform the implementation on his or her own, that needs to be considered when reviewing the matrix. Once the training needs have been determined, create a roadmap for each individual, keeping future technologies in mind. Once the roadmaps have been created and the client accepts them, this step is finished.

Developing a Maintenance Plan

This step in the design phase is intended to plan and identify how maintenance and operations will take place once the network is operational. The *maintenance plan* should cover all pieces of the network including operations and management. Also, the plan needs to take the skill set and training needs into consideration. Once a maintenance plan is developed and the client agrees to it, the implementation plan can be developed.

Developing an Implementation Plan

The *high-level implementation plan* should be an overview of the major steps required to implement the design. It should be comprehensive and it should highlight all steps from the design. Things to include in this step should be timelines, impact on existing network, and cost. The implementation plan and the detailed design documents will be the basis for the next phase: implementing the network design.

Creating the Detailed Design Documents

The *detailed design documents* should be a summarized section of all of the documents from the entire design phase, as well as the architecture and plan deliverables. As with the architecture deliverable, we recommend that you present this information in several forms, including (but not limited to) CD-ROM, a single design document, or a dedicated Web site. Once this step is complete, the design phase of the project is finished. The next step is to move on to the implementation phase and install the new network. The details for the implementation phase are specific to each design.

Now that you have been through a detailed examination of the how and why of network design, let's look at some design principles specific to wireless networking.

Understanding Wireless Network Attributes from a Design Perspective

In traditional short-haul microwave transmission (that is, line-of-sight microwave transmissions operating in the 18 GHz and 23 GHz radio bands), RF design engineers typically are concerned with signal aspects such as fade margins, signal reflections, multipath signals, and so forth. Like an accountant seeking to balance a financial spreadsheet, an RF design engineer normally creates an RF budget table, expressed in decibels (dB), in order to establish a wireless design. Aspects like transmit power and antenna gain are registered in the assets (or plus) column, and free space attenuation, antenna alignment, and atmospheric losses are noted in the liabilities (or minus) column. The goal is to achieve a positive net signal strength adequate to support the wireless path(s) called for in the design.

As we continue to build a holistic view of the design process, it is important to take into account those signal characteristics unique to wireless technologies from several design perspectives. We will explore both sides of the spectrum, so to speak, examining characteristics that are unique and beneficial to implementation—as well as those that make this medium cumbersome and awkward to manage. Equally important is the ability to leverage these attributes and apply them to meet your specific needs. Ultimately, it is from this combined viewpoint of understanding RF signal characteristics as well as exploiting those wireless qualities that we approach this next section.

For the sake of clarity, however, it is worth reiterating that the wireless characteristics described in the following sections are not focused on traditional short-haul licensed microwave technologies. Furthermore, it is not our intent to delve deeply into radio frequency theory or the historical applications of line-of-sight Point-to-Point Microwave. Rather, the purpose at this juncture is to entice you into exploring the possibilities of unlicensed wireless technologies by examining their characteristics from several design perspectives.

Application Support

Interest in wireless LAN technologies has skyrocketed dramatically over the last few years. Whether the increase in popularity stems from the promise of mobility or the inherent ability to enable a network with minimal intrusion, interest in wireless LAN technologies remains high. However, these aspects by themselves do not validate the need to embrace a wireless network—or any other network for that matter. To understand the real cause for adopting a network, wireless or otherwise, we must look to the intrinsic value of the network itself. What is the purpose of the network? How will the network enhance my current processes? Does the overall benefit of the network outweigh all operational, administrative, and maintenance (OAM) costs associated with deploying it?

In our search to find that intersection between cost and benefit, we ultimately come to the realization that it is the applications and services that are supported over the network that bring value to most end users. Except for those truly interested in learning how to install, configure, or support wireless or wireline networks, most users find the value of a given network to be in the applications or services derived from what is on the network. So then, how do unlicensed wireless technologies enhance user applications, and what are some of the associated dependencies that should be considered to support these applications or services?

It is undisputed that one of the key aspects of wireless technology is the inherent capability to enable mobility. Although wireless applications are still largely under development, services that accommodate demands for remote access are emerging rapidly. From *web clipping*, where distilled information requested on behalf of a common user base is posted for individual consumption upon request, to e-mail access and retrieval from remote locations within the network footprint, wireless personal information services are finding their place in our mobile society.

At this point, it should be realized that one wireless application dependency is found in the supporting form factor or device. Speculation is rampant as to what the ultimate "gadget" will look like. Some believe that the ultimate form factor will incorporate data and voice capabilities, all within a single handheld device. There is movement in the marketplace that suggests corporations and service providers are embracing a single device solution. We only need to look at their own cellular phones or newly released products like the Kyocera QCP 6035 that integrate PDA functionality with cellular voice to see this trend taking hold.

On the other hand, technologies like Bluetooth point to, perhaps, a model whereby applications and services are more easily supported by a two-form factor approach. Although still in the early development stage, with a Bluetooth enabled wireless headset communicating to a supporting handheld device or wristwatch, both voice and data communications may be supported without compromising session privacy or ergonomic function. As a result, from an applications perspective, knowing what physical platform will be used to derive or deliver your application or service is an important design consideration.

Power consumption and operating system efficiency are two more attributes that should be considered when planning applications and services over wireless LAN technologies. Many of us are aware of the importance of battery life, whether that battery is housed in a cellular telephone, laptop, or even the TV remote control. However, it should not go without mention that these two factors play a significant role in designing applications and services for wireless networking.

Unlike normal desktop operations, whereby the PC and supporting peripherals have ready access to nearby wall outlets to supply their power budget, developers that seek to exploit the mobile characteristics of wireless LAN are not afforded the same luxury. As a result, power consumption, heat dissipation, and operating system efficiencies are precious commodities within the mobile device that require preservation whenever the opportunity exists. Companies like Transmeta Corporation understand these relationships and their value to the mobile industry, and have been working diligently to exploit the operating system efficiencies of Linux in order to work beyond these constraints. Nevertheless, applications and service developers should take into account these characteristics in order to maintain or preserve service sessions.

Beyond these immediate considerations, the design developer may be limited in terms of what types of services, including supporting operating systems and plug-ins, are readily available. Synchronous- or isochronous-dependent services may prove difficult to support, based on the wireless transport selected. Therefore, take caution as you design your wireless service or application.

Subscriber Relationships

Unlike wired LAN topologies, where physical attachment to the network is evidenced merely by tracing cables to each respective client, physical connectivity in a wireless network is often expressed in decibels (dB) or decibel milliwatts (dBm). Simply put, these are units of measure that indicate signal strength expressed in terms of the signal levels and noise levels of a given radio channel, relative to 1 watt or 1 milliwatt, respectively. This ratio is known as a signal-to-noise (S/N) ratio, or SNR.

As a point of reference, for the Orinoco RG1000 gateway, the SNR level expressed as a subjective measure is shown in Figure 6.1.

Figure 6.1 SNR Levels for the Orinoco RG1000

From a wireless design perspective, subscriber relationships are formed, not only on the basis of user authentication and IP addressing, as is common within a wired network, but also on the signal strength of a client and its location, a secure network ID, and corresponding wireless channel characteristics. Taking into account, as an example, the wireless channel plan defined in the 802.11b specification, remember that Lucent Technologies AP1000 access point affords the user with a total of 11 useable channels to transport data. It is imperative that the network design engineer understands the subscriber relationships to be supported and develops a channel plan accordingly. Let's take a closer look.

Like traditional short-haul microwave technologies, 802.11 direct sequence spread spectrum (DSSS) wireless technology requires frequency diversity between different radios. Simply stated, user groups on separate access points within a wireless LAN must be supported on separate and distinct channels within that wireless topology. Similarly, adjacent channel spacing and active channel separation play an important role when planning and deploying a wireless network. These aspects refer to the amount of space between contiguous or active channels used in the wireless network. From a design perspective, the integrity and reliability of the network is best preserved when the channels assigned to access points in the same wireless network are selected from opposite ends of the wireless spectrum whenever possible. Failure to plan in accordance with these attributes most likely will lead to cochannel interference, an RF condition in which channels within the wireless spectrum interfere with one another. In turn, this may cause your service session to lock up, or it may cause severe network failure or total network collapse. Other attributes that depend on subscriber relationships involve network security (we reserve discussion on this characteristic separately in order to consider this wireless attribute more carefully; see the Network Security section later in this chapter).

Physical Landscape

Even if adequate channel spacing, sound channel management, and RF design principles are adhered to, other wireless attributes associated with the given environment must be taken into account. As mentioned at the onset of this section, antennas are constructed with certain gain characteristics in order to transmit and receive information. This attribute of the antenna serves to harness wireless information for transmission or reception; through the use of modulation and demodulation techniques, the transmitted signal ultimately is converted into useable information. However, the propensity of antennas to transmit and receive a signal is regulated largely by the obstructions, or lack thereof, between the transmit antenna and the receive antenna.

Make no mistake, although radio-based spread spectrum technologies do not require line-of-sight between the transmitter and corresponding receiver, signal strength is still determined by the angle in which information is received. The following diagnostic screens in Figures 6.2 (Screen A) and 6.3 (Screen B) show impacts to data when the angle of reception from the emitted signal is changed by less than five degrees.

Figure 6.2 Diagnostic Screen A

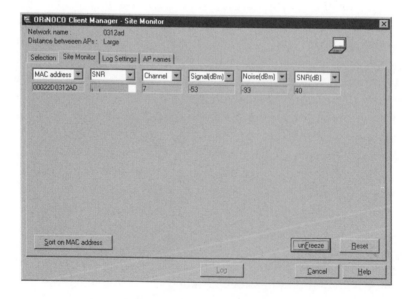

From a physical landscape perspective, we can easily see how physical obstructions may affect signal quality and overall throughput. As such, placement of antennas, angles of reception, antenna gain and distance to the radio should be considered carefully from a design perspective.

Obviously, with each type of antenna, there is an associated cost that is based on the transport characteristics of the wireless network being used. Generally speaking, wireless radios and corresponding antennas that require support for more physical layer interfaces will tend to cost more, due to the additional chipset integration within the system. However, it might also be that the benefit of increased range may outweigh the added expense of integrating more radios to your design.

Figure 6.3 Diagnostic Screen B

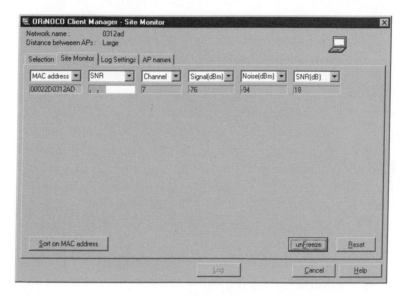

Beyond the physical environment itself, keep in mind that spectral capacity, or available bits per second (bps), of any given wireless LAN is not unlimited. Couple this thought of the aggregate bandwidth of a wireless transport with the density of the users in a given area, and the attribute of *spatial density* is formed. This particular attribute, spatial density, undoubtedly will be a key wireless attribute to focus on and will grow in importance proportionate to the increase in activity within the wireless industry. The reason for this is very clear. The wireless industry is already experiencing congestion in the 2.4 GHz frequency range. This has resulted in a "flight to quality" in the less congested 5 GHz unlicensed spectrum. Although this frequency range will be able to support more channel capacity and total aggregate bandwidth, designers should be aware that, as demand increases, so too will congestion and bandwidth contention in that spectrum. Because of the spectral and spatial attributes of a wireless LAN, we recommend that no more than 30 users be configured on a supporting radio with a 10BaseT LAN interface. However, up to 50 users may be supported comfortably by a single radio with a 100BaseT LAN connection.

Network Topology

Although *mobility* is one of the key attributes associated with wireless technologies, a second and commonly overlooked attribute of wireless transport is the *ease of access*. Let's take a moment to clarify. Mobility implies the ability of a client on a particular network to maintain a user session while roaming between different environments or different networks. The aspect of roaming obviously lends itself to a multitude of services and applications, many yet to be developed. Is mobility the only valuable attribute of wireless technology?

Consider that market researchers predict that functional use of appliances within the home will change dramatically over the next few years. With the emergence of the World Wide Web, many companies are seizing opportunities to enhance their products and product features using the Internet. Commonly referred to as IP appliances, consumers are already beginning to see glimmers of this movement. From IP-enabled microwave ovens to Internet refrigerators, manufacturers and consumers alike are witnessing this changing paradigm. But how do I connect with my refrigerator? Does the manufacturer expect there to be a phone jack or data outlet behind each appliance? As we delve into the details of the wiring infrastructure of a home network, it becomes apparent that the value of wireless technology enables more than just mobility. It also provides the ease of access to devices without disrupting the physical structure of the home.

Whether these wireless attributes are intended for residential use via HomeRF, or are slated for deployment in a commercial environment using 802.11b, mobility and ease of access are important considerations from a design perspective and have a direct impact on the wireless network topology. From a network aspect, the wireless designer is faced with how the wireless network, in and of itself, should function. As stated earlier in this book, wireless LANs typically operate in either an ad-hoc mode or an infrastructure mode. In an ad-hoc configuration, clients on the network communicate in a peer-to-peer mode without necessarily using an access point via the Distributed Coordination Function (DCF) as defined in the 802.11b specification. Alternatively, users may prescribe to the network in a client/server relationship via a supporting access point through the Point Coordination Function (PCF) detailed in the 802.11b specification. It should be determined early in the design process how each client should interact with the network. However, beyond a client's immediate environment, additional requirements for roaming or connectivity to a disparate sub-network in another location may be imposed. It is precisely for these reasons that mobility and wireless access must be factored in from the design perspective early in the design process and mapped against the network topology.

Finally, wireless access should also be viewed more holistically from the physical point of entry where the wireless network integrates with the existing wired infrastructure. As part of your planned network topology, once again, the impacts to the overall network capacity—as well as the physical means of integrating with the existing network—should be considered. The introduction of wireless clients, whether in whole or in part, most likely will impact the existing network infrastructure.

Network Security

It is frequently said that an individual's greatest strengths are often their greatest weaknesses. The same can be said when examining the attributes of a wireless network. Both mobility and ease of access are touted as some of the greatest characteristics available when using a wireless LAN. Unfortunately, these same attributes give cause for the greatest concerns when deploying a wireless network.

Undoubtedly, it is in the best interest of all users on any given network, wired or wireless, to protect the integrity of the network. As a result, corporate network administrators that utilize both wired and wireless networks for corporate traffic normally employ high-level security measures like password authentication and secure login IDs in order to maintain network integrity. Lower level security measures, like installing corporate firewalls, are also commonly deployed in order to discourage or prevent undesirables from entering into both networks. It is at this point (that is, Layer 3 or the network layer of the OSI model) that security practices between a wired network and a wireless network typically traverse down different paths.

In a typical wired network, where Layers 1 and 2 (the physical and data-link layers) are regulated by supplying cable runs and network interfaces to known clients on the network, whereas wireless network emissions are distributed freely across numbers of users, in some cases unbeknownst to others in the same environment. However, because of the general availability of signals to users within the wireless footprint, wireless network providers counter the lack of physical control with additional security measures, namely encryption.

Within the Lucent product set, for example, where 802.11b is utilized, 64-bit key encryption, optional 128-bit key encryption schemes, and a secure network ID serve to counter unauthorized network entry. HomeRF standards leverage the inherent capabilities of FHSS, standard 128-bit encryption, and a user-specified secure ID to counterbalance unauthorized network intrusion. In both cases, encryption mechanisms are deployed over their wired network counterparts.

Many will argue the security merits of one wireless technology over another wireless technology. These arguments stem over ease of symbol rate conversion and unauthorized encrypted packet insertion. Still others may argue the merits of nonencrypted data over wired networks versus encrypted data communicated over a wireless network. Many US government agencies mandate TEMPEST-ready conditions, in which wired emissions are regulated to avoid intrusion. In either case, from a network design perspective, it is vital that the wireless network designer takes appropriate measures to ensure the security and stability of the wireless network. At a minimum, ensure that the logical placement of your wireless access points, if required, are placed appropriately in front of your network firewall. Finally, take into account the value of the information being transmitted and secure it accordingly.

Summary

Designing a wireless network is not an easy task. Many wireless attributes should be considered throughout the design process. In the preliminary stages of your design, it is important to query users in order to accommodate their needs from a design perspective. Keep in mind that with wireless networks, attributes such as mobility and ease of access can impact your network in terms of cost and function.

The methodology used in this chapter incorporates elements of Lucent's Network Engagement Methodology (NEM). The design methodology is broken down into several parts, one being *execution and control*. This part has been categorized to include many of the most common types of projects; the category presented here is based on the service-provider methodologies. The execution and control part is broken down in this chapter into planning, architecture, and design.

The planning phase contains several steps responsible for gathering all information and documenting initial ideas regarding the design. The plan consists mostly of documenting and conducting research about the needs of the client. At the conclusion of the planning phase, documents that provide information such as competitive practices, gap analysis, and risk analysis can be presented to the client.

The architecture phase is responsible for taking the results of the planning phase and marrying them with the business objectives or client goals. The architecture is a high-level conceptual design. At the conclusion of the architecture phase, the client will have documents that provide information such as a high-level topology, a high-level physical design, a high-level operating model, and a collocation architecture.

The design phase takes the architecture and makes it reality. It identifies specific details necessary to implement the new design and is intended to provide all information necessary to create the new network. At the conclusion of the design phase, the design documents provided to the client will include a detailed topology, detailed physical design, detailed operations design, and maintenance plan.

Solutions Fast Track

Exploring the Design Process

- ☑ The design process consists of six major phases: preliminary investigation, analysis, preliminary design, detailed design, implementation, and documentation.

- ☑ In the early phases of the design process, the goal is to determine the cause or impetus for change. As a result, you'll want to understand the existing network as well as the applications and processes that the network is supporting.

☑ Because access to your wireless network takes place "over the air" between the client PC and the wireless access point, the point of entry for a wireless network segment is critical in order to maintain the integrity of the overall network.

☑ PC mobility should be factored into your design as well as your network costs. Unlike a wired network, users may require network access from multiple locations or continuous presence on the network between locations.

Identifying the Design Methodology

☑ Lucent Worldwide Services has created a network lifecycle methodology, called the Network Engagement Methodology (NEM), for its consultants to use when working on network design projects. The design methodology contains the best-of-the-best samples, templates, procedures, tools, and practices from their most successful projects.

☑ The NEM is broken down into several categories and stages; the category presented in this chapter is based on the execution and control category, for a service provider methodology. The execution and control category is broken down into planning, architecture, design, implementation, and operations.

☑ The planning phase contains several steps that are responsible for gathering all information and documenting initial ideas regarding the design. The plan consists mostly of documenting and conducting research about the needs of the client, which produces documents outlining competitive practices, gap analysis, and risk analysis.

☑ The architecture phase is responsible for taking the results of the planning phase and marrying them with the business objectives or client goals. The architecture is a high-level conceptual design. At the conclusion of the architecture phase, a high-level topology, a high-level physical design, a high-level operating model, and a collocation architecture will be documented for the client.

☑ The design phase takes the architecture and makes it reality. It identifies specific details necessary to implement the new design and is intended to provide all information necessary to create the new network, in the form of a detailed topology, detailed physical design, detailed operations design, and maintenance plan.

Understanding Wireless Network Attributes from a Design Perspective

☑ It is important to take into account signal characteristics unique to wireless technologies from several design perspectives. For example, power consumption and operating system efficiency are two attributes that should be considered when planning applications and services over wireless LAN technologies.

☑ Spatial density is a key wireless attribute to focus on when planning your network due to network congestion and bandwidth contention.

Frequently Asked Questions

The following Frequently Asked Questions, answered by the authors of this book, are designed to both measure your understanding of the concepts presented in this chapter and to assist you with real-life implementation of these concepts. To have your questions about this chapter answered by the author, browse to **www.syngress.com/solutions** and click on the **"Ask the Author"** form.

Q: Several customers want me to give them up-front costs for designing and installing a network. When is the most appropriate time to commit to a set price for the job?

A: Try to negotiate service charges based on deliverables associated with each phase of the design process. In doing so, you allow the customer to assess the cost prior to entering into the next phase of the design.

Q: I'm very confused by all the different home network standards. Is there any way that I can track several of the different home networking standards from a single unbiased source?

A: Yes. There are several means of tracking various home network standards and initiatives. For comprehensive reports in the home network industry, I would suggest contacting Parks Associates at www.parksassociates.com. The Continental Automated Buildings Association (CABA) at www.caba.org is another good source for learning about home network technologies from a broad and unbiased perspective.

Q: I am trying to create a design of a wireless campus network and I keep finding out new information, causing me to change all of my work. How can I prevent this?

A: If you have done a thorough job in the planning phase you should already have identified all of the requirements for the project. Once you identify all of the requirements, you need to meet with the client and make sure that nothing was overlooked.

Q: How can I learn more about the Network Engagement Methodology (NEM)?

A: Lucent has a considerable amount of information available on NEM and all of their professional services on their Web site, www.networkcare.com/consulting. From there you can learn more about the various services offered by Lucent ESS, see a live demo of NEM, and read about some of the successful engagements that Lucent has recently completed.

Chapter 7

Wireless Network Architecture and Design

Solutions in this chapter:

- **Fixed Wireless Technologies**

- **Developing WLANs through the 802.11 Architecture**

- **Developing WPANs through the 802.15 Architecture**

- **Mobile Wireless Technologies**

- **Optical Wireless Technologies**

- **Exploring the Design Process**

- **Creating the Design Methodology**

- **Understanding Wireless Network Attributes from a Design Perspective**

☑ **Summary**

☑ **Solutions Fast Track**

☑ **Frequently Asked Questions**

Fixed Wireless Technologies

The basic definition of a fixed wireless technology is any wireless technology where the transmitter and the receiver are at a fixed location such as a home or office, as opposed to mobile devices such as cellular phones. Fixed wireless devices normally use utility main power supplies (AC power), which will be discussed later in more detail. The technologies under fixed wireless can be MMDS connectivity models, LMDS, encompassing WLL, Point-to-Point Microwave, or WLAN.

Fixed wireless technologies provide advantages to service providers in several areas. First, just by nature of the wireless technology, fixed wireless systems provide the ability to connect to remote users without having to install costly copper cable or optical fiber over long distances. The service provider can deploy a fixed wireless offering much quicker and at a much lower cost than traditional wireline services. Also, the service provider can provide services via fixed wireless access without having to use the local service provider's last mile infrastructure. The disadvantages to fixed wireless vary, depending on which technology is being used, but some of the issues include line-of-sight and weather issues as well as interference from various sources, and licensing issues. After we discuss service provider implementations of fixed wireless, we will discuss how fixed wireless benefits the home and enterprise users.

Multichannel Multipoint Distribution Service

Allocated by the Federal Communications Commission (FCC) in 1983 and enhanced with two-way capabilities in 1998, Multichannel Multipoint Distribution Service is a licensed spectrum technology operating in the 2.5 to 2.7 GHz range, giving it 200 MHz of spectrum to construct cell clusters. Service providers consider MMDS a complimentary technology to their existing digital subscriber line (DSL) and cable modem offerings by providing access to customers not reachable via these wireline technologies (see Figure 7.1 for an example of a service provider MMDS architecture).

MMDS provides from 1 to 2 Mbps of throughput and has a relative range of 35 miles from the radio port controller (RPC) based on signal power levels. It generally requires a clear line of sight between the radio port (RP) antenna and the customer premise antenna, although several vendors are working on MMDS offerings that don't require a clear line of sight. The *fresnel* zone of the signal (the zone around the signal path that must be clear of reflective surfaces) must be clear from obstruction as to avoid absorption and reduction of the signal energy. MMDS is also susceptible to a condition known as *multipath reflection*. Multipath reflection or interference happens when radio signals reflect off surfaces such as water or buildings in the fresnel zone, creating a condition where the same signal arrives at different times. Figure 7.2 depicts the fresnel zone and the concept of absorption and multipath interference.

Figure 7.1 MMDS Architecture

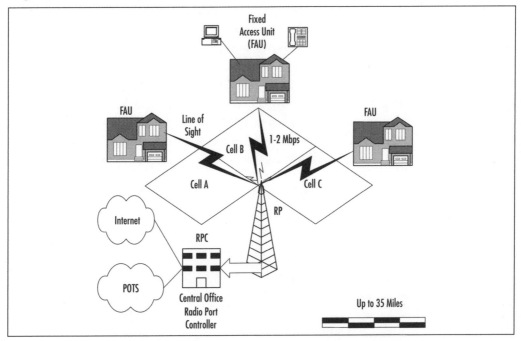

Figure 7.2 Fresnel Zone: Absorption and Multipath Issues

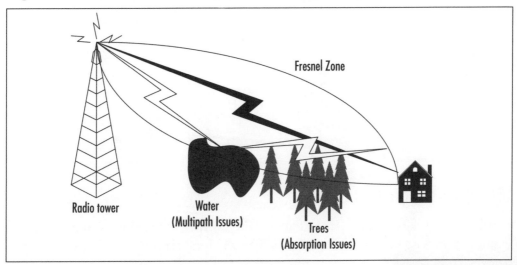

Local Multipoint Distribution Service

Local Multipoint Distribution Service (LMDS) is a broadband wireless point-to-multipoint microwave communication system operating above 20 GHz (28–31 GHz in the US). It is similar in its architecture to MMDS with a couple of exceptions. LMDS provides very high-speed bandwidth (upwards of 500 Mbps) but is currently limited to a relative maximum range of 3 to 5 miles of coverage. It has the same line-of-sight issues that MMDS experiences, and can be affected by weather conditions, as is common among line-of-sight technologies.

LMDS is ideal for short-range campus environments requiring large amounts of bandwidth, or highly concentrated urban centers with large data/voice/video bandwidth requirements in a relatively small area. LMDS provides a complementary wireless architecture for the wireless service providers to use for markets that are not suited for MMDS deployments. Figure 7.3 illustrates a generic LMDS architecture.

Figure 7.3 Local Multipoint Distribution Service (LMDS) Architecture

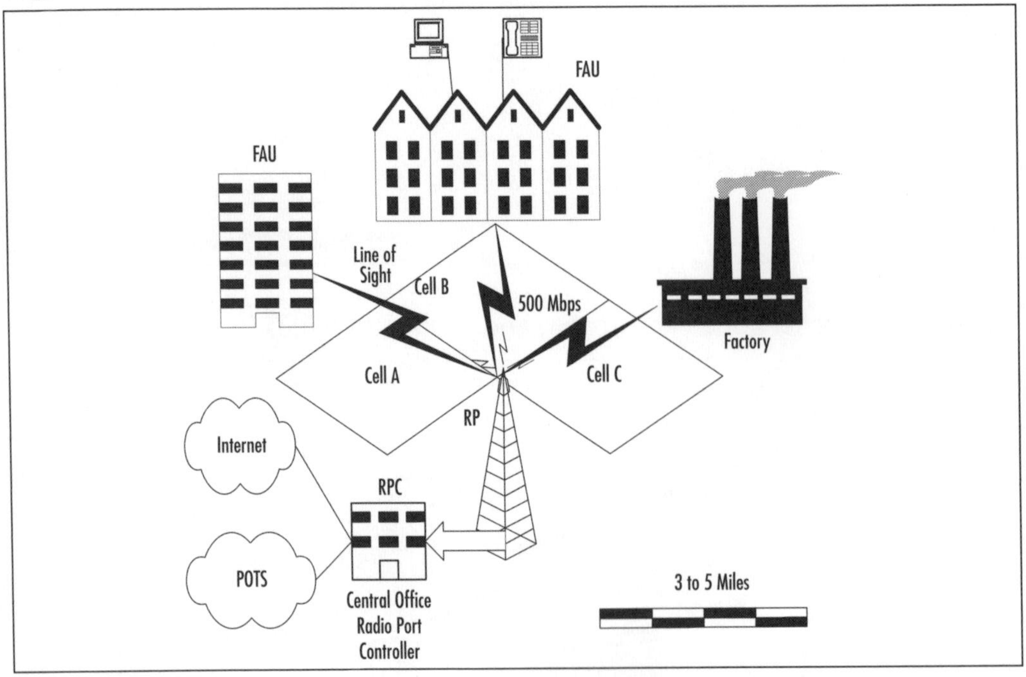

Wireless Local Loop

Wireless Local Loop (WLL) refers to a fixed wireless class of technology aimed at providing last-mile services normally provided by the local service provider over a wireless medium. This includes Plain Old Telephone Service (POTS) as well as broadband offerings such as DSL service. As stated earlier, this technology provides service without the laying of cable or use of the Incumbent Local Exchange Carrier (ILEC), which in layman's terms is the Southwestern Bells of the world.

The generic layout involves a point-to-multipoint architecture with a central radio or radio port controller located at the local exchange (LE). The RPC connects to a series of base stations called radio ports (RPs) via fixed access back to the LE. The RPs are mounted on antennas and arranged to create coverage areas or sectored cells. The radios located at the customer premise, or fixed access unit (FAU), connects to an external antenna optimized to transmit and receive voice/data from the RPs. The coverage areas and bandwidth provided vary depending on the technology used, and coverage areas can be extended through the use of repeaters between the FAU and the RPs. Figure 7.4 provides a generic depiction of a wireless local loop architecture.

Figure 7.4 Wireless Local Loop Architecture

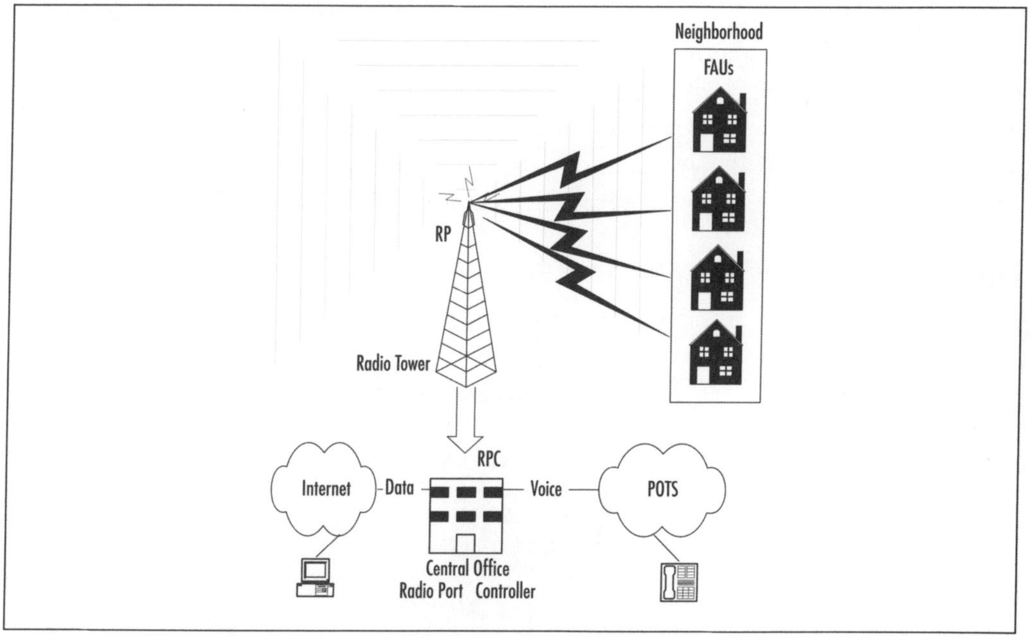

Point-to-Point Microwave

Point-to-Point (PTP) Microwave is a line-of-sight technology, which is affected by multipath and absorption much like MMDS and LMDS. PTP Microwave falls into two categories: licensed and unlicensed, or spread spectrum. The FCC issues licenses for individuals to use specific frequencies for the licensed version. The advantage with the licensed PTP Microwave is that the chance of interference or noise sources in the frequency range is remote. This is critical if the integrity of the traffic on that link needs to be maintained. Also, if the link is going to span a long distance or is in a heavily populated area, the licensed version is a much safer bet since the probability of interference is greater in those cases. The drawback to licensed PTP Microwave is that it may take a considerable amount of time for the FCC to issue the licenses, and there are fees associated with those licenses. Unlicensed PTP Microwave links can be used when a licensed PTP Microwave is not necessary and expediency is an issue.

Since PTP can span long distances, determined mostly by the power of the transmitter and the sensitivity of the receiver, as well as by traditional weather conditions, many different aspects need to be considered in designing a PTP Microwave link. First, a site survey and path analysis need to be conducted. Obstructions and curvature of the earth (for links over six miles) determine the height of the towers or the building required to build the link in a line-of-sight environment. As stated earlier, the fresnel zone must be clear of obstructions and reflective surfaces to avoid absorption and multipath issues. Predominant weather conditions can limit the distance of the PTP Microwave link since the signal is susceptible to a condition called *rain fade*. The designers must take the predicted amount of signal degradation in a projected area and factor that into the design based on reliability requirements for the PTP Microwave link. Figure 7.5 gives a basic depiction of a PTP Microwave link.

Figure 7.5 Point-to-Point Microwave

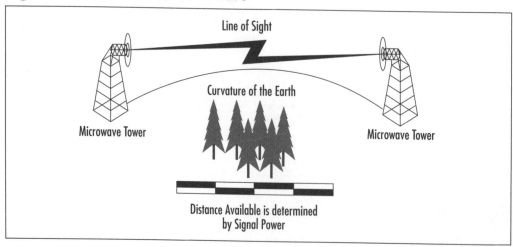

Wireless Local Area Networks

Benefits of fixed wireless can also provide value to the enterprise and home networks. This is where wireless capabilities get exciting for the end user. The benefits are literally at your fingertips. Imagine sitting at your desk when your boss calls announcing an emergency meeting immediately—there is a document on its way to you via e-mail that will be the focus of the meeting. Before wireless, you would first have to wait for your computer to receive the e-mail, and then perhaps print the document before traveling to the meeting; with a laptop, you would have to consider cords, batteries, and connections. After the meeting, you would go back to your desk for any document changes or further correspondence by e-mail. In a wireless environment, you can receive the e-mail and read the document while you are on your way to the meeting, and make changes to the document and correspond with other attendees real-time during the meeting.

Why the Need for a Wireless LAN Standard?

Prior to the adoption of the 802.11 standard, wireless data-networking vendors made equipment that was based on proprietary technology. Wary of being locked into a relationship with a specific vendor, potential wireless customers instead turned to more standards-based wired technologies. As a result, deployment of wireless networks did not happen on a large scale, and remained a luxury item for large companies with large budgets.

The only way wireless local area networks (WLANs) would be generally accepted would be if the wireless hardware involved had a low cost and had become commodity items like routers and switches. Recognizing that the only way for this to happen would be if there were a wireless data-networking standard, the Institute of Electrical and Electronics Engineers' (IEEE's) 802 Group took on their eleventh challenge. Since many of the members of the 802.11 Working Group were employees of vendors making wireless technologies, there were many pushes to include certain functions in the final specification. Although this slowed down the progress of finalizing 802.11, it also provided momentum for delivery of a feature-rich standard left open for future expansion.

On June 26, 1997, the IEEE announced the ratification of the 802.11 standard for wireless local area networks. Since that time, costs associated with deploying an 802.11-based network have dropped, and WLANs rapidly are being deployed in schools, businesses, and homes.

In this section, we will discuss the evolution of the standard in terms of band-width and services. Also, we will discuss the WLAN standards that are offshoots of the 802.11 standard.

> **NOTE**
>
> The IEEE (www.ieee.org) is an association that develops standards for almost anything electronic and /or electric. Far from being limited to computer-related topics, IEEE societies cover just about any technical practice, from automobiles to maritime, from neural networks to super-conductors. With 36 Technical Societies covering broad interest areas, more specific topics are handled by special committees. These other committees form Working Groups (WGs) and Technical Advisory Groups (TAGs) to create operational models that enable different vendors to develop and sell products that will be compatible. The membership of these committees and groups are professionals who work for companies that develop, create, or manufacture with their technical practice. These groups meet several times a year to discuss new trends within their industry, or to continue the process of refining a current standard.

What Exactly Does the 802.11 Standard Define?

As in all 802.x standards, the 802.11 specification covers the operation of the media access control (MAC) and physical layers. As you can see in Figure 7.6, 802.11 defines a MAC sublayer, MAC services and protocols, and three physical (PHY) layers.

Figure 7.6 802.11 Frame Format

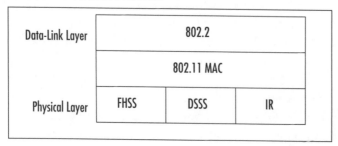

Data-Link Layer	802.2		
	802.11 MAC		
Physical Layer	FHSS	DSSS	IR

The three physical layer options for 802.11 are infrared (IR) baseband PHY and two radio frequency (RF) PHYs. Due to line-of-sight limitations, very little development has occurred with the Infrared PHY. The RF physical layer is composed of Frequency Hopping Spread Spectrum (FHSS) and Direct Sequence Spread Spectrum (DSSS) in the 2.4 GHz band. All three physical layers operate at either 1 or 2 Mbps. The majority of 802.11 implementations utilize the DSSS method.

FHSS works by sending bursts of data over numerous frequencies. As the name implies, it hops between frequencies. Typically, the devices use up to four frequencies simultaneously to send information and only for a short period of time before hopping to new frequencies. The devices using FHSS agree upon the frequencies being used. In fact, due to the short time period of frequency use and device agreement of these frequencies, many autonomous networks can coexist in the same physical space.

DSSS functions by dividing the data into several pieces and simultaneously sending the pieces on as many different frequencies as possible, unlike FHSS, which sends on a limited number of frequencies. This process allows for greater transmission rates than FHSS, but is vulnerable to greater occurrences of interference. This is because the data is spanning a larger portion of the spectrum at any given time than FHSS. In essence, DHSS floods the spectrum all at one time, whereas FHSS selectively transmits over certain frequencies.

Designing and Planning…

Additional Initiatives of the 802 Standards Committee

802.1 LAN/MAN Bridging and Management 802.1 is the base standard for LAN/MAN Bridging, LAN architecture, LAN management, and protocol layers above the MAC and LLC layers. Some examples would include 802.1q, the standard for virtual LANs, and 802.1d, the Spanning Tree Protocol.

802.2 Logical Link Control Since Logical Link Control is now a part of all 802 standards, this Working Group is currently in hibernation (inactive) with no ongoing projects.

802.3 CSMA/CD Access Method (Ethernet) 802.3 defines that an Ethernet network can operate at 10 Mbps, 100 Mbps, 1 Gbps, or even 10 Gbps. It also defines that category 5 twisted pair cabling and fiber optic cabling are valid cable types. This group identifies

Continued

how to make vendors' equipment interoperate despite the various speeds and cable types.

802.4 Token-Passing Bus This Working Group is also in hibernation with no ongoing projects.

802.5 Token Ring Token Ring networks operate at 4 Mbps or 16 Mbps. Currently, there are Working Groups proposing 100 Mbps Token Ring (802.5t) and Gigabit Token Ring (802.5v). Examples of other 802.5 specs would be 802.5c, Dual Ring Wrapping, and 802.5j, fiber optic station attachment.

802.6 Metropolitan Area Network (MAN) Since Metropolitan Area Networks are created and managed with current internetworking standards, the 802.6 Working Group is in hibernation.

802.7 Broadband LAN In 1989, this Working Group recommended practices for Broadband LANs, which were reaffirmed in 1997. This group is inactive with no ongoing projects. The maintenance effort for 802.7 is now supported by 802.14.

802.8 Fiber Optics Many of this Working Group's recommended practices for fiber optics get wrapped into other Standards at the Physical Layer.

802.9 Isochronous Services LAN (ISLAN) Isochronous Services refer to processes where data must be delivered within certain time constraints. Streaming media and voice calls are examples of traffic that requires an isochronous transport system.

802.10 Standard for Interoperable LAN Security (SILS) This Working Group provided some standards for Data Security in the form of 802.10a, Security Architecture Framework, and 802.10c, Key Management. This Working Group is currently in hibernation with no ongoing projects.

802.11 Wireless LAN (WLAN) This Working Group is developing standards for Wireless data delivery in the 2.4 GHz and 5.1 GHz radio spectrum.

802.12 Demand Priority Access Method This Working Group provided two Physical Layer and Repeater specifications for the development of 100 Mbps Demand Priority MACs. Although they were accepted as ISO standards and patents were received for their operation, widespread acceptance was overshadowed by Ethernet. 802.12 is currently in the process of being withdrawn.

802.13 This standard was intentionally left blank.

802.14 Cable-TV Based Broadband Comm Network

Continued

This Working Group developed specifications for the Physical and Media Access Control Layers for Cable Televisions and Cable Modems. Believing their work to be done, this Working Group has no ongoing projects.

802.15 Wireless Personal Area Network (WPAN) The vision of Personal Area Networks is to create a wireless interconnection between portable and mobile computing devices such as PCs, peripherals, cell phones, personal digital assistants (PDAs), pagers, and consumer electronics, allowing these devices to communicate and interoperate with one another without interfering with other wireless communications.

802.16 Broadband Wireless Access The goal of the 802.16 Working Group is to develop standards for fixed broadband wireless access systems. These standards are key to solving "last-mile" local-loop issues. 802.16 is similar to 802.11a in that it uses unlicensed frequencies in the unlicensed national information infrastructure (U-NII) spectrum. 802.16 is different from 802.11a in that Quality of Service for voice/video/data issues are being addressed from the start in order to present a standard that will support true wireless network backhauling.

Does the 802.11 Standard Guarantee Compatibility across Different Vendors?

As mentioned earlier, the primary reason WLANs were not widely accepted was the lack of standardization. It is logical to question whether vendors would accept a nonproprietary operating standard, since vendors compete to make unique and distinguishing products. Although 802.11 standardized the PHY, MAC, the frequencies to send/receive on, transmission rates and more, it did not absolutely guarantee that differing vendors' products would be 100 percent compatible. In fact, some vendors built in backward compatibility features into their 802.11 products in order to support their legacy customers. Other vendors have introduced proprietary extensions (for example, bit-rate adaptation and stronger encryption) to their 802.11 offerings.

To ensure that consumers can build interoperating 802.11 wireless networks, an organization called the Wireless Ethernet Compatibility Alliance (WECA) tests and certifies 802.11 devices. Their symbol of approval means that the consumer can be assured that the particular device has passed a thorough test of interoperations with devices from other vendors. This is important when considering devices to be implemented into your existing network, because if the devices cannot communicate, it complicates the management of the network—in fact, essentially you will have to deal with two autonomous networks. It is also important when building a new network because you may be limited to a single vendor.

Since the first 802.11 standard was approved in 1997, there have been several initiatives to make improvements. As you will see in the following sections, the 802.11 standard has and will continue to improve WLAN technologies that will boast throughput, strengthen security, and provide better interoperability.

802.11b

The 802.11b amendment to the original standard was ratified in 1999. It uses an extension of the DSSS modulation technique (used by the original standard) called Complementary code keying (CCK). CCK is a modulation scheme that can transfer more data per unit time than the DSSS modulation scheme. Data rates for CCK are 5.5 and 11 Mbps. The increased throughput of 802.11b (11 Mbps) compared to the original standard (1-2 Mbps) led to the wide acceptance of the 802.11b WLAN technology by both home users and corporations.

The 802.11b security mechanism, Wired Equivalent Privacy (WEP) was designed to provide a level of protection equivalent to that provided on a wired network. It utilizes an RC4-based encryption scheme, and it is not intended for end-to-end encryption or as a sole method of securing data. Its design was proven to have security weaknesses and is superseded by WPA and WPA2.

802.11g

To further higher-speed physical layer extension using the 2.4 GHz band, in June 2003, the 802.11 standard was amended to include 802.11g. 802.11g improved upon 802.11b WLAN technologies in the 2.4 GHz radio spectrum which increased throughput to 54 Mbps. 802.11g operates within the same 2.4 GHz band as 802.11b; however, it uses a different modulation scheme called Orthogonal Frequency Division Multiplexing (OFDM). OFDM allows data rates of 6, 9, 12, 18, 24, 36, 48, and 54 Mbps.

In addition to speed enhancements, 802.11g hardware is backward compatible with 802.11b hardware. The backward compatibility feature allows interoperability between the two technologies, but does significantly reduce the speed of an 802.11g network when using 802.11b hardware. When using a mixture of 802.11b and 802.11g hardware, the 802.11b (DSSS) modulation scheme is used reducing your data rate from between 5.5 to 11 Mbps.

The range of 802.11g devices is better than 802.11b devices, however, the range that you can achieve the maximum data rate (54 Mbps) is much shorter than of 802.11b devices.

The 802.11b/g standard uses any one of 14 center-frequency channels in the 2.4 GHz Industrial, Scientific, and Medical (ISM) radio band. As Table 7.1 shows, North America allows 11 channels; Europe allows 13, the most channels allowed. Japan has only one channel reserved for 802.11, at 2.483 GHz.

Table 7.1 802.11b/g Channels and Participating Countries

Channel Number			Frequency			
GHz	North America		Europe	Spain	France	Japan
1	2.412	X	X			
2	2.417	X	X			
3	2.422	X	X			
4	2.427	X	X			
5	2.432	X	X			
6	2.437	X	X			
7	2.442	X	X			
8	2.447	X	X			
9	2.452	X	X			
10	2.457	X	X	X	X	
11	2.462	X	X	X	X	
12	2.467		X		X	
13	2.472		X		X	
14	2.483					X

There are many different devices competing for airspace in the 2.4 GHz radio spectrum. Unfortunately, most of the devices that cause interference are especially common in the home environment, such as microwaves and cordless phones.

One of the more recent entrants to the 802.11b/g airspace comes in the form of the emerging Bluetooth wireless standard. Though designed for short-range transmissions, Bluetooth devices utilize FHSS to communicate with each other. Cycling through thousands of frequencies a second, this looks as if it poses the greatest chance of creating interference for 802.11. Further research will determine exactly what—if any—interference Bluetooth will cause to 802.11b networks. Many companies are concerned with over saturating the 2.4 GHz spectrum, and are taking steps to ensure that their devices "play nicely" with others in this arena.

802.11a

Due to the overwhelming demand for more bandwidth and the growing number of technologies operating in the 2.4 GHz band, the 802.11a standard was created for WLAN use in North America as an upgrade from the 802.11b standard. 802.11a provides 25 to 54 Mbps bandwidth in the 5 GHz spectrum (the unlicensed national information infrastructure [U-NII] spectrum). Since the 5 GHz band is currently mostly clear, chance of interference is reduced. However, that could change since it is still an unlicensed portion of the spectrum. 802.11a still is designed mainly for the enterprise, providing Ethernet capability.

802.11a is one of the physical layer extensions to the 802.11 standard. Abandoning spread spectrum completely, 802.11a uses an encoding technique called Orthogonal Frequency Division Multiplexing (OFDM), also used in 802.11g. Although this encoding technique is similar to the European 5-GHz HiperLAN physical layer specification, which will be explained in greater detail later in the chapter, 802.11a currently is specific to the United States.

As shown in Table 7.2, three 5-GHz spectrums have been defined for use with 802.11a. Each of these three center-frequency bands covers 100 MHz.

Table 7.2 802.11a Channels Usable in the 5-GHz U-NII Radio Spectrum

Regulatory Area	Frequency Band	Channel Number	Center Frequencies
USA	U-NII Lower Band 5.15 - 5.25 GHz	36	5.180 GHz
		40	5.200 GHz
		44	5.220 GHz
		48	5.240 GHz
USA	U-NII Middle Band 5.25 - 5.35 GHz	52	5.260 GHz
		56	5.280 GHz
		60	5.300 GHz
		64	5.320 GHz
USA	U-NII Upper Band 5.725 - 5.825 GHz	149	5.745 GHz
		153	5.765 GHz
		157	5.785 GHz
		161	5.805 GHz

802.11e

The IEEE 802.11e is providing enhancements to the 802.11 standard while retaining compatibility with 802.11b/g, 802.11a and 802.11i. The enhancements include multimedia capability made possible with the adoption of quality of service (QoS) functionality as well as security improvements. What does this mean for a service provider? It means the ability to offer video on demand, audio on demand, high-speed Internet access and Voice over IP (VoIP) services. What does this mean for the home or business user? It allows high-fidelity multimedia in the form of MPEG2 video and CD quality sound, and redefinition of the traditional phone use with VoIP.

QoS is the key to the added functionality with 802.11e. It provides the function-ality required to accommodate time-sensitive applications such as video and audio. QoS includes queuing, traffic shaping tools, and scheduling. These characteristics allow priority of traffic. For example, data traffic is not time sensitive and therefore has a lower priority than applications like streaming video. With these enhancements, wireless networking has evolved to meet the demands of today's users.

802.11i

802.11i, also known as WPA2 provides enhanced security mechanisms for 802.11 beyond the capabilities of the wired equivalent privacy (WEP) method used in the original standard. The new security features of 802.11i are considered upgrades to the original security specification, WEP, which was proven to have security weak-nesses. WPA2 retains WEP features for backward compatibility with existing 802.11 devices. Most new WLAN hardware has out of the box support for WPA2. Usually, support is available for existing wireless cards using a firmware upgrade or patch, available at the manufacturer's website.

Developing WLANs through the 802.11 Architecture

The 802.11 architecture can best be described as a series of interconnected cells, and consists of the following: the wireless device or station, the Access Point (AP), the wireless medium, the distribution system (DS), the Basic Service Set (BSS), the Extended Service Set (ESS), and station and distribution services. All of these working together providing a seamless mesh gives wireless devices the ability to roam around the WLAN looking for all intents and purposes like a wired device.

The Basic Service Set

The core of the IEEE 802.11 standard is the Basic Service Set (BSS). As you can see in Figure 7.7, this model is made up of one or more wireless devices communicating with a single Access Point in a single radio cell. If there are no connections back to a wired network, this is called an *independent Basic Service Set*.

Figure 7.7 Basic Service Set

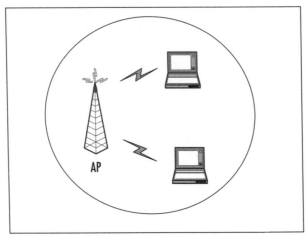

If there is no Access Point in the wireless network, it is referred to as an *ad-hoc network*. This means that all wireless communications is transmitted directly between the members of the ad-hoc network. Figure 7.8 describes a basic ad-hoc network.

Figure 7.8 Ad-Hoc Network

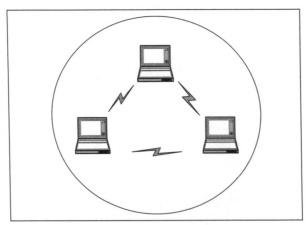

When the BSS has a connection to the wired network via an AP, it is called an *infrastructure BSS*. As you can see in the model shown in Figure 7.9, the AP bridges the gap between the wireless device and the wired network.

Figure 7.9 802.11 Infrastructure Architecture

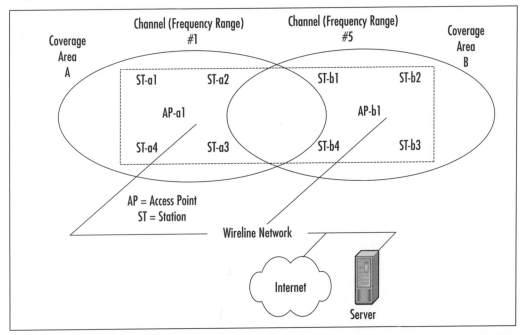

Since multiple Access Points exist in this model, the wireless devices no longer communicate in a peer-to-peer fashion. Instead, all traffic from one device destined for another device is relayed through the AP. Even though it would look like this would double the amount of traffic on the WLAN, this also provides for traffic buffering on the AP when a device is operating in a low-power mode.

The Extended Service Set

The compelling force behind WLAN deployment is the fact that with 802.11, users are free to move about without having to worry about switching network connections manually. If we were operating with a single infrastructure BSS, this moving about would be limited to the signal range of our one AP. Through the Extended Service Set (ESS), the IEEE 802.11 architecture allows users to move between multiple infrastructure BSSs. In an ESS, the APs talk amongst themselves forwarding traffic from one BSS to another, as well as switch the roaming devices from one BSS to another. They do this using a medium called the distribution system (DS). The distribution system forms the spine of the WLAN, making the decisions whether to forward traffic from one BSS to the wired network or back out to another AP or BSS.

What makes the WLAN so unique, though, are the invisible interactions between the various parts of the Extended Service Set. Pieces of equipment on the wired network have no idea they are communicating with a mobile WLAN device, nor do they see the switching that occurs when the wireless device changes from one AP to another. To the wired network, all it sees is a consistent MAC address to talk to, just as if the MAC was another node on the wire.

Services to the 802.11 Architecture

There are nine different services that provide behind-the-scenes support to the 802.11 architecture. Of these nine, four belong to the *station services* group and the remaining five to the *distribution services* group.

Station Services

The four station services (*authentication*, *de-authentication*, *data delivery*, and *privacy*) provide functionality equal to what standard 802.3 wired networks would have.

The authentication service defines the identity of the wireless device. Without this distinct identity, the device is not allowed access to the WLAN. Authentication can also be made against a list of MACs allowed to use the network. This list of allowable MAC addresses may be on the AP or on a database somewhere on the wired network. A wireless device can authenticate itself to more than one AP at a time. This sort of "pre-authentication" allows the device to prepare other APs for its entry into their airspace.

The de-authentication service is used to destroy a previously known station identity. Once the de-authentication service has been started, the wireless device can no longer access the WLAN. This service is invoked when a wireless device shuts down, or when it is roaming out of the range of the Access Point. This frees up resources on the AP for other devices.

Just like its wired counterparts, the 802.11 standard specifies a data delivery service to ensure that data frames are transferred reliably from one MAC to another. This data delivery will be discussed in greater detail in following sections.

The privacy service is used to protect the data as it crosses the WLAN. The original mechanism utilizes an RC4-based encryption scheme, it is not intended for end-to-end encryption or as a sole method of securing data. Its design was to provide a level of protection equivalent to that provided on a wired network, Wireless

Equivalency Privacy (WEP). As new standards were introduced stronger encryption schemes are available for use such as WPA and WPA2.

Distribution Services

Between the Logical Link Control (LLC) sublayer and the MAC, five distribution services make the decisions as to where the 802.11 data frames should be sent. As we will see, these distribution services make the roaming handoffs when the wireless device is in motion. The five services are *association, reassociation, disassociation, integration,* and *distribution.*

The wireless device uses the association service as soon as it connects to an AP. This service establishes a logical connection between the devices, and determines the path the distribution system needs to take in order to reach the wireless device. If the wireless device does not have an association made with an Access Point, the DS will not know where that device is or how to get data frames to it. As you can see in Figure 7.10, the wireless device can be authenticated to more than one AP at a time, but it will never be associated with more than one AP.

Figure 7.10 Wireless Authentication through the Association Service

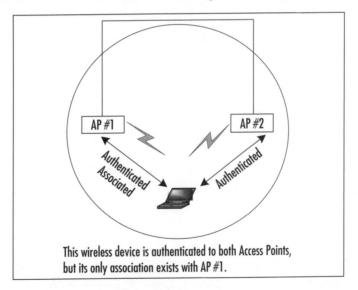

This wireless device is authenticated to both Access Points, but its only association exists with AP #1.

As we will see in later sections dealing with roaming and low-power situations, sometimes the wireless device will not be linked continuously to the same AP. To keep from losing whatever network session information the wireless device has, the reassociation service is used. This service is similar to the association service, but includes current information about the wireless device. In the case of roaming, this information tells the current AP who the last AP was. This allows the current AP to contact the previous AP to pick up any data frames waiting for the wireless device and forward them to their destination.

The disassociation service is used to tear down the association between the AP and the wireless device. This could be because the device is roaming out of the AP's area, the AP is shutting down, or any one of a number of other reasons. To keep communicating to the network, the wireless device will have to use the association service to find a new AP.

The distribution service is used by APs when determining whether to send the data frame to another AP and possibly another wireless device, or if the frame is destined to head out of the WLAN into the wired network.

The integration service resides on the APs as well. This service does the data translation from the 802.11 frame format into the framing format of the wired network. It also does the reverse, taking data destined for the WLAN, and framing it within the 802.11 frame format.

The CSMA-CA Mechanism

The basic access mechanism for 802.11 is carrier sense multiple access collision avoidance (CSMA-CA) with binary exponential backoff. This is very similar to the carrier sense multiple access collision detect (CSMA-CD) that we are familiar with when dealing with standard 802.3 (Ethernet), but with a couple of major differences.

Unlike Ethernet, which sends out a signal until a collision is detected, CSMA-CA takes great care to not transmit unless it has the attention of the receiving unit, and no other unit is talking. This is called *listening before talking* (LBT).

Before a packet is transmitted, the wireless device will listen to hear if any other device is transmitting. If a transmission is occurring, the device will wait for a randomly determined period of time, and then listen again. If no one else is using the medium, the device will begin transmitting. Otherwise, it will wait again for a random time before listening once more.

The RTS/CTS Mechanism

To minimize the risk of the wireless device transmitting at the same time as another wireless device (and thus causing a collision), the designers of 802.11 employed a mechanism called Request To Send/Clear To Send (RTS/CTS).

For example, if data arrived at the AP destined for a wireless node, the AP would send a RTS frame to the wireless node requesting a certain amount of time to deliver data to it. The wireless node would respond with a CTS frame saying that it would hold off any other communications until the AP was done sending the data.

Other wireless nodes would hear the transaction taking place, and delay their transmissions for that period of time as well. In this manner, data is passed between nodes with a minimal possibility of a device causing a collision on the medium.

This also gets rid of a well-documented WLAN issue called *the hidden node*. In a network with multiple devices, the possibility exists that one wireless node might not know all the other nodes that are out on the WLAN. Thanks to RST/CTS, each node hears the requests to transmit data to the other nodes, and thus learns what other devices are operating in that BSS.

Acknowledging the Data

When sending data across a radio signal with the inherent risk of interference, the odds of a packet getting lost between the transmitting radio and the destination unit are much greater than in a wired network model. To make sure that data transmissions would not get lost in the ether, *acknowledgment* (ACK) was introduced. The acknowledgement portion of CSMA-CA means that when a destination host receives a packet, it sends back a notification to the sending unit. If the sender does not receive an ACK, it will know that this packet was not received and will transmit it again.

All this takes place at the MAC layer. Noticing that an ACK has not been received, the sending unit is able to grab the radio medium before any other unit can and it resends the packet. This allows recovery from interference without the end user being aware that a communications error has occurred.

Configuring Fragmentation

In an environment prone to interference, the possibility exists that one or more bits in a packet will get corrupted during transmission. No matter the number of corrupted bits, the packet will need to be re-sent.

When you are operating in an area where interference is not a possibility, but a reality, it makes sense to transmit smaller packets than those traditionally found in wired networks. This allows for a faster retransmission of the packet to be accomplished.

The disadvantage to doing this is that in the case of no corrupted packets, the cost of sending many short packets is greater than the cost of sending the same information in a couple of large packets. Thankfully, the 802.11 standard has made this a configurable feature. This way, a network administrator can specify short packets in some areas and longer packets in more open, noninterfering areas.

Using Power Management Options

Because the whole premise of wireless LANs is mobility, having sufficient battery power to power the communications channel is of prime concern. The IEEE recognized this and included a power management service that allows the mobile client to go into a sleep mode to save power without losing connectivity to the wireless infrastructure.

Utilizing a 20-byte Power Save Poll (PS-Poll) frame, the wireless device sends a message to its AP letting it know that is going into power-save mode, and the AP needs to buffer all packets destined for the device until it comes back online. Periodically, the wireless device will wake up and see if there are any packets waiting for it on the AP. If there aren't, another PS-Poll frame is sent, and the unit goes into a sleep mode again. The real benefit here is that the mobile user is able to use the WLAN for longer periods of time without severely impacting the battery life.

Multicell Roaming

Another benefit to wireless LANs is being able to move from wireless cell to cell as you go around the office, campus, or home without the need to modify your network services. Roaming between Access Points in your ESS is a very important portion of the 802.11 standard. Roaming is based on the ability of the wireless device to determine the quality of the wireless signal to any AP within reach, and decide to switch communications to a different AP if it has a stronger or cleaner signal. This is based primarily upon an entity called the signal-to-noise (S/N) ratio. In order for wireless devices to determine the S/N ratio for each AP in the network, Access Points send out *beacon* messages that contain information about the AP as well as link measurement data. The wireless device listens to these beacons and determines which AP has the clearest and cleanest signal. After making this determination, the wireless device sends authentication information and attempts to reassociate with the new AP. The reassociation process tells the new AP which AP the device just came from. The new AP picks up whatever data frames that might be left at the old AP, and notifies the old AP that it no longer needs to accept messages for that wireless device. This frees up resources on the old AP for its other clients.

Even though the 802.11 standard covers the concepts behind the communications between the AP and the DS, it doesn't define exactly how this communication should take place. This is because there are many different ways this communication can be implemented. Although this gives a vendor a good deal of flexibility in AP/DS design, there could be situations where APs from different vendors might not be able to interoperate across a distribution system due to the differences in how those vendors implemented the AP/DS interaction. Currently, there is an 802.11 Working Group (802.11f) developing an Inter-Access Point Protocol. This protocol will be of great help in the future as companies who have invested in one vendor's products can integrate APs and devices from other vendors into their ESSes.

Security in the WLAN

One of the biggest concerns facing network administrators when implementing a WLAN is data security. In a wired environment, the lack of access to the physical wire can prevent someone from wandering into your building and connecting to your internal network. In a WLAN scenario, it is impossible for the AP to know if the person operating the wireless device is sitting inside your building, passing time in your lobby, or if they are seated in a parked car just outside your office. Acknowledging that passing data across an unreliable radio link could lead to possible snooping, the IEEE 802.11 standard provides three ways to provide a greater amount of security for the data that travels over the WLAN. Adopting the following mechanisms will decrease the likelihood of an accidental security exposure.

The first method makes use of the 802.11 Service Set Identifier (SSID). The SSID is a code attached to all packets on the WLAN used to identify each packet as part of that network. All WLAN devices must share the same SSID to communicate with one another. A very weak form of network security is to turn off the SSID broadcast on the access point. To the average user, there is not a network to authenticate to, but by using a wireless network sniffer such as kismet or netstumbler you can easily uncloak the SSID and authenticate to the access point.

As mentioned earlier in the station services section, the AP also can authenticate a wireless device against a list of MAC addresses. This list could reside locally on the AP, or the authentication could be checked against a database of allowed MACs located on the wired network. This typically is a very low level of security but none the less can help keep your neighbor off you network. Again, by using a wireless network sniffer you can identify wireless clients connected to an access point and get information such as their MAC address. So again, MAC filtering is again a trivial security mechanism.

The third mechanism 802.11 offers to protect data traversing the WLAN is by using encryption features such as WPA2, WPA, or at the minimum WEP. WPA2 provides enhanced security mechanisms for 802.11 beyond the capabilities of WEP. These features not only provide encryption for the data in transit from the wireless client to the access point but they provide authentication to the wireless network.

Some network designers consider WLANs to be in the same crowd as Remote Access Service (RAS) devices, and claim the best protection is to place the WLAN architecture behind a firewall or Virtual Private Network (VPN) device. This would make the wireless client authenticate to the VPN or firewall using third-party software (on top of WPA2, WPA, or WEP). The benefit here is that the bulk of the authenticating would be up to a non-WLAN device, and would not require additional AP maintenance.

While disabling the SSID broadcast or providing MAC filtering alone is a weak security practice, using them combined with encryption and authentication can provide adequate protection for your WLAN.

The uses of 802.11 networks can range from homes to public areas like schools and libraries, to businesses and corporate campuses. The ability to deploy a low-cost network without the need to have wires everywhere is allowing wireless networks to spring up in areas where wired networks would be cost prohibitive. The 802.11 services allow the wireless device the same kind of functionality as a wired network, yet giving the user the ability to roam throughout the WLAN.

Next, we will discuss another wireless technology breakthrough, appealing to the truly free-spirited. This emerging technology is capable of providing a personal network that moves along with you wherever you go. Let's say you receive a text message on your cellular and personal communications services (PCS) phone and would like to transfer the contents into your PDA. No problem—with the 802.15 standard, this is possible no matter where you are. And if you happen to be in a public place and someone near you is using the same technology, there is no need to worry, because your information is encrypted.

Developing WPANs through the 802.15 Architecture

Wireless personal area networks (WPANs) are networks that occupy the space surrounding an individual or device, typically involving a 10m radius. This is referred to as a personal operating space (POS). This type of network adheres to an ad-hoc system requiring little configuration. The devices in a WPAN find each other and communicate with little effort by the end user.

WPANs generally fall under the watchful eyes of the IEEE 802.15 working group (technically, 802.15 networks are defined as *short-distance wireless networks*). The growing trend toward more "smart" devices in the home and the increasing number of telecommuters and small office/home office (SOHO) users is driving the demand for this section of the wireless industry. Another driving requirement for this segment is the need for simplistic configuration of such a network. As this segment grows, the end users involved are not the technically elite, early technology adopters, but the average consumer. The success of this segment is rooted in its ability to simplify its use while maintaining lower costs. In addition, various efforts are under way to converge the 802.11 and 802.15 standards for interoperability and the reduction of interference in the 2.4 GHz space. Since this is the same unlicensed range shared by numerous wireless devices such as garage door openers, baby monitors, and cordless phones, 802.15 devices must be able to coexist. They fall under two categories. The first is the collaborative model where both standards not only will coexist with interference mitigated, but also will interoperate. The second is the noncollaborative model, where the interference is mitigated but the two standards do not interoperate.

Bluetooth

Bluetooth technology was named after Harold Blaatand (Bluetoothe) II, who was the King of Denmark from 940–981 and was generally considered a "unifying figurehead" in Europe during that period. The unification of Europe and the unification of PDAs and computing devices is the parallelism that the founders of this technology sought to create when they chose the name *Bluetooth*. Bluetooth began in 1994 when Ericsson was looking for inexpensive radio interfaces between cell phones and accessories such as PDAs. In 1998, Ericsson, IBM, Intel, Nokia, and Toshiba formed the Bluetooth Special Interest Group (SIG) and expanded to over 1000 members by 1999, including Microsoft. However, the Bluetooth technology is currently behind schedule and the projected cost of $5 per transceiver is not being realized. This combined with the expansion and success of the 802.11 standard may threaten the survivability of this technology.

Bluetooth is primarily a cable replacement WPAN technology that operates in the 2.4 GHz range using FHSS. One of the main drivers for the success of the Bluetooth technology is the proposition of low-cost implementation and size of the wireless radios. Bluetooth networks are made up of *piconets*, which are loosely fashioned or *ad-hoc* networks. Piconets are made up of one master node and seven simultaneously active slaves or an almost limitless number of virtually attached but not active (standby) nodes. Master nodes communicate with slaves in a hopping pattern determined by a 3-bit Active Member Address (AMA). Parked nodes are addressed with an 8-bit Parked Member Address, (PMA). Up to 10 piconets can be collocated and linked into what is called *scatternets*. A node can be both a master in one piconet and a slave in another piconet at the same time, or a slave in both piconets at the same time. The range of a Bluetooth standard piconet is 10 meters, relative to the location of the master. Bluetooth signals pass through walls, people, and furniture, so it is not a line-of-sight technology. The maximum capacity of Bluetooth is 740 Kbps per piconet (actual bit rate) with a raw bit rate of 1 Mbps.

Figure 7.11 provides a logical depiction of several piconets linked together as a scatternet.

Figure 7.11 Bluetooth Piconet and Scatternet Configuration

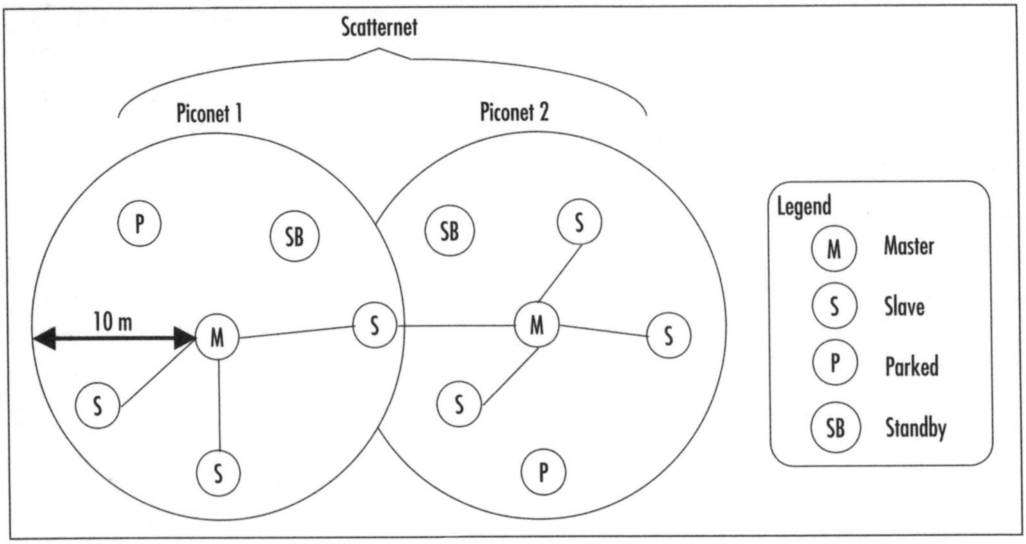

Since Bluetooth shares the 2.4 GHz frequency range with 802.11b, there is a possibility for interference between the two technologies if a Bluetooth network is within ten meters of an 802.11b network. Bluetooth was designed to be a complementary technology to the 802.11 standard and the IEEE Task Group f (TGf) is chartered with proposing interoperability standards between the two technologies. Bluetooth has also been working with the FCC and FAA to provide safe operation on aircraft and ships. Figure 7.12 gives a broad view of the envisioned uses of Bluetooth as a technology (more information on Bluetooth can be obtained at www.bluetooth.com).

HomeRF

HomeRF is similar to Bluetooth since it operates in the 2.4 GHz spectrum range and provides up to 1.6 Mbps bandwidth with user throughput of about 650 Kb/s. HomeRF has a relative range of about 150 feet as well. Home RF uses FHSS as its physical layer transmission capability. It also can be assembled in an ad hoc architecture or be controlled by a central connection point like Bluetooth. Differences between the two are that HomeRF is targeted solely towards the residential market— the inclusion of the Standard Wireless Access Protocol (SWAP) within HomeRF gives it a capability to handle multimedia applications much more efficiently.

Figure 7.12 Bluetooth Uses

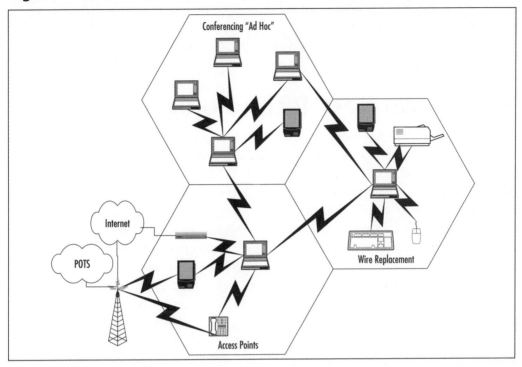

SWAP combines the data beneficial characteristics of 802.11's CSMA–CA with the QoS characteristics of the Digital Enhanced Cordless Telecommunications (DECT) protocol to provide a converged network technology for the home. SWAP 1.0 provides support for four DECT toll quality handsets within a single ad-hoc network. SWAP 1.0 also provides 40-bit encryption at the MAC layer for security purposes.

SWAP 2.0 will extend the bandwidth capabilities to 10 Mbps and provide roaming capabilities for public access. It also provides upward scalability for support of up to eight toll quality voice handsets based on the DECT protocol within the same ad-hoc network. The QOS features are enhanced by the addition of up to eight prioritized streams supporting multimedia applications such as video. SWAP 2.0 extends the security features of SWAP 1.0 to 128 bits encryption. For more information on HomeRF, go to www.homerf.com.

High-Performance Radio LAN

High-Performance Radio LAN (HiperLAN) is the European equivalent of the 802.11 standard. HiperLAN Type 1 supports 20 Mbps of bandwidth in the 5 GHz range.

HiperLAN Type 2 (HiperLAN2) also operates in the 5 GHz range but offers up to 54 Mbps bandwidth. It also offers many more QoS features and thus currently supports many more multimedia applications that its 802.11a counterpart. HiPerLAN2 is also a connection-oriented technology, which, combined with its QoS and bandwidth, gives it applications outside the normal enterprise networks.

Mobile Wireless Technologies

The best way to describe *mobile wireless* is to call it your basic cellular phone service. The cell phone communications industry has migrated along two paths; the United States has generally progressed along the Code Division Multiple Access (CDMA) path, with Europe following the Global System for Mobile Communications (GSM) path. However, both areas' cellular growth has progressed from analog communications to digital technologies, and both continents had an early focus on the voice communication technology known as 1G and 2G (the G stands for *generations*). Emerging technologies are focused on bringing both voice and data as well as video over the handheld phones/devices. The newer technologies are referred to as 2.5G and 3G categorically. A linear description of the evolution of these two technologies is presented in the following sections.

Figure 7.13 illustrates a generic cellular architecture. A geographic area is divided into cells; the adjacent cells always operate on different frequencies to avoid interference—this is referred to as *frequency reuse*. The exact shape of the cells actually vary quite a bit due to several factors, including the topography of the land, the anticipated number of calls in a particular area, the number of man-made objects (such as the buildings in a downtown area), and the traffic patterns of the mobile users. This maximizes the number of mobile users.

Figure 7.13 Basic Cell Architecture

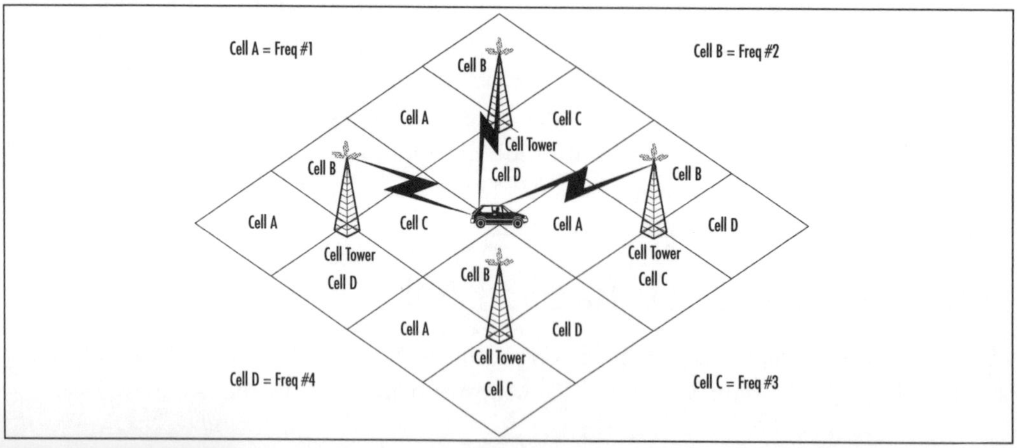

A lower powered antenna is placed at a strategic place, but it is not in the center of the cell, as you might think. Instead, the transmitter is located at a common point between adjacent cells. For example, in Figure 4.13, a base station is built at the intersection of cells A, B, C, and D. The tower then uses directional antennas that point inward to each of the adjacent cells. Other transmitters subsequently are placed at other locations through the area. By using the appropriately sized transmitter, frequencies in one particular cell are also used in nearby cells. The key to success is making sure cells using the same frequency cannot be situated right next to each other, which would result in adverse effects. The benefit is that a service provider is able to reuse the frequencies allotted to them continually so long as the system is carefully engineered. By doing so, more simultaneous callers are supported, in turn increasing revenue.

As a cell phone moves through the cells, in a car for example, the cell switching equipment keeps track of the relative strength of signal and performs a handoff when the signal becomes more powerful to an adjacent cell site. If a particular cell becomes too congested, operators have the ability to subdivide cells even further. For example, in a very busy network, the operator may have to subdivide each of the cells shown in Figure 4.12 into an even smaller cluster of cells. Due to the lower powered transmitters, the signals do not radiate as far, and as we mentioned, the frequencies are reused as much as we desire as long as the cells are spaced apart appropriately.

Mobile technology has developed with various protocols associated with each generation. These protocols will be explained in greater detail in the following sections, after we introduce the migration scheme.

First Generation Technologies

The introduction of semiconductor technology and the smaller microprocessors made more sophisticated mobile cellular technology a reality in the late 1970s and early 1980s. The *First Generation* (1G) technologies started the rapid growth of the mobile cellular industry. The most predominant systems are the Advanced Mobile Phone System (AMPS), Total Access Communication System (TACS), and the Nordic Mobile Telephone (NMT) system. However, analog systems didn't provide the signal quality desired for a voice system. These systems provided the foundation for the growth of the industry into the digital systems characterized by 2G.

Second Generation Technologies

The need for better transmission quality and capacity drove the development of the *Second Generation* (2G) systems and brought about the deployment of digital systems in the mobile industry. The U.S. companies like Sprint PCS predominantly gravitated towards the CDMA systems; most of the rest of the world embraced the GSM systems. Dual band mobile phones were created to allow roaming between digital 2G coverage areas through analog 1G areas. The CDMA and GSM 2G technologies are currently incompatible. The globalization of the world economy and the market for mobile data capabilities fueled the development of the 2.5G and 3G technologies. Both provide a migration path towards convergence of the two standards (GSM and CDMA) toward a globally interoperable mobile system. Both 2.5G and 3G also provide a migration path for a fully converged mobile voice/data/video system.

2.5G Technology

With the beginning of convergence came the development of new protocols created to optimize the limited bandwidth of mobile systems. The Wireless Access Protocol (WAP) was one of the first specifications for protocols created to meet these challenges by creating more efficient applications for the mobile wireless environment. The General Packet Radio Service (GPRS) was created to provide a packet-switched element (classical data) to the existing GSM voice circuit-switched architecture. In addition, GPRS seeks to increase the relative throughput of the GSM system fourfold, using a permanent IP connection from the handset to the Internet. Enhanced Data Rates for GSM Evolution (EDGE) was created as a further extension to the GSM data rates but is not limited to the time division multiple access (TDMA)-based GSM systems. EDGE's acceptance in the market to date is limited, and as with any technology, may be affected by the low acceptance rate. Many mobile service providers may migrate directly from existing GSM/GPRS systems directly to 3G systems.

Third Generation Technologies

The promise of the *Third Generation* (3G) mobile wireless technologies is the ability to support applications such as full motion video that require much larger amounts of bandwidth. This capability is known as Broadband and generally refers to bandwidths in excess of 1 Mbps. Wideband CDMA and cdma2000 are two versions of systems designed to meet this demand; however, they still are not globally compatible. A global group of standards boards called the Third-Generation Partnership Project (3GPP) has been created to develop a globally compatible 3G standard so the global interoperability of mobile systems can be a reality. The standard this group has developed is named the Universal Mobile Telecommunications System (UMTS). For more information on 3G and UMTS, go to www.umts.com.

Figure 7.14 illustrates the progression of the mobile wireless industry.

Figure 7.14 Mobile Wireless Progression

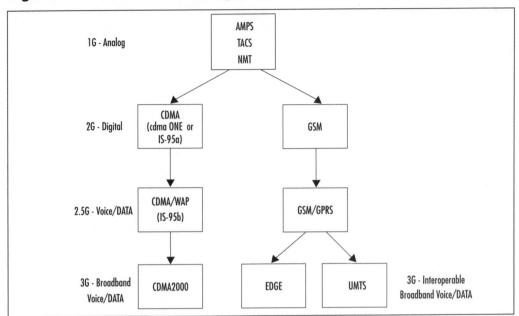

Wireless Application Protocol

The *Wireless Application Protocol* (WAP) has been implemented by many of the carriers today as the specification for wireless content delivery. WAP is an open specification that offers a standard method to access Internet-based content and services from wireless devices such as mobile phones and PDAs. Just like the OSI reference model, WAP is nonproprietary. This means anyone with a WAP-capable device can utilize this specification to access Internet content and services. WAP is also not dependent on the network, meaning that WAP works with current network architectures as well as future ones.

WAP as it is known today is based on the work of several companies that got together in 1997 to research wireless content delivery: Nokia, Ericsson, Phone.com, and Motorola. It was their belief at that time that the success of the wireless Web relied upon such a standard. Today, the WAP Forum consists of a vast number of members including handset manufacturers and software developers.

WAP uses a model of accessing the Internet very similar in nature to the standard desktop PC using Internet Explorer. In WAP, a browser is embedded in the software of the mobile unit. When the mobile device wants to access the Internet, it first needs to access a WAP gateway. This gateway, which is actually a piece of software and not a physical device, optimizes the content for wireless applications. In the desktop model, the browser makes requests from Web servers; it is the same in wireless. The Web servers respond to URLs, just like the desktop model, but the difference is in the formatting of the content. Because Internet-enabled phones have limited bandwidth and processing power, it makes sense to scale down the resource-hungry applications to more manageable ones. This is achieved using the Wireless Markup Language (WML). A WML script is used for client-side intelligence.

Global System for Mobile Communications

The *Global System for Mobile Communications* (GSM) is an international standard for voice and data transmission over a wireless phone. Utilizing three separate components of the GSM network, this type of communication is truly portable. A user can place an identification card called a Subscriber Identity Module (SIM) in the wireless device, and the device will take on the personal configurations and information of that user. This includes telephone number, home system, and billing information. Although the United States has migrated toward the PCS mode of wireless communication, in large part the rest of the world uses GSM.

The architecture used by GSM consists of three main components: a *mobile station*, a *base station subsystem*, and a *network subsystem*. These components work in tandem to allow a user to travel seamlessly without interruption of service, while offering the flexibility of having any device used permanently or temporarily by any user.

The mobile station has two components: mobile equipment and a SIM. The SIM, as mentioned, is a small removable card that contains identification and connection information, and the mobile equipment is the GSM wireless device. The SIM is the component within the mobile station that provides the ultimate in mobility. This is achieved because you can insert it into any GSM compatible device and, using the identification information it contains, you can make and receive calls and use other subscribed services. This means that if you travel from one country to another with a SIM, and take the SIM and place it into a rented mobile equipment device, the SIM will provide the subscriber intelligence back to the network via the mobile GSM compatible device. All services to which you have subscribed will continue through this new device, based on the information contained on the SIM. For security and billing purposes, SIM and the terminal each have internationally unique identification numbers for independence and identification on the network. The SIM's identifier is called the International Mobile Subscriber Identity (IMSI). The mobile unit has what is called an International Mobile Equipment Identifier (IMEI). In this way a user's identity is matched with the SIM via the IMSI, and the position of the mobile unit is matched with the IMEI. This offers some security, in that a suspected stolen SIM card can be identified and flagged within a database for services to be stopped and to prevent charges by unauthorized individuals.

The base station subsystem, like the mobile station, also has two components: the base transceiver station and the base station controller. The base transceiver station contains the necessary components that define a cell and the protocols associated with the communication to the mobile units. The base station controller is the part of the base station subsystem that manages resources for the transceiver units, as well as the communication with the mobile switching center (MSC). These two components integrate to provide service from the mobile station to the MSC.

The network subsystem is, in effect, the networking component of the mobile communications portion of the GSM network. It acts as a typical class 5 switching central office. It combines the switching services of the core network with added functionality and services as requested by the customer. The main component of this subsystem is the MSC. The MSC coordinates the access to the POTS network, and acts similarly to any other switching node on a POTS network. It has the added ability to support authentication and user registration. It coordinates call hand-off with the Base Station Controller, call routing, as well as coordination with other subscribed services. It utilizes Signaling System 7 (SS7) network architecture to take advantage of the efficient switching methods. There are other components to the network subsystem called *registers*: visitor location register (VLR) and home location register (HLR). Each of these registers handles call routing and services for mobility when a mobile customer is in their local or roaming calling state. The VLR is a database consisting of visitor devices in a given system's area of operation. The HLR is the database of registered users to the home network system.

General Packet Radio Service

General Packet Radio Service (GPRS), also called GSM-IP, sits on top of the GSM networking architecture offering speeds between 56 and 170 Kbps. GPRS describes the bursty packet-type transmissions that will allow users to connect to the Internet from their mobile devices. GPRS is nonvoice. It offers the transport of information across the mobile telephone network. Although the users are always on like many broadband communications methodologies in use today, users pay only for usage. This provides a great deal of flexibility and efficiency. This type of connection, coupled with the nature of packet-switched delivery methods, truly offers efficient uses of network resources along with the speeds consumers are looking for. The data rates offered by GPRS will make it possible for users to partake in streaming video applications and interact with Web sites that offer multimedia, using compatible mobile handheld devices. GPRS is based on Global System for Mobile (GSM) communication and as such will augment existing services such as circuit-switched wireless phone connections and the Short Message Service (SMS).

Short Message Service

Short Message Service (SMS) is a wireless service that allows users to send and receive short (usually 160 characters or less) messages to SMS-compatible phones. SMS, as noted earlier, is integrated with the GSM standard. SMS is used either from a computer by browsing to an SMS site, entering the message and the recipient's number, and clicking **Send**, or directly from a wireless phone.

Optical Wireless Technologies

The third wireless technology we'll cover in this chapter is *optical*, which marries optical spectrum technology with wireless transmissions.

An optical wireless system basically is defined as any system that uses modulated light to transmit information in open space or air using a high-powered beam in the optical spectrum. It is also referred to as *free space optics* (FSO), *open air photonics*, or *infrared broadband*. FSO systems use low-powered infrared lasers and a series of lenses and mirrors (known as a telescope) to direct and focus different wavelengths of light towards an optical receiver/telescope. FSO is a line-of-sight technology and the only condition affecting its performance besides obstruction is fog, and to a lesser degree, rain. This is due to the visibility requirements of the technology. Fog presents a larger problem than rain because the small dense water particles deflect the light waves much more than rain does. The technology communicates bi-directionally (that is, it is full duplex) and does not require spectrum licensing. Figure 7.15 represents a common FSO implementation between buildings within a close proximity, which is generally within 1000 feet, depending on visibility conditions and reliability requirements. Some FSO vendors claim data rates in the 10Mbps to 155Mbps range with a maximum distance of 3.75 kilometers, as well as systems in the 1.25 Gbps data rate range with a maximum distance of 350 meters. The optical sector is growing in capability at a rapid rate, so expect these data rates and distance limits to continue to increase.

Figure 7.15 Free Space Optical Implementation

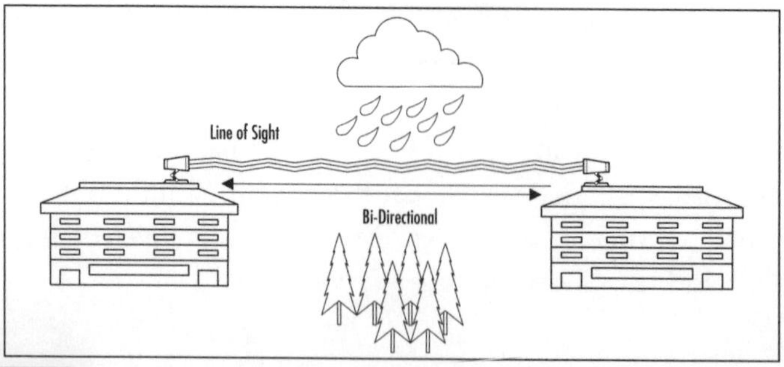

Summary

This chapter provides an overview of differences and purposes of the emerging technologies in the wireless sector. The three primary areas of discussion are *fixed wireless*, *mobile wireless*, and *optical wireless* technology.

We began with a discussion of the fixed wireless technologies that include Multichannel Multipoint Distribution Service (MMDS), Local Multipoint Distribution Service (LMDS), Wireless Local Loop (WLL) technologies, and the Point-to-Point Microwave technology. The primary definition of a fixed wireless technology is that the transmitter and receiver are both in a fixed location. Service providers consider MMDS a complimentary technology to their existing digital subscriber line (DSL) and cable modem offerings; LMDS is similar, but provides very high-speed bandwidth (it is currently limited in range of coverage). Wireless Local Loop refers to a fixed wireless class of technology aimed at providing last mile services normally provided by the local service provider over a wireless medium. Point-to-Point (PTP) Microwave is a line-of-sight technology that can span long distances. Some of the hindrances of these technologies include line of sight, weather, and licensing issues.

In 1997, the Institute of Electrical and Electronics Engineers (IEEE) announced the ratification of the 802.11 standard for wireless local area networks. The 802.11 specification covers the operation of the media access control (MAC) and physical layers; the majority of 802.11 implementations utilize the DSSS method that comprises the physical layer. The introduction of the standard came with 802.11b. Then along came 802.11a, which provides up to five times the bandwidth capacity of 802.11b. 802.11g also provides the same bandwidth as 802.11a but on the same 2.4 GHz as 802.11b, and provides interoperability for both. 802.11i provides enhanced security features. Now, accompanying the ever-growing demand for multimedia services is the development of 802.11e.

The 802.11 architecture can be best described as a series of interconnected cells, and consists of the following: the wireless device or station, the Access Point (AP), the wireless medium, the distribution system (DS), the Basic Service Set (BSS), the Extended Service Set (ESS), and station and distribution services. All these working together providing a seamless mesh allows wireless devices the ability to roam around the WLAN looking for all intents and purposes like a wired device.

High Performance Radio LAN (HiperLAN) is the European equivalent of the 802.11 standard. Wireless personal area networks (WPANs) are networks that occupy the space surrounding an individual or device, typically involving a 10m radius. This is referred to as a personal operating space (POS). This type of network adheres to an ad-hoc system requiring little configuration. Various efforts are under way to converge the 802.11 and 802.15 standards for interoperability and the reduction of interference in the 2.4 GHz space.

Bluetooth is primarily a cable replacement WPAN technology that operates in the 2.4 GHz range using FHSS. One of the main drivers for the success of the Bluetooth technology is the proposition of low-cost implementation and size of the wireless radios. HomeRF is similar to Bluetooth but is targeted solely toward the residential market.

The second category of wireless technology covered in the chapter is *mobile wireless*, which is basically your cell phone service. In this section we described the evolution of this technology from the analog voice (1G) to the digital voice (2G) phases. We continued with a discussion of the next generation technologies including the digital voice and limited data phase (2.5G) to the broadband multimedia (3G) phase, which supports high data rate voice, video, and data in a converged environment.

An *optical wireless* system basically is defined as any system that uses modulated light to transmit information in open space or air using a high-powered beam in the optical spectrum. It is also referred to as free space optics (FSO); it has growing capabilities in the infrared arena for bi-directional communication. It does not require licensing.

Designing a wireless network is not an easy task. Many wireless attributes should be considered throughout the design process. In the preliminary stages of your design, it is important to query users in order to accommodate their needs from a design perspective. Keep in mind that with wireless networks, attributes such as mobility and ease of access can impact your network in terms of cost and function.

The architecture phase is responsible for taking the results of the planning phase and marrying them with the business objectives or client goals. The architecture is a high-level conceptual design. At the conclusion of the architecture phase, the client will have documents that provide information such as a high-level topology, a high-level physical design, a high-level operating model, and a collocation architecture.

The design phase takes the architecture and makes it reality. It identifies specific details necessary to implement the new design and is intended to provide all information necessary to create the new network. At the conclusion of the design phase, the design documents provided to the client will include a detailed topology, detailed physical design, detailed operations design, and maintenance plan.

Hopefully this chapter has provided you with enough basic understanding of the emerging wireless technologies to be able to differentiate between them. The information in this chapter affords you the ability to understand which technology is the best solution for your network design. Evaluate the advancements in these technologies and see how they may impact your organization.

Solutions Fast Track

Fixed Wireless Technologies

☑ In a fixed wireless network, both transmitter and receiver are at fixed locations, as opposed to mobile. The network uses utility power (AC). It can be point-to-point or point-to-multipoint, and may use licensed or unlicensed spectrums.

☑ Fixed wireless usually involves line-of-sight technology, which can be a disadvantage.

☑ The *fresnel* zone of a signal is the zone around the signal path that must be clear of reflective surfaces and clear from obstruction, to avoid absorption and reduction of the signal energy. *Multipath reflection* or interference happens when radio signals reflect off surfaces such as water or buildings in the fresnel zone, creating a condition where the same signal arrives at different times.

☑ Fixed wireless includes Wireless Local Loop technologies, Multichannel Multipoint Distribution Service (MMDS) and Local Multipoint Distribution Service (LMDS), and also Point-to-Point Microwave.

Developing WLANs through the 802.11 Architecture

☑ The North American wireless local area network (WLAN) standard is 802.11, set by the Institute of Electrical and Electronics Engineers (IEEE); HiperLAN is the European WLAN standard.

☑ The three physical layer options for 802.11 are infrared (IR) baseband PHY and two radio frequency (RF) PHYs. The RF physical layer is comprised of Frequency Hopping Spread Spectrum (FHSS) and Direct Sequence Spread Spectrum (DSSS) in the 2.4 GHz band.

☑ WLAN technologies are not line-of-sight technologies.

☑ The standard has evolved through various initiatives from 802.11b, to 802.11a, which provides up to five times the bandwidth capacity of 802.11b. 802.11g also provides the same bandwidth as 802.11a but on the same 2.4 GHz as 802.11b, and provides interoperability for both. 802.11i provides enhanced security features. Now, accompanying the ever-growing demand for multimedia services is the development of 802.11e.

- ☑ 802.11b provides 11 Mbps raw data rate in the 2.4 GHz transmission spectrum.

- ☑ 802.11a provides 25 to 54 Mbps raw data rate in the 5 GHz transmission spectrum.

- ☑ 802.11g provides up to 54 Mbps raw data rate in the 2.4 GHz transmission spectrum, and provides backward compatibility for 802.11a/b devices.

- ☑ 802.11i or WPA2 provides additional security features and utilized AES encryption.

- ☑ HiperLAN type 1 provides up to 20 Mbps raw data rate in the 5 GHz transmission spectrum.

- ☑ HiperLAN type 2 provides up to 54 Mbps raw data rate and QOS in the 5 GHz spectrum.

- ☑ The IEEE 802.11 standard provides three ways to provide a greater amount of security for the data that travels over the WLAN: use of the 802.11 Service Set Identifier (SSID); authentication by the Access Point (AP) against a list of MAC addresses; and the use of encryption technologies.

Developing WPANs through the 802.15 Architecture

- ☑ Wireless personal area networks (WPANs) are networks that occupy the space surrounding an individual or device, typically involving a 10m radius. This is referred to as a personal operating space (POS). WPANs relate to the 802.15 standard.

- ☑ WPANs are characterized by short transmission ranges.

- ☑ Bluetooth is a WPAN technology that operates in the 2.4 GHz spectrum with a raw bit rate of 1 Mbps at a range of 10 meters. It is not a line-of-sight technology. Bluetooth may interfere with existing 802.11 technologies in that spectrum.

- ☑ HomeRF is similar to Bluetooth but targeted exclusively at the home market. HomeRF provides up to 10 Mbps raw data rate with SWAP 2.0.

Mobile Wireless Technologies

☑ Mobile wireless technology is basic cell phone technology; it is not a line-of-sight technology. The United States has generally progressed along the Code Division Multiple Access (CDMA) path, with Europe following the Global System for Mobile Communications (GSM) path.

☑ Emerging technologies are known in terms of *generations*: 1G refers to analog transmission of voice; 2G refers to digital transmission of voice; 2.5G refers to digital transmission of voice and limited bandwidth data; 3G refers to digital transmission of multimedia at broadband speeds (voice, video, and data).

☑ The Wireless Application Protocol (WAP) has been implemented by many of the carriers today as the specification for wireless content delivery. WAP is a nonproprietary specification that offers a standard method to access Internet-based content and services from wireless devices such as mobile phones and PDAs.

☑ The Global System for Mobile Communications (GSM) is an international standard for voice and data transmission over a wireless phone. A user can place an identification card called a Subscriber Identity Module (SIM) in the wireless device, and the device will take on the personal configurations and information of that user (telephone number, home system, and billing information).

Optical Wireless Technologies

☑ Optical wireless is a line-of-sight technology in the infrared (optical) portion of the spread spectrum. It is also referred to as free space optics (FSO), open air photonics, or infrared broadband.

☑ Optical wireless data rates and maximum distance capabilities are affected by visibility conditions, and by weather conditions such as fog and rain.

☑ Optical wireless has very high data rates over short distances (1.25 Gbps to 350 meters). Full duplex transmission provides additional bandwidth capabilities. The raw data rate available is up to a 3.75 kilometer distance with 10 Mbps.

☑ There are no interference or licensing issues with optical wireless, and its data rate and distance capabilities are continuously expanding with technology advances.

Frequently Asked Questions

The following Frequently Asked Questions, answered by the authors of this book, are designed to both measure your understanding of the concepts presented in this chapter and to assist you with real-life implementation of these concepts. To have your questions about this chapter answered by the author, browse to **www.syngress.com/solutions** and click on the **"Ask the Author"** form.

Q: What does the G stand for in 1G, 2G, 2.5G, and 3G mobile wireless technologies?

A: It stands for *generation* and the use of it implies the evolutionary process that mobile wireless is going through.

Q: What are the primary reasons that service providers use a Wireless Local Loop (WLL)?

A: The primary reasons are speed of deployment, deployment where wireline technologies are not practical, and finally, for the avoidance of the local exchange carrier's network and assets.

Q: Why is digital transmission better than analog in mobile wireless technologies?

A: Digital transmissions can be reconstructed and amplified easily, thus making it a cleaner or clearer signal. Analog signals cannot be reconstructed to their original state.

Q: Why does fog and rain affect optical links so much?

A: The tiny water particles act as tiny prisms that fracture the light beam and minimize the power of the signal.

Q: What is the difference between an ad-hoc network and an infrastructure network?

A: Ad-hoc networks are ones where a group of network nodes are brought together dynamically, by an Access Point (AP), for the purpose of communicating with each other. An infrastructure network serves the same purpose but also provides connectivity to infrastructure such as printers and Internet access.

Q: Several customers want me to give them up-front costs for designing and installing a network. When is the most appropriate time to commit to a set price for the job?

A: Try to negotiate service charges based on deliverables associated with each phase of the design process. In doing so, you allow the customer to assess the cost prior to entering into the next phase of the design.

Q: I'm very confused by all the different home network standards. Is there any way that I can track several of the different home networking standards from a single unbiased source?

A: Yes. There are several means of tracking various home network standards and initiatives. For comprehensive reports in the home network industry, I would suggest contacting Parks Associates at www.parksassociates.com. The Continental Automated Buildings Association (CABA) at www.caba.org is another good source for learning about home network technologies from a broad and unbiased perspective.

Q: I am trying to create a design of a wireless campus network and I keep finding out new information, causing me to change all of my work. How can I prevent this?

A: If you have done a thorough job in the planning phase you should already have identified all of the requirements for the project. Once you identify all of the requirements, you need to meet with the client and make sure that nothing was overlooked.

Chapter 8

Monitoring and Intrusion Detection

Solutions in this chapter:

- **Designing for Detection**
- **Defensive Monitoring Considerations**
- **Intrusion Detection Strategies**
- **Conducting Vulnerability Assessments**
- **Incident Response and Handling**
- **Conducting Site Surveys for Rogue Access Points**

☑ Summary

☑ Solutions Fast Track

☑ Frequently Asked Questions

Introduction

Network monitoring and intrusion detection have become an integral part of network security. The monitoring of your network becomes even more important when introducing wireless access, because you have added a new, openly available entry point into your network. Security guards patrol your building at night. Even a small business, if intent on retaining control of its assets, has some form of security system in place—as should your network. Monitoring and intrusion detection are your security patrol, and become the eyes and ears of your network, alerting you to potential vulnerabilities, and intrusion attempts. Designing secure wireless networks will rely on many of the standard security tools and techniques but will also utilize some new tools.

In this chapter, you'll learn about the planning and deployment issues that must be addressed early on in order to make monitoring and intrusion detection most effective when the system is fully operational.

You'll also learn how to take advantage of current intrusion principles, tools, and techniques in order to maximize security of your wireless network. Specialized wireless tools such as NetStumbler and AirSnort will also be used to provide a better overall picture of your wireless security.

Intrusion Prevention (IP) systems may offer an additional layer to detection. We'll discuss the pros and cons of their use, and their relationship to conventional intrusion detection. You'll also learn how to respond to incidents and intrusions on a wireless network, as well as conduct site surveys to identify the existence of rogue Access Points (APs).

Designing for Detection

In this section, we will discuss how to design a wireless network with an emphasis on monitoring, focusing on the choice of equipment, physical layout and radio interference. The decision-making involved in the design, deployment, and installation of a wireless local area network (WLAN), combined with the choice of product vendor, can play a key role in later efforts to monitor the network for intrusions. *Designing for detection* occurs when you build a network with monitoring and intrusion detection principles in mind from the start. For example, when a bank is built, many of the security features, such as the vault security modules, closed circuit cameras, and the alarm are part of the initial design. Retrofitting these into a building would be much more expensive and difficult than including them in the beginning. The same idea is true with a network. Designing your network for detection, having made the decisions about monitoring strategies and the infrastructure to support them, will save you time and money in the long run.

If you've followed the design and configuration advice given in this book, you should be able to identify certain false alarms. Knowledge of your building's layout and physical obstacles, as discussed earlier, will strengthen your ability to identify red herrings. Additionally, understanding sources of radio interference and having an idea of the limits of your network signal can also help avoid potential headaches from false alarms and misleading responses when patrolling the network for intruders. Keeping these points in mind, laying out your wireless network for the most appropriate detection should be no problem.

Starting with a Closed Network

The choice of vendor for your wireless gear can dramatically alter the visible footprint of your wireless network. After an Access Point is installed, it will begin emitting broadcasts, announcing, among other things, its Service Set Identifier (SSID). This is a very useful function for clients to be able to connect to your network. It makes discovery and initial client configuration very easy, and quick. The ease of contact, however, has some security implications. The easily available nature of the network is not only available for your intended users, but for anyone else with a wireless card. The easier any system is to find, the easier it is to exploit.

In order to counteract some of the troubles with openly available and easily discoverable wireless networks, some vendors have developed a system known as closed network. With closed network functionality enabled, the wireless AP no longer broadcasts its SSID to the world; rather it waits for a client to connect with the proper SSID and channel settings. This certainly makes the network more difficult to find, as programs such as NetStumbler and dstumbler will not see it. The network is now much more secure, because it is much more difficult for an attacker to compromise a network he or she can't see. The potential disadvantage, however, is that clients must now know the SSID and settings of your network in advance in order to connect. This process can be difficult for some users, as card configuration will be required. From a security standpoint, however, a closed network system is the ideal foundation from which to begin designing a more secure wireless network solution. A closed network–capable AP is recommended for all but those who wish to have an openly available wireless network (in such a scenario, security concerns are generally not primary).

Ruling Out Environmental Obstacles

Another important design consideration is the physical layout. A knowledge of the obstacles you are designing around is vital for determining the number of APs that will be required to provide adequate coverage for your wireless network. Many installations have suffered from administrators failing to take notice of trees, indoor waterfalls, and even the layout and construction materials of the building. Features such as large indoor fountains and even translucent glass walls can be a barrier to proper signal path. Fixing a broken network is much more of a burden than making sure everything is set up properly from the beginning. Before starting, learn as much as you can about the building in which you're planning to deploy. If the building is concrete with a steel frame, the 802.11 signal will be much more limited than if it were passing through a wood/drywall frame building. When placing the initial 802.11 AP, design from the inside-out. Place the AP toward the center of your user base and take advantage of the fact that the signal will radiate outwards. The goal of this placement is to provide the best quality of signal to your users, while limiting the amount and strength of the signal that passes outside of your walls. Remember, potential attackers will be looking for a signal from your network, and the weaker the signal is when it leaves your premises, the less likely an attacker can safely snoop on your network. *Safely*, in this case, means that an attacker doesn't need to worry about being seen in an unusual place with a laptop. For example, an attacker sitting in your lobby with a wireless card is suspicious, but, someone sipping coffee in a coffee shop with their laptop isn't. Of course, signal strength alone isn't a security measure, but is part of a whole secure security package you will want to have built into your wireless network.

The second physical consideration that should be kept in mind when designing a wireless network is the building floor plan. Using the inside-out method of AP placement, place the AP as far from possible from external windows and doors. If the building layout is a square, with cubicles in all directions, place the AP in the center. If the building is a set of long corridors and rooms, then it will be best to experiment with placement. Try putting the APs at different locations, and then scout the location with NetStumbler or other tools to determine where the signal is strongest, and whether or not it can be seen from outside of your facility. We'll talk more about using NetStumbler and other site evaluation tools a bit later.

Another consideration should be your neighbors. In most environments, there will be other companies or businesses operating nearby. Either from the floors above, below, or right next door, your signal may be visible. If you have competitors, this may be something which you wish to avoid, because they will be able to join your network, and potentially exploit it. Close proximity means that an attacker could

easily and discreetly begin deciphering your wireless encryption keys. Proper placement and testing of your APs before deployment can help you gain a better understanding of your availability to those around you.

SECURITY ALERT

Remember that good design requires patience and testing. Avoid at all
costs the temptation to design around obstacles simply by throwing
more APs at the situation, or increasing the signal strength. While providing more signal and availability, this potentially dangerous scenario
adds more points of entry to your network, and can increase your
chance of compromise.

Ruling Out Interference

Thought should also be given to whether or not there are external or internal
sources of radio interference present in your building. Potential problems can come
from microwave ovens, 2.4GHz wireless phones, wireless video security monitors,
and other 802.11b wireless networks. If these are present in large numbers in your
environment, it may be necessary to do some experimentation with AP placement
and settings to see which combination will provide the most available access. We'll
discuss interference in more detail in the next section, but be aware that these
devices may create holes, or weaken your range. Having properly identified these
sources and potential problems can help you diagnose future problems, and realize
that an outage may not necessarily be an attacker but rather a hungry employee
warming lunch.

Defensive Monitoring Considerations

Monitoring wireless networks for intrusion attempts requires attention to some
newer details, which many security administrators have not encountered in the past.
The use of radio for networking introduces new territory for security administrators
to consider. Issues such as signal strength, distortion by buildings and fixtures, inter-
ferences from local and remote sources, and the mobility of users are some of these
new monitoring challenges not found in the wired world. Any attempt to develop
an intrusion detection regime must take into account these new concepts. Security

administrators must make themselves familiar with radio technology and the direct impact the environment will have on networks using these technologies.

Security monitoring is something that should be built into your initial wireless installation. Many devices have logging capabilities and these should be fully utilized in order to provide the most comprehensive overall picture possible of what is happening on your network. Firewalls, routers, internal Web servers, Dynamic Host Configuration Protocol (DHCP) servers, and even some wireless APs will provide log files, which should be stored and reviewed frequently. Simply collecting the logs isn't enough; they should be thoroughly reviewed by security administrators. This is something that should be built into every security procedures guide, but is often overlooked. A firewall log is worthless if it's never reviewed! Having numerous methods and devices in place to review traffic and usage on your network will provide critical insight into any type of attack, either potential or realized.

Availability and Connectivity

Obviously the most important things in building and operating a wireless network are availability and connectivity. A wireless network that users cannot connect to, while very secure, is completely useless. Interference, signal strength and denial of service (DoS) attacks can all dramatically affect your availability. In the past, for an attacker to perform a denial of service attack against your internal network, they would have needed to gain access to it, not always a trivial task. Now, however, an attacker with a grudge against your organization needs only to know that a wireless network is present in order to attack. We'll discuss the possibilities of denial of service attacks later in this section. Even if the network has been designed securely, simply the fact that the network is radio-based means these issues must be considered.

Interference and Noise

Identifying potential sources of interference during the design phase can help you identify potentially malicious sources of interference within your environment once you undertake your monitoring activities.

For example, during one wireless deployment, we were experiencing a major denial of service in one group. Users in one group were either unable to connect to the AP at all, or suffered from diminished bandwidth. It was suspected there was a potentially malicious source of activity somewhere, but after reviewing our initial design notes about the installation, we remembered a kitchen near these users. At the time of deployment, there was no known source of interference in the kitchen, but upon investigating further, we discovered the group had just installed a new commercial grade, high wattage microwave oven. As you can see, when deploying a wireless network, it's important to explore all possible solutions of interference before suspecting foul play. If your organization uses noncellular wireless phones, or any other type of wireless devices, be certain you check whether or not they are operating in the 2.4GHz spectrum. While some devices like telephones won't spark a complete outage, they can cause intermittent problems with connections. Other devices like wireless video monitors can cause serious conflicts, and should be avoided at all costs. Identified potential problems early can be very useful when monitoring for interference and noise in your wireless network environment.

It should be noted that some administrators may have few, if any, problems with microwave ovens, phones, or other wireless devices, and tests have been performed on the World Wide Web supporting this. A simple Web search for microwave ovens and 802.11b will give you plenty of information. However, do realize that while some have had few problems, this is no guarantee you will be similarly blessed. Instead, be thorough. Having an idea of potential problems can save you time identifying later connectivity issues.

As mentioned earlier, knowledge of your neighbors is a good idea when building a wireless network. If you are both running a wireless network with similar settings, you will be competing on the same space with your networks, which is sure to cause interference problems. Given this, it's best to monitor what your neighbors are doing at all times to avoid such problems. Notice that conflicts of this kind are generally inadvertent. Nevertheless, similar situations can be used to create a denial of service, which we'll discuss later.

Signal Strength

From a monitoring standpoint, signal strength is one of the more critical factors to consider. First, it is important to monitor your signal regularly in order to know the extent to which it is available. Multiple APs will require multiple investigations in order to gain a complete picture of what a site looks like externally. Site auditing discovery tools should be used to see how far your signal is traveling. It will travel much farther than most manufacturer claims, so prepare to be surprised. If the signal is adequate for your usage, and you'd like to attempt to limit it, some APs will allow you to fine-tune the signal strength. If your AP supports this feature, experiment with it to provide the best balance between internal and external availability.

Whether you can fine-tune your signal strength or not, during initial design you should have noted points externally where the signal was available. Special attention should have been paid to problematic areas, such as cafes, roadways or parking lots. These areas are problematic because it is difficult, or impossible to determine whether or not an attacker is looking at your wireless network specifically. When monitoring, those areas should be routinely investigated for potential problems. If you are facing an intrusion, knowledge of places like these, with accessibility to your network could help lead you to your attacker.

Detecting a Denial of Service

Monitoring the wireless network for potential denial of service attacks should be part of your security regime. Surveying the network, checking for decreases in signal strength, unauthorized APs, and unknown Media Access Control (MAC) addresses, are all ways to be proactive about denial of service.

Denial of service attacks can be incredibly destructive. Often times, however, their severity is overlooked because a DoS attack doesn't directly put classified data at risk. While this attitude may be acceptable at certain organizations, at others it can cost a tremendous amount of money both in lack of employee productivity and lost customer revenue. One only needs to look back at the DoS attacks conducted in February 2000 against several major E-commerce companies to realize the threat from such attacks.

On an Internet level, this type of attack can be devastating, but at the wireless networking level, they may not be as severe. The largest possible loss could come from lost employee productivity. The availability of a wired alternative can help mitigate the risks from a wireless DoS, but as networking moves toward the future, and away from wires, this may become less of a possibility.

As mentioned earlier, the radio-based nature of 802.11b makes it more susceptible to denial of service. In the wired world, an attacker generally needed access to your internal network in order to cause a DoS outage. Since many wireless installations offer instant access into this network, it can be much easier for an attacker to get in and start shutting things down. There are two main ways an attacker can conduct a DoS against your wireless LAN. The first method would be fairly traditional. They would connect to the network, and simply start blasting packets to any of your internal machines—perhaps your DNS servers or one of your routers. Either scenario is likely to cause connectivity outages on the network. A second method of denying service to wireless LANs wouldn't even require a wireless LAN card, but rather just a knowledge of how the technology works. An attacker with a device known to cause interference could place it in the path of your wireless network. This is a very crude, but potentially effective method of performing a DoS attack. A third way to conduct a DoS against a wireless LAN is similar to the scenario we've just discussed, but requires a wireless AP. In this scenario, an attacker would configure a wireless AP to mimic the settings on your AP, but not connect the AP to the network. Therefore, users connecting to this AP would not be able to communicate on the LAN. And, if this AP were placed in an area with many of your users, since their cards are generally configured to connect to the strongest signal, the settings would match, making detection potentially difficult. A good way to save yourself from this scenario is to identify the MAC addresses of all your wireless APs, and then routinely do surveys for any nonmatching APs. This type of situation closely mirrors what we will discuss later when talking about rogue APs.

Monitoring for Performance

Keeping an eye on the performance of your network is always a good idea. Knowing your typical baseline usage, the types of traffic that travel on your network, as well as the odd traffic patterns that might occur will not only help you keep an eye on capacity, but clue you in to potential intrusions. This type of monitoring is generally part of a good security regime in the wired world, but should be adopted to cover traffic on your wireless network as well.

Knowing the Baseline

Knowing the baseline usage that your network generally sees can help you identify potential problems. Over time, you should be watching the network to get an idea of how busy it gets throughout the day. Monitoring baseline performance will give you a good idea of your current capacity, and help provide you with a valuable picture of how your network generally operates. Let's say, for example, your network generally sees its peak usage at 9AM at which point it generally sees a load of 45 percent. Then, in monitoring your performance logs you notice usage peaks at 3AM with much higher bandwidth consumed—you have an anomaly that should be investigated. Additionally, if, when monitoring, you find that massive amounts of bandwidth are being consumed, and you only have four or five users with minimal usage needs, this should be a red flag as well. A common attack motive for intruders is to gain access to bandwidth.

Monitoring Tools of the Trade

There are many performance-monitoring tools, with diverse prices and levels of functionality. Commercially available tools such as Hewlett-Packard's OpenView have great amounts of market share. OpenView can be configured to watch just about any aspect of your network, your servers, bandwidth, and even traffic usage patters. It is a very powerful tool that is also customizable and can be made to monitor just about anything imaginable. Being a solution designed for enterprise type organizations, it does come with a hefty price tag, but is generally considered one of the best monitoring tools available. There are some downsides to

OpenView, however. It isn't security friendly, in that it requires the use of the User Datagram Protocol (UDP), which is something that is sometimes not allowed through firewalls due to the fact that it is a connectionless protocol. Connectionless protocols do not allow firewalls to verify that all transmissions are requested by the initiating party. In other words, there is no connection handshake like with the Transport Control Protocol (TCP). OpenView also has some problems working in a Network Address Translation (NAT) environment. Implementing OpenView into a secure environment can also be a real challenge, and may require some security requirement sacrifices. Proceed with caution.

If you are looking for something with a lower price tag, and potentially easier integration, SNIPS (formerly known as NOCOL) is an excellent monitoring package. It is very flexible in what it can do, but one particularly useful function is that it can be used to watch your Ethernet bandwidth. Watching bandwidth, as mentioned earlier, is a good idea because it can help you spot potential excess usage. SNIPS can also be configured to generate alarms when bandwidth reaches a certain level above what is considered normal use in your environment. Notification of this kind could alert you early to network intrusion, and when combined with specially designed detection software can be a very powerful combination. The screenshot in Figure 8.1 shows the different alert levels SNIPS features, and how they are sorted.

Figure 8.1 SNIPS: A Freely Available Monitoring Package

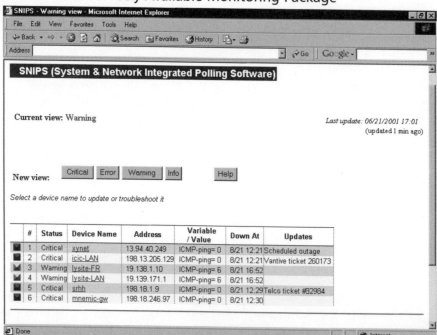

Another excellent tool for watching bandwidth on your network is called EtherApe. It provides an excellent graphical view of what bandwidth is being consumed, and where. With breakdowns by IP or MAC address, and protocol classifications, it is one tool that should be explored. It is freely available at http://etherape.sourceforge.net. For example, if you were detecting great slowdowns on your network, and you needed to quickly see what was consuming your resources, start EtherApe. It listens to your network and identifies traffic, protocols, and network load. Additionally, it traces the source and destination of the traffic, and provides a nice visual picture of the network. It's a great tool for identifying problems with the network, and can assist in explaining bandwidth and traffic issues to nontechnical people. Figure 8.2 shows EtherApe in action, illustrating how the traffic is displayed, graphically. The hosts are presented in a ring, with connections shown as lines drawn between them. The more intense the traffic, the larger the connection lines. Traffic can also be sorted by color, which makes it instantly easier to distinguish between types.

Figure 8.2 EtherApe for Linux

Intrusion Detection Strategies

Until now, we've primarily discussed monitoring in how it relates to intrusion detection, but there's more to an overall intrusion detection installation than monitoring alone. Monitoring can help you spot problems in your network, as well as identify performance problems, but watching every second of traffic that passes through your network, manually searching for attacks, would be impossible. This is why we need specialized network intrusion detection software. This software inspects all network traffic, looking for potential attacks and intrusions by comparing it to a predefined list of attack strings, known as *signatures*. In this section, we will look at different intrusion detection strategies and the role monitoring plays. We'll learn about different strategies designed for wireless networks, which must take into account the nature of the attacks unique to the medium. These include a lack of centralized control, lack of a defined perimeter, the susceptibility to hijacking and spoofing, the use of rogue APs, and a number of other features that intrusion detection systems were not designed to accommodate. Only a combination of factors we've discussed earlier, such as good initial design and monitoring, can be combined with traditional intrusion detection software to provide an overall effective package.

Integrated Security Monitoring

As discussed earlier, having monitoring built in to your network will help the security process evolve seamlessly. Take advantage of built-in logging-on network devices such as firewalls, DHCP servers, routers, and even certain wireless APs. Information gathered from these sources can help make sense of alerts generated from other intrusion detection sources, and will help augment data collected for incidents. Additionally, these logs should help you to manually spot unauthorized traffic and MAC addresses on your network.

Tools & Traps...

Beware of the Auto-responding Tools!

When designing your intrusion detection system, you will likely come across a breed of tools, sometimes known as Intrusion Prevention Systems. These systems are designed to automatically respond to incidents. One popular package is called PortSentry. It will, upon detection of a port scan, launch a script to react. Common reactions include dropping the route to the host that has scanned you, or adding firewall rules to block it. While this does provide instant protection from the host that's scanning you, and might seem like a great idea at first, it creates a very dangerous denial of service potential. Using a technique known as IP spoofing, an attacker who realizes PortSentry is being used can send bogus packets that appear to be valid port scans to your host. Your host will, of course, see the scan and react, thinking the address that its coming from is something important to you, such as your DNS server, or your upstream router. Now, network connectivity to your host is seriously limited. If you do decide to use auto-responsive tools, make sure you are careful to set them up in ways that can't be used against you.

Watching for Unauthorized Traffic and Protocols

As a security or network administrator, it is generally a good idea to continuously monitor the traffic passing over your network. It can give you an idea of the network load, and more importantly, you can get an idea of what kinds of protocols are commonly used. For most corporate networks, you are likely to see SMTP (e-mail), DNS lookups, Telnet or SSH, and, of course, Web traffic. There is also a good chance if you are using Hewlett-Packard printers, there will be JetDirect traffic on port 9100. If you have Microsoft products such as Exchange server, look for traffic on a number of other ports, with connections to or from your mail servers. After several sample viewings of network traffic, you should start to notice some patterns as to what is considered normal usage. It is from these samples that you can start looking for other unknown and possibly problematic traffic. IRC, Gnutella, or heavy FTP traffic can be a sign that your network is being used maliciously. If this is the case, you should be able to track the traffic back to its source, and try to identify who is using the offending piece of software. There are many Gnutella clients today, and it has become the most heavily used peer-to-peer networking system available. It is advised you become familiar with a few Gnutella clients, so they can be quickly identified and dealt with. BearShare, Gnotella, and LimeWire are some of the more popular ones. LimeWire, shown in Figure 8.3, provides an easy-to-use interface for Gnutella and offers lots of information about clients. Another point of caution about peer-to-peer client software should be the fact that it is often bundled with spyware—software which shares information about the user and their computer, often without their knowledge.

Figure 8.3 LimeWire: A Popular Gnutella Peer-to-peer File Sharing Program

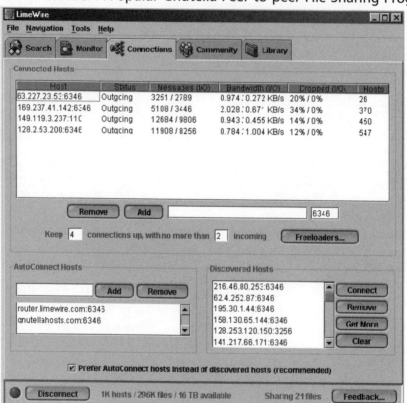

Within your security policy, you should have defined which types of applications are not considered acceptable for use in your environment. It is advisable to ban peer-to-peer networking software like Napster, Gnutella, and Kazaa. Constant monitoring is essential because the list grows larger each day and current policies may not prohibit the latest peer-to-peer software. Aside from possibly wasting company bandwidth, these tools allow others on the Internet to view and transfer files from a shared directory. It is very easy to misconfigure this software to share an entire hard drive. If shared, any other user on the peer-to-peer network would potentially have access to password files, e-mail files, or anything else that resides on the hard disk. This is more common than one would expect. Try a search on a peer-to-peer network for a sensitive file name like archive.pst, and you might be surprised by what you find.

Internet Relay Chat (IRC) traffic can also be a sign that something fishy is happening on your network. There are legitimate uses for IRC on an internal network. It makes a great team meeting forum for large groups separated by distances, or for those who require a common real-time chat forum. It should be kept in mind though that attackers commonly use IRC to share information or illegally copied software. If you are using IRC on your network, make sure you have a listing of your authorized IRC servers, and inspect IRC traffic to insure it is originating from one of those hosts. Anything else should be treated as suspect. If you aren't using IRC on your network, any IRC traffic (generally found on TCP port 6666 or 6667) should be treated as suspect.

A good way to automate this kind of scanning is generally available in intrusion detection packages. Snort, the freely available IDS has a signature file that identifies Gnutella, Napster, IRC, and other such types of traffic. Network Flight Recorder has similar filters, and supports a filter writing language that is incredibly flexible in its applications. We'll discuss some of the IDS packages a bit later in this chapter.

Unauthorized MAC Addresses

MAC address filtering is a great idea for wireless networks. It will only allow wireless cards with specified MAC addresses to communicate on the network. Some APs have this capability built in, but if yours doesn't, DHCP software can often be configured to do the same. This could be a major headache for a large organization, because there could simply be too many users to keep track of all of the MAC addresses. One possible way around this is to agree upon the same vendor for all of your wireless products. Each wireless card vendor has an assigned OUI or organizationally unique identifier, which makes up the first part of an Ethernet card's MAC address. So, if you chose Lucent wireless cards, you could immediately identify anything that wasn't a Lucent card just by noting the first part of the MAC address. This type of system could be likened to a company uniform. If everyone wore orange shirts to work, someone with a blue shirt would be easily spotted. This is not foolproof, however. An attacker with the same brand of wireless card would slide thorough unnoticed. In a more complicated vein, it is possible for attackers to spoof their MAC addresses, meaning they can override the wireless network card's MAC address. A system based solely on vendor OUIs alone wouldn't provide much protection, but it can make some intrusions much easier to identify.

Popular Monitoring Products

The number of available intrusion detection packages has increased dramatically in the past few years. There are two main types of intrusion detection software: host-based and network-based. Host-based intrusion detection is generally founded on the idea of monitoring a system for changes to its file system. It doesn't generally inspect network traffic. For that functionality, you'll need a network intrusion detection system (IDS), which looks specifically at network traffic, and will be our focus for this section.

Signature files are what most Intrusion Detection Systems use to identify attacks. Therefore, an IDS is generally only as good as its signature files. Using just a small snippet from an attack, the IDS compares packets from captured traffic to the signature file, searching for the specified attack string. If there's a match, an alert is triggered. This is why it's important to have control and flexibility with your signature files. When spotting new attacks, time is always of the essence. New attacks occur daily, and the ability to add your own signature files to your IDS sensor can save you the wait for a vendor to release a new signature file. Another thing to keep in mind with signature files is that, if they are written too generically, false alarms will become the norm. The downfall of any IDS system, false alarms can desensitize administrators to warnings, thus allowing attacks to sneak through—a perfect real-life example of "crying wolf."

Of all of the commercially available IDS products, one of the most flexible and adaptable is Network Flight Recorder, from NFR Security. Its sensors are run from a CD-ROM based on an OpenBSD kernel. Its greatest flexibility comes with the specially developed N-Code system for filter writing. N-Code can be used to grab any type of packet and dissect it to the most minimal of levels, then log the output. This is particularly useful when searching for attack strings, but can also be used to identify unknown network protocols, or to learn how certain software communicates over the network. Having the ability to write your own filters can be very helpful as well. For example, if your company has a specially developed piece of software, and you would like to identify its usage and make sure it isn't being utilized outside your network, a filter could be written to identify traffic from that specific program—a task which would be impossible with a hard-coded signature file system. Another excellent use of N-Code is in developing custom attack signatures. We'll discuss why having custom signatures can be important in the next section. NFR also supports the use of multiple sensors distributed throughout an environment, with a central logging and management server. Configurations and N-Code additions are done via a GUI, through a Windows-based program. Changes are centrally done, then pushed out to all remote sensors, eliminating the need to manually update each remote machine. This can be a huge timesaver in big environments.

A free alternative to NFR is a program called Snort, which is an excellent and freely available tool (downloadable from www.snort.org). Snort is a powerful and lightweight IDS sensor that also makes a great packet sniffer. Using a signature file or rule set (essentially a text file with certain parameters to watch the traffic it is inspecting), it generates alerts to a text file or database. We'll take a more in-depth look at writing rules in the next section. Snort has a large community of developers, so it is continually being updated to stay current with the latest changes in security. It is also now more able to deal with tools like Stick and Snot, which were designed to fool IDS sensors. One potential downside to Snort, however, is that because it is freeware, the group that writes it does not offer technical support. For home or small business use this might not be a problem, but for larger companies who require support when using Snort, a company called Silicon Defense offers commercial support and also sells a hardware, ready-to-go Snort sensor.

Signatures

It isn't uncommon for a sophisticated attacker to know the signature files of common IDS sensors, and use that knowledge to confuse the system. For a very simplistic example of this, let's say a particular attack contains the string "Hacked by hAx0r." A default filter might therefore search specifically for the string "hAx0r." Countering, an attacker with knowledge of the default signature files could send benign packets to your network containing only the string "hAx0r." This technically wouldn't be an attack, but it could fool the IDS. By sending a large series of packets all with "hAx0r" in them, the sensor could become overwhelmed, generating alerts for each packet, and causing a flurry of activity. An attacker could use this to their advantage in one of two ways. They could either swamp the IDS with so many packets it can't log them any more, or they could swamp it with alerts in order to hide a real attack. Either strategy spells trouble.

A custom signature could be defined to look for "by hAx0r," therefore defeating this type of attack strategy. Again, this scenario is a very simplistic example of custom signature writing. In reality, there is much more in the way of actual analysis of attacks and attack strings that must be done. Simple signatures can be very easy to write or modify, but the more complex the attack, the more difficult it is to write the signature. The best way to learn how to write signatures is to investigate already written ones included with the system. In the case of NFR, there are many N-Code examples that ship with the software, and many more can be found on the Web. A comprehensive N–Code guide is also available, which gives a detailed explanation of all the features and abilities of N–Code.

Snort, on the other hand, as we earlier described, just uses a text file with rules. A sample rule file for snort looks like this:

```
alert tcp $HOME_NET 21 -> !$HOME_NET any (msg:"FTP-bad-login";flags:PA;
    content:"530 Login incorrect";)
alert tcp !$HOME_NET any -> $HOME_NET 21 (msg:"FTP-shosts";flags:PA;
    content:".shosts";)
alert tcp !$HOME_NET any -> $HOME_NET 21 (msg:"FTP-user-root";flags:PA;
    content:"user root |0d|";)
alert tcp !$HOME_NET any -> $HOME_NET 21 (msg:"FTP-user-warez";flags:PA;
    content:"user warez |0d|";)
alert tcp !$HOME_NET any -> $HOME_NET 21 (msg:"IDS213 - FTP-Password
    Retrieval"; content:"passwd"; flags: AP;)
alert icmp !$HOME_NET any -> $HOME_NET any (msg:"IDS118 - MISC-
    Traceroute ICMP";ttl:1;itype:8;)
```

From this example, the format is easily readable. To create a simple signature, one only needs to specify the port number, an alert string, which is written to the file, and a search string, which is compared to the packets being inspected. As an example, we'll write a rule to search for Xmas tree scans, or a port–scan where strange packets are sent with the FIN, PSH, and URG TCP flags set. Most port scanning software, like Nmap will perform these scans. To begin, we can run some test Xmas tree scans just to watch what happens. Using a packet sniffer like Snort or Ethereal, we can see exactly which flags are set in our scan. Once we have that information gathered, the next step is to actually write the rule. So, our sample rule looks like this:

```
alert tcp !$HOME_NET any -> $HOME_NET any (msg:"SCAN
    FullXMASScan";flags: FPU;)
```

All alert rules start with the word "alert." The next three fields tell Snort to look for Transmission Control Protocol (TCP) packets coming from outside of our network on any port. The other side of the arrow specifies the destination of the traffic. In this case, it is set to anything defined as our home network, on any port. Next, we set our message, which is logged to the alerts file. It's generally a good idea to make the message as descriptive as possible, so you know what you're logging. The final two parts of the rule are where we fill in the information gathered from our sniffer. We know that the TCP flags were set to FPU, so we enter that in the flags field. This way, from start to finish the rule reads "make an alert if there is any TCP packet that comes from outside of our network, on any port, to anywhere on our home network, on any port with the flags FPU." Try reading through some of the rules listed previously and see if they begin to make sense. The first rule would read "Make an alert if anything on our network tries to connect to an FTP server outside of our network, and fails." Snort rules are fairly straightforward to read and write. For more complex rules, and a better definition of all the features that can be included with Snort rule writing, see the Snort project's home page.

Damage & Defense…

Keep Your Signatures Up to Date!

Most IDS sensors work by comparing traffic to a predefined list of signatures. When a match is found, an alert is triggered. This system has worked well in the past, but a new type of tool has been developed to mimic authentic signatures. One common tool is called Stick, and can be used to generate thousands of "attacks" per second, all from spoofed IP addresses. An attacker could use this to cause a denial of service to your IDS sensors, or to provide cover for his or her specific attack to your network. Some IDS vendors claim to now be able to distinguish between these fake attacks and real ones. Nevertheless, proceed with caution. And don't forget to update your signatures often!

Conducting Vulnerability Assessments

Ini Chapter 12 of this book, we will cover in detail how to perform a wireless penetration test using the Auditor Security Collection. In this chapter, we'll cover the basics of a wireless vulnerability assessment. Being aware of changes in your network is one of the keys to detecting problems. Performing this kind of an assessment on a wireless network will be a fairly new exercise for most administrators. There are a number of new challenges that will arise from a radio transmission-based network, such as the mobility of clients and the lack of network boundaries.

When beginning a wireless vulnerability assessment, it's important to identify the extent of the network signal. This is where tools like NetStumbler, and the ORiNOCO client software will be very handy, because they will alert you to the presence of wireless connectivity. A good place to start the assessment is near the wireless AP. Start the monitoring software and then slowly walk away from the AP, checking the signal strength and availability as you move. Check out the entire perimeter of your area to make note of signal strength, taking special notice of the strong and weak points. Once you have a good idea about the signal internally, try connecting to your network from outside your facility. Parking lots, sidewalks, any nearby cafes, and even floors above and below yours should be investigated to analyze the extent of your signal. Anyplace where the signal is seen should be noted as a potential trouble area, and scrutinized in the future. If your signal is available far outside your premises, it might be a good idea to rethink the locations of your APs. If you can see your network, so can an attacker. Try to lower the signal strength of your AP by either moving it or making adjustments to its software, if possible. If limiting signal strength isn't an option, more emphasis should be placed on constant monitoring, as well as looking into other security devices.

If you have a signal from your network, externally, you'll now want to look at the visibility of your network resources from your wireless network. A good security design would isolate the wireless AP from the rest of the network, treating it as an untrusted device. However, more often than not, the AP is placed on the network with everything else, giving attackers full view of all resources. Generally, the first step an attacker takes is to gain an IP address. This is generally done via DHCP, which works by assigning an IP address to anyone who asks. Once an IP address has been handed out, the attacker becomes part of the network. They can now start looking around on the network just joined. In conducting a vulnerability assessment, become the attacker, and follow these steps to try to discover network resources. The next step is to perform a ping scan, or a connectivity test for the network, to see what else on the network is alive and responding to pings. Using Nmap, one of the best scanning tools available, a ping scan is performed like this:

```
# nmap -sP 10.10.0.1-15

Starting nmap V. 2.54BETA7 ( www.insecure.org/nmap/ )
Host  (10.10.0.1) appears to be up.
Host  (10.10.0.5) appears to be up.
Nmap run completed — 15 IP addresses (2 hosts up) scanned
    in 1 second
#
```

With this scan, we've checked all the hosts from 10.10.0.1 through 10.10.0.15 to see if they respond to a ping. From this, we gain a list of available hosts, which is essentially a Yellow Page listing of potentially vulnerable machines. In this case, .1

and .5 answered. This means they are currently active on the network. The next step is to see what the machines are, and what they run, so an exploit can be found to compromise them. An OS detection can also be done with Nmap like this:

```
# nmap -sS -O 10.10.0.1

Starting nmap V. 2.54BETA7 ( www.insecure.org/nmap/ )
Interesting ports on   (10.10.0.1):
(The 1530 ports scanned but not shown below are in state:
    closed)
Port        State        Service
22/tcp      open         ssh
25/tcp      open         smtp
53/tcp      open         domain
110/tcp     open         pop-3

TCP Sequence Prediction: Class=random positive increments
                         Difficulty=71574 (Worthy
                              challenge)
Remote operating system guess: OpenBSD 2.6-2.7

Nmap run completed — 1 IP address (1 host up) scanned in
    34 seconds
#
```

With this information, we now know that there is a machine with OpenBSD v2.6 or 2.7, running the services listed. We could now go and look for possible remote exploits that would allow us to gain access to this machine. If this were a real attack, this machine could have been compromised, giving the attacker a foothold into your wired network, and access to the rest of your network as well.

Snooping is another angle to consider when performing your vulnerability assessment. It can be every bit as dangerous as the outright compromising of machines. If confidential data or internal company secrets are being sent via wireless connection, it is possible for an attacker to capture that data. While 802.11b does support the Wired Equivalent Privacy (WEP) encryption scheme, it has been cracked, and can be unlocked via AirSnort or WEPcrack. These programs use the WEP weakness described by Scott Fluhrer, Itsik Mantin, and Adi Shamir in their paper "Weaknesses in the Key Scheduling Algorithm of RC4," which can be found at numerous Internet sites by searching for either the authors' or the paper's name. WEP does make it more difficult for an attacker to steal your secrets by adding one more obstacle: time. In some cases, it could take up to a week for an attacker to break your encryption. However, the busier the network, the faster the key will be discovered. To insure the best data privacy protection, have all wireless users connect to the internal network through a virtual private network (VPN) tunnel.

There are many opportunities for an attacker to gain access to a wireless network, simply because of their radio-based nature. After performing a vulnerability analysis, you should be able to spot some potential weaknesses in your security infrastructure. With these weakness identified, you can develop a plan of action to either strengthen your defenses, or increase your monitoring. Both are recommended.

Incident Response and Handling

Incidents happen. If your company has a network connection, there will eventually be some sort of incident. Therefore, an incident response and handling procedure is a critical component when it comes to protecting your network. This policy should be the definitive guide on how to handle any and all security incidents on your network. It should be clearly written and easy to understand, with steps on how to determine the level of severity of any incident. Let's take, for example, wireless intrusion attempts on two different networks, one without a good incident response policy, and one with more thorough policies in place.

Imagine one company without a formal security policy. As the company's network was built, the emphasis was placed on superior deployment, speed, and availability. While the network matured, and wireless access was added, there was little done in the way of documentation—they simply didn't afford it the time. There was still no security policy in place after adding wireless access, and no particular plans for how to handle an incident. Several weeks after deploying their companywide wireless network, the network administrators began to receive complaints of poor performance across the network. They investigated, based on what the various network administrators deemed necessary at that time. It was eventually concluded that perhaps one of the wireless Access Points was not functioning properly, and so they replaced it. After several more weeks, law enforcement officials visited the company—it seemed that a number of denial of service attacks had been originating from the company's network. Having had no formal security policy or incident handling process, the company was unable to cooperate with the officials, and could not produce any substantial evidence. Without this evidence, investigators could not locate the culprit. Not only was the company unable to help with the investigation, they had no idea they had even been attacked, nor did they know to what extent their internal data had been compromised. This left them with many more hours of work, rebuilding their network and servers, than if they had taken the time at the beginning to create a security and incident handling policy.

Next, imagine another company, one that attempted to balance performance and security considerations, and noticed some suspicious activity on their network from within their internal network. Through routine monitoring, the administrators detected some unusual traffic on the network. So, when their IDS sent an alarm message, they were ready to investigate. Within their security policy, guidelines as to how to handle the incidents were clearly detailed. The administrators had forms and checklists already prepared, so they were immediately able to start sleuthing. Using a number of steps outlined in their policy, they were able to determine that the traffic was coming from one of their wireless APs. They found this to be strange, as policy dictated that all APs were to have been configured with WEP. Further investigation found that this particular AP was mistakenly configured to allow non-WEP encrypted traffic.

In this case, having a good policy in place, the administrators were quickly able to track down the problem's source, and determine the cause. They were then able to systematically identify and reconfigure the problem Access Point.

Having an incident response policy is one thing, but the additional complexity posed by a wireless network introduces new challenges with forensics and information gathering. Let's investigate some of those new challenges, and consider some suggestions on how to contend with them.

Policies and Procedures

Wireless networking makes it easy for anyone to poke a gaping hole in any network, despite security measures. Simply putting a wireless AP on the internal network of the most secure network in the world would instantly bypass all security, and could make it vulnerable to anyone with a $100 wireless access card. It is for that reason that a provision to ban the unauthorized placement of any kind of wireless device should be drafted into a company's policy. This should be made to cover not just wireless APs, but the cards themselves. A user connected to your internal network could potentially be connected to an insecure wireless network, and bridging between the two interfaces on that machine would be very simple. The consequences of this to your network could be detrimental. Enforcing this policy can be difficult, however, as some popular laptop makers, such as Toshiba, have imbedded wireless access cards in their new notebooks. It should be considered a very severe infraction to place a wireless AP on the network—possibly one of the most severe—due to the level of risk involved. Having a wireless access card should also be treated seriously. Though this poses less of a risk than the AP, it should still be classified

accordingly. Excellent sample policies are available on the SANS Web site at
www.sans.org/newlook/resources/policies/policies.htm.

Reactive Measures

Knowing how to react to an incident is always a question of balance. On one hand,
it would be tempting to close everything down and pull the plug on the whole network. That would certainly give you ample time to investigate the incident without
further risk of compromise, but it would make your systems unavailable to your
users. Some balance must be reached. When dealing with a wireless network compromise, it might be a good idea to disable wireless access until you can identify the
entry point for the intrusion. Since wireless access is more of a luxury than a crucial
business need, this may be possible. Of course, in organizations where wireless is critical, this isn't feasible. In either case, the WEP keys should be immediately changed,
and if WEP isn't enabled, it should be. This will lock out the attacker for a limited
time, hopefully giving you more of an opportunity to deal properly with the intrusion. In a secure and well-designed network, the scenario of a user joining a wireless
network and immediately compromising it isn't as likely because more safeguards are
in effect. If your network has been compromised through its wireless network, it's
probably time to take some additional security measures.

While your network has been locked down, or at least had new keys installed,
make sure to gather evidence of the intrusion. If the attacker was just passively listening to the network, there will be little evidence available, and not much taken as
a result. However, if there were compromises into other network machines, it is critical to follow your company security policy guidelines to properly document the
intrusion and preserve the evidence for the proper authorities. As mentioned in the
introduction, covering how to handle evidence collection and performing forensics
on a hacked machine is a book of its own!

Reporting

A wireless intrusion should be reported in the same manner as any other type of
intrusion or incident. In most cases though, a wireless intrusion can be more severe,
and difficult to document. Reporting a serious intrusion is a key part of maintaining
a responsible approach to security. This is where a complete logging and monitoring
system with IDS will be very useful. Having gathered and examined all log files from
security devices; try to gain an understanding of the severity of the intrusion. Were
any of the machines successfully attacked? From where were the attacks originating?
If you suspect a machine was compromised, shut it down immediately, running as

few commands as possible. Unless you really know what you are doing, and are familiar with computer forensics, the evidence should be turned over to investigators or forensics experts. The reason for this is that attackers will generally install a rootkit or backdoor system in a machine. These often feature booby traps, which can run and destroy critical information on the server. The primary places for booby traps like these are in the shutdown scripts, so it is possible you will have to unplug the machine, rather than use a script to power it down. Once that has been done, it's best to make two copies of the infected machine's disk for evidence purposes. If the authorities have been notified and will be handling the case, they will ask for the evidence, which should now be properly preserved for further forensics and investigation.

Cleanup

Cleaning up after an incident can pose a huge challenge to an organization. Once the level and extent of the intrusion has been determined, and the proper evidence gathered, one can begin rebuilding network resources. Generally, servers can be rebuilt from tape backup, but in some cases it may be necessary to start again from scratch. This is the type of decision that should be made after determining the extent of the intrusion. It is critical that when restoring from tape, you don't restore a tape of the system, post-intrusion—the same problems and intrusion will still exist. Some administrators feel there is no need to rebuild an infected machine, but simply to patch the security hole that allowed the intrusion. This is a particularly bad idea, because of the problem we mentioned with backdoors. The most advisable solution is to begin from scratch, or a known-to-be-safe backup. From there, the machines should be updated with the latest verified patches from the vendor.

Assuming the compromise did come from a wireless source, the wireless network should be re-examined. It may be difficult to determine exactly which AP was used for the compromise, but if you have an AP in a location that makes it easily accessible externally, you should probably consider moving it.

Prevention

As we've emphasized throughout this chapter, the best way to prevent an attack to your wireless network is to be secure from the start. This means designing a secure installation, maintaining firewalls and server logs, and continually patrolling your network for possible points of attack.

A secure wireless network is one which takes as many precautions as possible. Combining a properly secured AP with a firewall will provide a minimum level of security. Several steps that can be taken to help secure the network are adding a VPN to provide data privacy protection to your network. This is a critical step for organizations that require their data not be captured or altered in transmission. Isolation of network APs by a firewall is another often-overlooked step which should be implemented. Finally, simply making sure that WEP is enabled and enforced in all of your wireless APs can be just enough of a deterrent to save you from an intrusion. This may sound like quite a bit of extra work, which it is, but in order to remain secure, precautions must be taken.

Conducting Site Surveys for Rogue Access Points

Even if you don't have a wireless network installed, it's a good idea to perform scans of your area for wireless traffic. The low cost and ease of setup makes installing unauthorized or rogue APs very appealing. Whether installed by well-intentioned users of your own network, or by malicious outsiders, making sure you routinely patrol for any wireless activity on your network is a sound idea.

In this section, we'll discuss some strategies for surveying your network and tracking down rogue wireless APs. Using tools like the ORiNOCO Client Manager and NetStumbler we'll describe how to locate unauthorized wireless access at your network site, and instruct you in how to see your network as an attacker would.

The Rogue Placement

There are really quite a few scenarios in which a rogue AP could be placed on the network. In this section, we'll take a look at two scenarios, one done without any bad intentions, and one placed by an attacker hoping to gain access to a network.

The Well-intentioned Employee

The first situation involves a well-meaning employee. This person has been looking at advertisements at computer shops that feature low cost wireless network equipment, and having just purchased a wireless networking installation for home, wants to bring that convenience to work. Believing that having a wireless network available for the other employees will provide a great service, this employee goes to the shop and brings back the $150 wireless AP on sale that particular week. After carefully following the instructions from the manufacturer, the AP is made available, and the user announces the availability of the AP to fellow employees. Wanting the configuration to be as simple as possible, the well-intentioned employee has configured the AP not to require a preconfigured SSID string, allowing anyone to connect to it. This now provides the freedom to other department employees to roam about freely with their wireless cards. Note that none of this was done with authorization, because the user had no idea of the security implications involved. As we've discussed earlier, this now provides an open point of entry to anyone within range of the signal.

Scenarios such as this demonstrate the need to educate users as to the dangers of adding wireless APs to the network. Visual demonstrations or real-world examples assist in providing powerful explanations detailing the repercussions of this kind of security breach. It should also be made known that there exists within the company security policy a provision banning any kind of wireless networking.

The Social Engineer

A determined attacker will stop at nothing to compromise a network, and the availability and low cost of wireless networking equipment has made this task slightly easier. In this scenario, an attacker who has either taken a position at your company as a nightly custodian or has managed to "social engineer" their way into your office space will place a rogue AP.

One often-overlooked possibility for intrusion comes from an attacker posing as a nightly custodian, or one that has officially obtained that position. Night custodial staff often have unsupervised access to many areas of an office space, and as such are in the position to place a rogue wireless AP. Given time to survey the surroundings and find an inconspicuous location for an AP, this type of attacker can establish an entry point into your network for later access. In this kind of situation, an attacker may try to disguise their AP both physically, and from the network side. If there are other wireless APs present in your environment, the attacker may choose to use the same vendor, and SSID naming schema, making it all the more necessary to keep listings of the MAC addresses of all your authorized wireless APs. Another possibility is that an attacker will enable WEP encryption on their AP, ensuring that only they are able to access it at a later date. Attackers often tend to feel very territorial towards their targets.

A similar scenario to this involves a technique known as social engineering. This generally involves representing oneself as someone else. A good way to social engineer a situation is to first know some inside information about the organization which you are targeting. If it's a large company, they may have a published org-chart which will have important names that the social engineer can quote from to seem legitimate. Other sources for names include the company's Web site and press releases. In one example, during a vulnerability assessment for a fairly large firm, we were generally unable to find easy access to the network, so we employed a social engineering tactic. Posing as a vendor replacing hardware, we were able to gain access to the Accounting department and were able to place an AP in the most suitable location we could find: a VP's hard-wall office, overlooking the parking garage across the street. With this AP in place, we were successfully able to demonstrate both the need for education about the dangers of social engineering, and the need for tightened security on the company's internal network.

Tracking Rogue Access Points

If after conducting a vulnerability assessment or site audit, you've spotted an AP that should not be present, it's time to begin tracking it down. It may be that your assessment found quite a few APs, in fact. In a city office environment this is to be expected, don't worry. There's a better than average chance that many organizations around yours are using wireless access, and their APs are showing up on your scan. Nevertheless, they should all be investigated. A clever attacker could give their AP on your network the name of a neighboring business.

Investigating APs can be a tricky proposition. Perhaps the first step is to try to rule out all those who aren't likely to be in your location. This can be done with signal testing tools like NetStumbler, or LinkManager from ORiNOCO. Signals that appear to be weak are less likely to be coming from your direct area. For example, let's say we're looking for an AP called *buzzoff* that turned up on our NetStumbler site survey.

In Figure 8.4, we can see on our NetStumbler screen that two APs have been spotted. The AP called covechannel has a pretty weak signal, when it's even visible, so it's probably not nearby, though we may want to check it again later. Instead, we'll look at buzzoff, because it's showing a very strong signal. A very useful tool for investigating signal strength is the ORiNOCO Site Monitor, which comes bundled with the ORiNOCO Client Manager. Bringing up the client manager software and clicking on the **Advanced** tab will reveal the Site Monitor option. In this example, the Site Monitor software reveals that the signal for buzzoff is still fairly weak.

From the information we've seen in Figure 8.5, it looks like we're still a bit far from the AP. The signal isn't all that strong, and that's not terribly surprising since we've just started looking. Now we need to find this AP. The signal is strong enough to assume that it's probably somewhere nearby, so we'll start walking around until we get a stronger signal. At this point, finding the AP becomes a lot like the children's game, "Hot and Cold." When we move out of range, the AP's signal becomes weaker or "cold," so we move back in until the signal strengthens. This process can be time-consuming and slow, but with patience you'll be able to close in on the signal (as seen in Figure 8.6).

With a signal this strong, we're very close to the AP. At this point, it's time for the grunt work of the physical search. Knowing where all the LAN jacks are is helpful, because the AP will be plugged into one. It wouldn't be much of a threat otherwise. So, by systematically checking all possible LAN connections, we are able to locate this rogue AP sitting on top of an employee's computer. In this particular instance, it appears we have found an AP that falls under the "well-intentioned employee" scenario. Though, since we don't know for sure that it was the employee who placed it there, the AP should be handled very carefully.

Figure 8.4 Network Stumbler: We've Found a Few Interesting APs

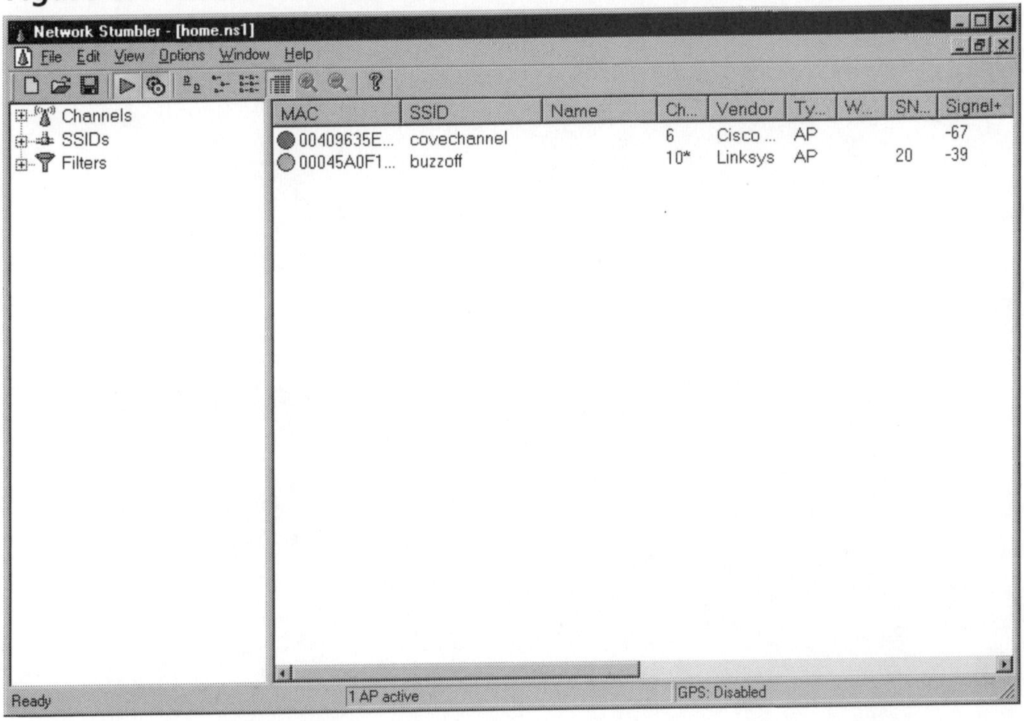

With the AP found, it would also be advisable to conduct more audits of system machines to see if there were any break-ins during the time the rogue AP was available. To do this, refer to the monitoring section earlier, and start watching traffic patterns on your network to see if anything out of the ordinary pops up. Another good area to watch is the CPU load average on machines around the network. A machine with an extraordinarily high load could be easily explained, but it could also be a warning sign.

Figure 8.5 ORiNOCO Site Monitor: Looks Like We're Not too Close Yet

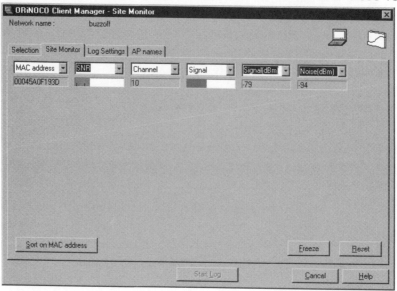

Figure 8.6 ORiNOCO Site Monitor: A Much Stronger Signal—We're Almost There

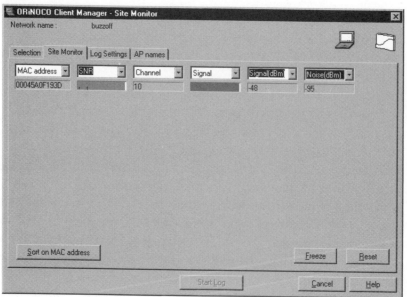

Summary

In this chapter, we've introduced some of the concepts of intrusion detection and monitoring, and discussed how they pertain to wireless networking. Beginning with the initial design for a wireless network, we've focused on the fact that security is a process that requires planning and activity, rather than just a product shrink-wrapped at the computer store. Through proper investigation of our site, we can build a wireless network in which we are aware of potential problems before they occur. Examples of this are noting potential sources of interference, and knowing which physical structures may be a barrier to the network.

After designing the network, we discussed the importance of monitoring. Using a combination of software designed for monitoring and the logs from our security devices, we can gain a valuable picture of how the network is supposed to look, and from there deduce potential problems as they occur. Knowing that the network is under a much heavier load can be a sign of an intrusion. Along with monitoring, dedicated intrusion detection software should be used in order to watch for specific attacks to the network. The software, using signature files that can be customized to look for specific attacks, will generate alerts when it finds a signature match in the traffic.

From there, we moved on to discussing how to conduct a vulnerability assessment. This is important to do regularly because it can help you learn to see your wireless network as an attacker does, hopefully before they do. Spotting problems early on can save time and money that would be wasted dealing with an intrusion.

Intrusions do happen, and adding a wireless network without proper security definitely increases that risk. That is why it is critical to have a security policy in place that not only prohibits the use of unauthorized wireless equipment, but also educates users to the dangers of doing so. Updating the security policy to handle wireless issues is key to maintaining a secure network in today's environment. However, should an intrusion occur through the wireless network, we discussed a few strategies on how to deal with the incident itself, and then how to contend with the cleanup afterward. We didn't delve into the realm of the actual computer forensics, however. That is a very complex and involved field of security, and is definitely a book of its own. Should you be interested in learning more about forensics, there are a number of excellent manuals available on the Internet that deal specifically with the forensics of Unix and Windows systems.

Lastly, we dealt with rogue Access Points (APs), possibly one of the greatest new threats to network security. Rogue APs can be placed by an attacker seeking access to your network, or placed by a well-meaning employee, trying to provide a new service. Either way, they offer attackers a direct and anonymous line into the heart of your network. After conducting a routine site audit, in our example, we discovered a rogue AP and tracked it down using a combination of the ORiNOCO Site Monitor and the NetStumbler tool. Once it was found, we handled it very carefully, in order to uncover where it came from, and why.

Intrusion detection and monitoring are one of the key building blocks in designing a secure network. Being familiar with the operations of your network, and knowing how to spot problems can be a huge benefit when an attack occurs. Proper intrusion detection software, monitored by a conscious administrator, as well as a combination of other security devices such as virtual private networks (VPNs) and firewalls, can be the key to maintaining a secure and functional wireless network.

Solutions Fast Track

Designing for Detection

☑ Get the right equipment from the start. Make sure all of the features you need, or will need, are available from the start.

☑ Know your environment. Identify potential physical barriers and possible sources of interference.

☑ If possible, integrate security monitoring and intrusion detection in your network from its inception.

Defensive Monitoring Considerations

☑ Define your wireless network boundaries, and monitor to know if they're being exceeded.

☑ Limit signal strength to contain your network.

☑ Make a list of all authorized wireless Access Points (APs) in your environment. Knowing what's there can help you immediately identify rogue APs.

Intrusion Detection Strategies

☑ Watch for unauthorized traffic on your network. Odd traffic can be a warning sign.

☑ Choose an intrusion detection software that best suits the needs of your environment. Make sure it supports customizable and updateable signatures.

☑ Keep your signature files current. Whether modifying them yourself, or downloading updates from the manufacturer, make sure this step isn't forgotten.

Conducting Vulnerability Assessments

☑ Use tools like NetStumbler and various client software to measure the strength of your 802.11b signal.

☑ Identify weaknesses in your wireless and wired security infrastructure.

☑ Use the findings to know where to fortify your defenses.

☑ Increase monitoring of potential trouble spots.

Incident Response and Handling

☑ If you already have a standard incident response policy, make updates to it to reflect new potential wireless incidents.

☑ Great incident response policy templates can be found on the Internet.

☑ While updating the policy for wireless activity, take the opportunity to review the policy in its entirety, and make changes where necessary to stay current. An out-of-date incident response policy can be as damaging as not having one at all.

Conducting Site Surveys for Rogue Access Points

☑ The threat is real, so be prepared. Have a notebook computer handy to use specifically for scanning networks.

☑ Conduct walkthroughs of your premises regularly, even if you don't have a wireless network.

☑ Keep a list of all authorized APs. Remember, Rogue APs aren't necessarily only placed by attackers. A well-meaning employee can install APs as well.

Frequently Asked Questions

The following Frequently Asked Questions, answered by the authors of this book, are designed to both measure your understanding of the concepts presented in this chapter and to assist you with real-life implementation of these concepts. To have your questions about this chapter answered by the author, browse to **www.syngress.com/solutions** and click on the **"Ask the Author"** form.

Q: I already have a wireless network installed, without any of the monitoring or intrusion detection you've mentioned. What can I do from here?

A: It's never too late to start. If you already have a network in place, start from the design phase anyway, and follow the steps we've listed. Adding to a currently in-production wireless network doesn't have to be difficult.

Q: I don't really think I know enough about security to perform a proper vulnerability assessment. What should I do?

A: You can always try. That's the best way to learn. However, until you're more comfortable, consider hiring an outside security vendor to perform a network vulnerability analysis for you. Even if you do know what you're doing, a second set of eyes on something can always be beneficial.

Q: I've bought an IDS system that says it is host-based. How can I make it start seeing the network traffic like you described in this chapter?

A: You can't. Host-based intrusion detection software is very different from network IDS. It mainly looks at the file system of the server on which it is installed, notices any changes to that system, and generates an alert from there. To watch the traffic, you need to look specifically for a network-based intrusion detection system.

Q: I can see a ton of APs from my office. How can I tell if any of them are on my network?

A: The first way would be to check the signal strength. If you're getting a faint signal that only appears intermittently, chances are it's not in your area. If you detect a strong signal, you can attempt to join the network and see if it assigns you an address from your network. Additionally, you could look at some of the traffic on the network to determine if it's yours, but that may introduce some legality questions, and is definitely not advised.

Q: I've found a rogue AP on my network. Now what?

A: First, start by determining who placed it. Was it an employee or an outside party? If it appears to be the work of an employee, question them about it to find out how long it has been present. The longer it has been around, the more likely an intrusion has taken place. In the case of it being put there by an attacker, handle it very carefully, and if necessary, be prepared to hand it over to the authorities. Also, consider having a professional system audit to see if any machines have been compromised.

Chapter 9

Designing a Wireless Enterprise Network: Hospital Case Study

Solutions in this chapter:

- Introducing the Enterprise Case Study
- Evaluating Network Requirements
- Designing a Wireless Solution
- Implementing and Testing the Wireless Solution
- Lessons Learned

☑ Summary

☑ Solutions Fast Track

☑ Frequently Asked Questions

Introduction

An *enterprise network*, sometimes called a *campus network*, is a network that spans across multiple buildings. The case study we'll explore in this chapter follows the process of planning, designing, and implementing a wireless network in a hospital and associated medical buildings on the hospital campus. We will also review the advantages and cost savings associated with the implementation of this type of wireless networking versus leased lines.

We will walk through the steps you must complete first, gathering the network requirements and analyzing the current network. At that time, you will perform a site survey to determine the building infrastructure and see first-hand whether there are any line-of-sight issues. Based on the analysis of the current network and the site survey, you then develop a high-level design, and based upon the results, select the wireless equipment and develop the detailed design. Implementation of the network goes forward according to the high-level design. After the equipment is installed and configured, you make sure to perform acceptance testing to verify that the wireless links are working correctly.

Since these buildings are medical in nature, you must take special care when deploying wireless communications. Medical equipment in particular may be sensitive to wireless devices and channels. You must pay special attention to limiting radio frequencies in and around the emergency room area to avoid any interference. When you are writing the design, make sure that you speak to the Information Systems (IS) department to determine any concerns you need to be aware of while operating wireless equipment in the 2.4GHz range.

Applying Wireless in an Enterprise Network

Today, the IEEE 802.11b and g standards make wireless networks possible. These standards enable you to implement frequencies around 2.4GHz in North America without having to go through licensing paperwork. Wireless Local Area Network (WLAN) equipment using the 802.11b protocol provides for wireless capabilities with up to 11 Mbps of bandwidth and the 802.11g protocol provides up to 54 Mbps of bandwidth.

Wireless devices help the campus networks evolve by providing users flexibility and mobility within the campus. Users are not limited to the locations or the number of available data ports in an office or conference room. They may now take their laptop computer and connect to the LAN virtually anywhere, even in hallways, within the access point range.

Introducing the Enterprise Case Study

Our case study puts us in partnership with Jones Hospital & Associates. The IS manager from Jones Hospital has hired us as consultants to provide a wireless LAN solution for their hospital network. The hospital employs just over 800 employees, working in a complex composed of seven buildings in an urban area. All of the buildings are close enough in proximity that employees frequently walk between buildings.

The management staff wants to deploy wireless technology in their enterprise network. They have heard that this type of network will save them money as opposed to the current leased line network. They believe that the added mobility will also dramatically increase productivity. Since the staff moves from building to building, it seems to be the right idea to provide them with computer equipment and data that will roam with them.

Assessing the Opportunity

Jones Hospital & Associates is composed of a group of affiliated primary care doctors and specialists. The hospital operates six leased satellite office buildings surrounding an eight-story central hospital tower. The IS department handles all LAN requirements. In preparation for this initiative, the IS department has gathered requirements with representatives from each medical area, including Administration, Pediatrics, Surgery, the Cancer Center, Heart Center, Pharmacy, Radiology, and Emergency departments. As consultants, we have been brought in to help the IS department assess the situation, gather requirements, and determine a wireless solution.

At this point we are in the planning phase of a wireless project. The first step in the planning phase is to gather the technical network requirements and current network architecture. To determine network requirements and current architecture, we will need to ask the following questions:

- What problems are you experiencing in the current network? Look for areas where wireless technology can help ease the IS burden as well as the departmental issues. By making the network easy to update, you've eliminated some issues.

- What is the architecture of the current network? Evaluate the current network to see how to integrate the wireless technology with the current network.

- What are the bandwidth-limiting areas of your network? Obviously, you must limit the bandwidth in and around the surgery areas. Are there any other areas in which wireless technology could interfere with other medical equipment?

- Are there any applications that are limited by the current network? Or is there additional functionality required that the current network is not capable of? You may be able to increase productivity by overcoming any current obstacles.

- What is the building floor plan or layout? This information can help you plan where the wireless hardware equipment can be installed.

- Is there any equipment that might be susceptible to wireless frequencies? This should tell you whether the network transmission could possibly interfere with medical equipment. This is one area where research is very important.

- Are there any existing monthly costs that can be replaced with wireless link? The hospital administration staff wants to review the costs of the wireless network versus the current leased line system. The hospital Board of Directors is very cost-conscious.

- Are there any constraints in placing wireless equipment and antennas in computer rooms, communication closets, walls, roofs, offices, or conference rooms? If you can identify constraints at this time, you can save a lot of time during the implementation phase.

- What are the hospital plans for adding or deleting any office buildings? How accurate are the plans? Changes to the physical structure should be identified early to save time and effort.

These questions are important in understanding the current environment. They also help you design for constraints or limitations.

Evaluating Network Requirements

After meeting with the hospital IS team, you determine the following requirements and constraints. It is important to define all requirements and constraints to ensure that your wireless solution falls within the expectations of your customer.

In the current physical network, all the satellite buildings use internal Category 5 Unshielded Twisted Pair (UTP) wiring. The Administration department expressed the need for LAN access in conference rooms in the main building. Since all employees are receiving laptop computers, all departments will require the ability to access the network from anywhere in the satellite buildings and in the conference room in the main hospital building. All users want the ability to use their laptops anywhere in any of the buildings. This will improve productivity because the medical staff often travels between the main hospital and the satellite buildings. The IS manager will provide building floor plans but foresees no physical limitations for placing wireless devices in the satellite buildings.

The satellite buildings are not owned by the hospital, they are leased. The satellite buildings connect to the main hospital via leased T-1 links. These leased links have become a recurring cost that the IS manager would like to reduce or eliminate. These links are highly utilized and the IS staff has received complaints from the different medical groups. There are no fiber lines to the satellite buildings and the cost to install fiber is not within the current budget. The IS staff wants to implement wireless technology to provide links from the main hospital to the satellite buildings. The wireless network must provide security and encryption. The IS staff requires some level of redundancy for the site links.

Hospital officers mention that there are immediate plans to increase the hospital employee population. However, they want the developed solution to be scalable to support additional buildings in the future. They also express concerns about using radio frequencies in main hospital building, so they want to limit the wireless devices to conference rooms and site links within the hospital building.

Assessing the Satellite Buildings' Physical Landscape

The IS manager has provided the floor plans. All of the satellite buildings were designed around the same floor plan, shown in Figure 9.1. Notice that there are communications and mechanical rooms in the center of the floor plan area; these rooms may generate noise and affect the wireless design. The rest of the floor plan is open-area space with offices and a conference room. You will perform a walk-through in each of the buildings to visually assess the environment.

Evaluating the Outside Physical Landscape

The hospital campus area topography is shown in Figure 9.2. The main hospital is eight stories high and is separated from all other buildings with roads located in an urban zone. There is a park with high trees at the south of the main hospital building. Each satellite building is four stories high. There is a clear line-of-sight from the hospital to each satellite building, and between the satellite buildings, except from Building 301 to Building 201. We must take this information into consideration to ascertain the redundancy requirements.

Figure 9.1 Satellite Building Floor Plan

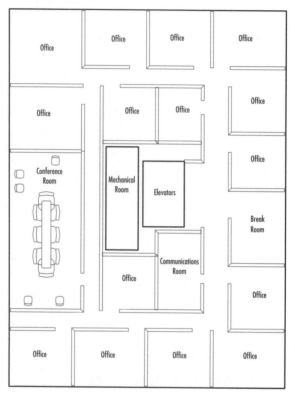

Figure 9.2 The Topology of the Jones Hospital Buildings

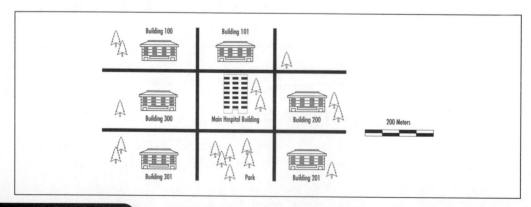

Evaluating the Current Network

The current campus network has a router located in the computer center of the main hospital building with six serial interfaces. Each serial interface is used to connect to each of the satellite buildings via T-1 lines. Since the outside infrastructure is not privately owned, the T-1 lines are leased via the local exchange carrier. There is no fiber infrastructure between the buildings.

Each building has category 5 UTP wired infrastructure terminating into closet patch panels. Each LAN drop is terminated into a LAN switch. Fiber cables connect the LAN switches to the building router. As shown in Figure 9.3, the floor LAN switches connect to the building router, which has the T-1 line to the main building. This architecture does not provide for any redundancy, and the hospital wants to overcome this problem in the current architecture.

Figure 9.3 Current Network Topology

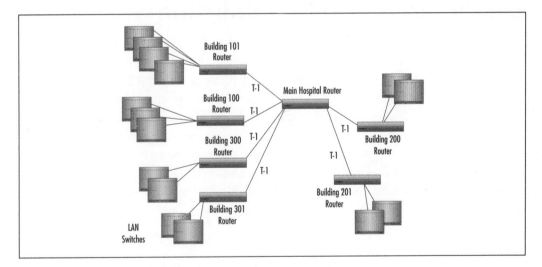

Evaluating the Hospital Conference Room Networking Landscape

An assessment of the conference rooms in the hospital building shows four existing LAN drops in each room. The conference room layout is shown in Figure 9.4. Because the network connection already exists, and there is easy access to the drop, one of these drops can be used to place the access point bridge. As an additional benefit, the building infrastructure permits wireless LAN access from nearby offices.

Figure 9.4 Conference Room Layout

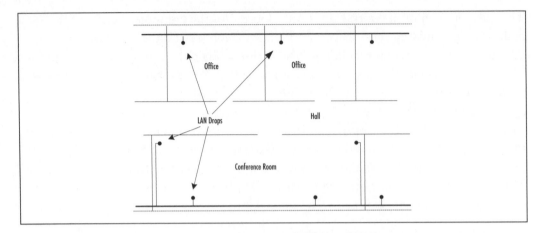

Designing a Wireless Solution

Once we define the requirements and baseline the current network, we can move from the planning phase to the design phase. For convenience, we'll divide this wireless design project into three smaller projects, compartmentalizing the large project into smaller ones to make it easier to provide a solution for each project on a smaller scale. The first project provides *wireless access in the satellite buildings.* The second project simply adds *wireless LAN connectivity to the conference room* in the main hospital building. This enables the employees with wireless connectivity through the wireless interface cards in their laptops. The third project adds the *wireless links from the satellite buildings to the main hospital* and then adds the redundant links between each pair of satellite buildings. Let's review the design requirements:

- Provide wireless access for laptops in all satellite buildings.
- Provide wireless connectivity in conference rooms in the main hospital building.
- Provide a replacement to the leased T-1s that connect the satellite buildings to the main hospital building.
- Provide increased bandwidth to the satellite buildings.
- Provide redundancy to the satellite buildings.
- Maintain a level of security and encryption for the links.

Project 1: Providing Satellite Building Access

When designing a wireless network in an enterprise building, you must determine the placement of antennas and access points for best coverage. In this example, the mechanical room, elevators, and communications room are sources of frequency interference that you need to consider. A single omnidirectional antenna might be capable of covering the office area in a satellite building (over 100 feet). However, with the interference items to consider, it would be better to place omnidirectional antennas (and access points) in each hallway, as shown in the Figure 9.5, to get better coverage. Also, each access point can provide redundancy. If one access point fails, the other provides access to all computers on the floor.

The access point wireless bridges will be placed on shelves near the antennas. The Ethernet ports of the access point bridges will be connected to the LAN switches that serve the floor. The LAN switch must be configured to permit multiple media access control (MAC) addresses on the data port.

Designing & Planning...

Other Antenna and Access Point Bridge Placements

There are several methods of placing antennas to have full coverage in a floor. Directional antennas could be placed in each of the four corners of the floor aiming at a 45-degree angle toward the center of the building. To verify antenna placement, place an access point bridge at each location and test its range with a laptop with a wireless card. It is helpful to perform this test using a roll cart, so you can roll around the hallways, offices, and conference rooms to verify coverage.

Project 2: Providing Wireless Technology to the Conference Rooms

For the conference room project, plan to install one access point wireless bridge. Users requiring wireless LAN connectivity will need to install wireless LAN network interface cards into their laptop computers. The access point will be configured as a bridge with the Ethernet port connecting to the LAN jack. An antenna will be installed in the conference room. This solution meets the requirement for access to the LAN from the hospital conference room.

Figure 9.5 Project 1: Placement of Access Point Antennas in Satellite Buildings

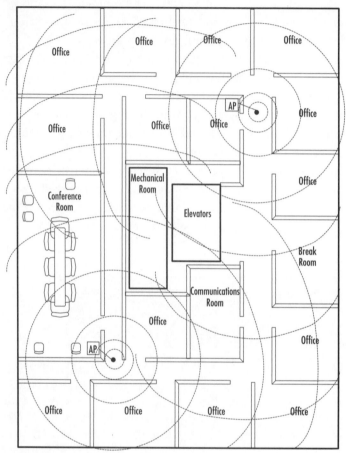

As shown in Figure 9.6, the access point is placed on a shelf on the corner of the conference room. A directional antenna is also placed on the corner providing access to the conference rooms and nearby offices. The LAN switch that serves the conference room drop must be configured to permit more that one MAC address on the LAN port. This same solution is applied to each conference room in each satellite building.

Figure 9.6 Project 2: Conference Room Solution

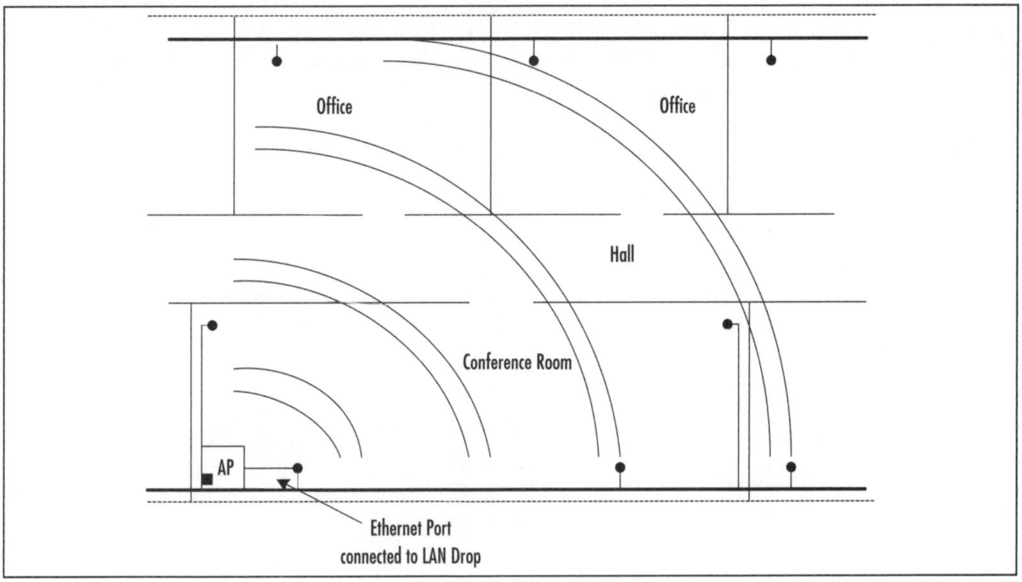

Project 3: Providing Building-to-Building Connectivity

There are many different ways to provide a wireless solution for Jones Hospital's network. A single wireless link could be implemented between the hospital building and each satellite building, but this solution would not provide redundancy. A full mesh could also be implemented, but it might be an overkill solution.

The solution presented here is one design approach that meets the requirements. Let's review the design requirements for this connectivity:

- Provide a replacement to the leased T-1s that connect the satellite buildings to the main hospital building.

- Provide increased bandwidth to the satellite buildings.

- Provide redundancy to the satellite buildings.

- Maintain a level of security and encryption for the links.

Based on the requirements, the existing lease lines will need to be replaced with wireless links from the main hospital building to each satellite building. Data encryption will be enabled to provide link security. The wireless links will provide increased bandwidth from 1.5 Mbps to 11 Mbps. To provide redundancy, we could link every building in a loop, but this would add additional cost to the solution. The redundancy goal can be accomplished by just adding wireless links between building pairs; for example, adding a wireless link between Buildings 100 and 101, Buildings 200 and 201, and Buildings 300 and 301. A high-level illustration of the proposed solution for Jones Hospital is shown in Figure 9.7.

Figure 9.7 Project 3: Proposed Building Wireless Connectivity

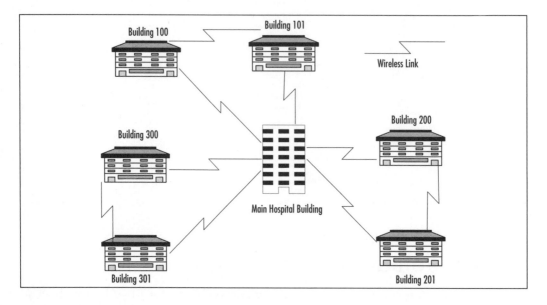

Describing the Detailed Design of the Building Links

As previously described, we want to create point-to-point wireless links between buildings. Some vendors have wireless devices called *outdoor routers* that can provide a solution for Jones Hospital. We will create an architecture using the existing routers in each building. The access-point outdoor routers will connect via Ethernet to the hospital router. We'll use each outdoor router to create point-to-point links to each satellite building. As Figure 9.8 shows, Building 100 will use two wireless outdoor routers to link with the main hospital router and to Building 101 for redundancy. Data will be encrypted using 64-bit Wired Equivalent Privacy (WEP) or 128-bit RC4.

Figure 9.8 Router-to-Router Wireless Connectivity

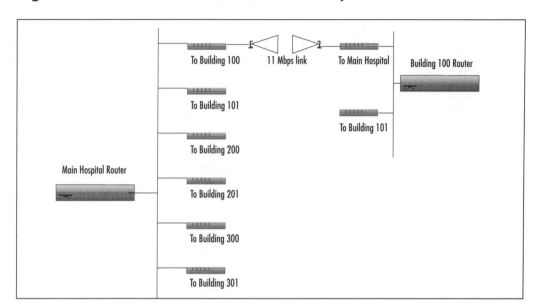

This solution provides for 11 Mbps of bandwidth between the buildings and the main hospital building, a significant increase from the 1.5 Mbps of bandwidth provided by the leased T-1 lines. Also, the hospital IS team will reduce costs by eliminating the monthly recurring costs for the leased lines.

Let's now look at how to add redundant links to provide backup connectivity in case of link or device failure. As shown in Figure 9.9, the main hospital router connects via Ethernet to the access-point outdoor routers. Each satellite building has two access-point outdoor routers to connect to the hospital and to the other building. If the link between Building 100 and the main hospital fails, the Building 100 router will still have access to the hospital via its link to Building 101. The same loop would be created for linking Buildings 200 and 201 to the main hospital route and for Buildings 300 and 301.

These designs will provide redundant connectivity for all satellite buildings. If there is a problem with any link or access point device, all traffic takes the redundant path to the main hospital router. These designs provide increased bandwidth to 11 Mbps. Also, users with wireless cards in their laptops will be able to meet in the conference room and access the local area network. Since the leased T-1 lines are not required, recurring costs also are eliminated.

Figure 9.9 Redundant Links: Hospital to Building 100 and Building 101

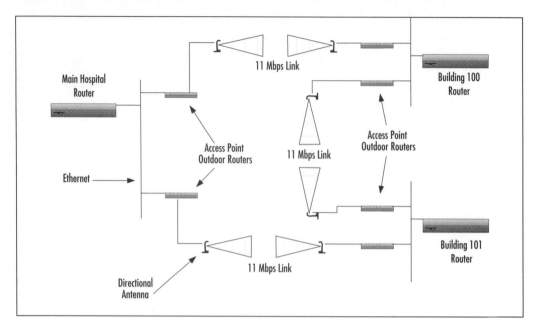

Part of planning and design will be to reserve network closet and computer room space for the placement of the access point devices. We will need to estimate the distance of the antenna cables. Also, we need to determine the necessary equipment and obtain the building owner's permission to place the antennas on the building roof.

Implementing and Testing the Wireless Solution

When the design phase is complete and all the equipment has been acquired, we can begin the implementation phase. The following sections describe the steps to follow when installing, configuring, and testing the wireless devices.

Project 1: Implementing the Satellite Building LAN Access

Install the access point devices and antennas in the building floors as described in the design:

1. Mount the antennas in the hallways, and connect cables to the access point devices in the nearby offices.

2. Connect the access points to the floor LAN switch.

3. Configure the access point frequencies, keeping configuration information available for laptop configuration.

4. Configure the access point for bridging, and enable multiple MAC addresses on the LAN switch.

At this point we are ready to test wireless access throughout the floor plan. We begin by verifying access from each office and the conference room. The hospital laptops can be equipped with the wireless PC Memory Card International Association (PCMCIA) cards and configured to connect to the LAN via the access points.

Project 2: Implementing the Hospital Conference Room

Implementation for the hospital conference room includes the same steps used in Project 1. For the conference rooms, install the access point and antenna at the corner of the room as described in the design diagram. Place the directional antenna so that the antenna energy covers the conference room completely. The access point is configured for bridging (no routing). Connect the Ethernet port of the access point to a LAN drop. Configure the building switch that serves the used LAN drop to permit multiple MAC addresses on that port. The following steps provide a review of this implementation:

1. Mount the access point and directional antenna in the conference room.

2. Connect the access points to the floor LAN switch.

3. Configure the access point frequencies, keeping configuration information available for laptop configuration.

4. Configure the access point for bridging, and enable multiple MAC addresses on the LAN switch.

Use a laptop to verify access to the LAN in the conference room and nearby offices. Make sure that the connection is reliable.

Project 3: Implementing the Building-to-Building Connectivity

The implementation of the wireless links between buildings is made in parallel to the current T-1 connectivity. No serial interfaces are used on the existing routers. In the server room of the main hospital, you connect the main router to six access-point outdoor routers. These routers reside in the server room, not outside. Install the directional antennas on the roof of the hospital, each pointing toward the direction of its respective satellite building. Install and configure the primary links between each satellite building and the hospital before installing the redundant lines. Figure 9.10 shows, at a high-level, the primary links. For implementation of the primary wireless links, follow these steps:

1. Install and configure the access–point outdoor routers.

2. Install the outdoor antennas and connect them to the outdoor routers.

3. Verify that the frequencies are configured and test the wireless link.

4. Verify that the received connection is strong enough to be a reliable connection.

Figure 9.10 Primary Wireless Links

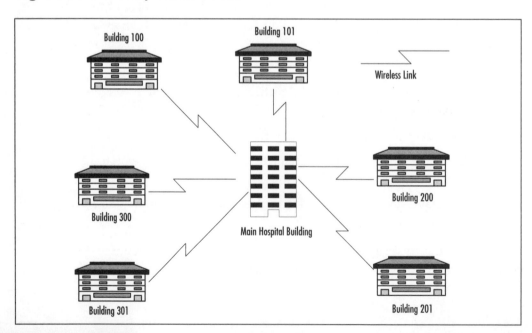

At this point we connect the outdoor routers to the existing building routers. Use the following steps to make these connections:

1. Connect the routers via their Ethernet ports.

2. Enable the encryption protocols for data security.

3. Configure the existing routers to forward packets via the wireless link.

4. Since the wireless link provides greater bandwidth, verify that the packets are getting forwarded via the wireless link over the leased T-1 lines.

When all six wireless links are installed and are passing traffic, install and configure the redundant links as shown in the steps for the primary wireless links listed at the beginning of this section. The next task is to test and verify the links:

1. Test the routing failover of wireless links by deactivating a primary link interface or an outdoor router.

2. Verify that the building still has access to the main hospital router.

3. Reenable the primary link or an outdoor router.

When all of the wireless devices and links are tested, the IS Manager can place cancellation orders for the leased T-1 links.

Configuring & Implementing…

Verify the Wireless Service

When installing the wireless antennas, use the testing capabilities of the access point devices to make sure that the wireless connection is reliable. If you encounter problems with the connection, try moving the antenna in one or the other direction to correct the problem. Verify that interference is not caused by trees or reflection from nearby buildings.

Reviewing the Hospital's Objectives

Hold a follow-up meeting with the IS Department and hospital teams to demonstrate the functionality of the new wireless network and to determine the successes or failures of the wireless project. This meeting will help you determine whether you need to expand the wireless access points further in the satellite buildings. In the meeting held with Jones Hospital, we hear that the feedback received from the hospital groups is good. They are satisfied with their new ability to access the LAN from the satellite buildings and the conference room in the main hospital building. The medical staff is very happy with their new mobility.

The IS manager is very pleased with the savings of leased line costs and the increased bandwidth to the satellite buildings. Since installing wireless devices is much faster than requesting that fiber cabling be installed, they are very pleased that the project was completed in a short time span.

Lessons Learned

From this case study we learned how to deploy wireless technologies to extend wired Ethernet LANs for office and conference room areas. We also learned how to use wireless outdoor routers to provide campus links to an enterprise hospital network.

We learned to follow the process of planning, designing, implementing, and testing a wireless network. Following this approach makes wireless projects easier to manage. We learned to consider interference when designing for antenna placement for best coverage of the floor area, and decided on placing two access points in the satellite building floors because of interference in the mechanical, elevator, and communications rooms. In addition, having two access points increases the range and redundancy of the wireless network.

We decided to pair up satellite buildings to provide redundancy for the building links. This solution was not as costly as compared to creating a full mesh of wireless links. We enabled encryption to provide data security to the wireless links. We also learned that verification testing is very important. When the wireless network is implemented, testing access and routing will help validate the solution.

Summary

In this case study chapter, we reviewed an enterprise network example involving a hospital. Wireless local area network (LAN) devices in this scenario were to help information services (IS) managers provide additional functionality and services. By installing wireless LAN access points, hospital personnel could access the LAN in the conference rooms and nearby offices. By installing IEEE 802.11b outdoor routers, the hospital was able to save leased line cost and increase bandwidth to 11 Mbps. Redundancy and security issues were also addressed in this case study.

We followed a wireless project approach of planning the project by gathering the requirements and baselining the current network before designing the wireless network. Implementation of the network was followed by testing and verification.

For Jones Hospital we gathered requirements for three projects broken out of the main challenge of providing a wireless solution: wireless access for laptops in satellite buildings, wireless access in the hospital building, and links between buildings. The requirements can be summarized as follows:

- Provide wireless access for laptops in all satellite buildings.

- Provide wireless connectivity in conference room in the main hospital building.

- Provide a replacement to the leased T-1 lines that connect the satellite buildings to the main hospital building.

- Provide increased bandwidth to the satellite buildings.

- Provide redundancy to the satellite buildings.

We designed a wireless access solution for the satellite buildings using two access points with omnidirectional antennas per floor in each satellite building. The design consisted of a wireless solution that contained an access point with a directional antenna in the hospital conference room. We designed a wireless solution for replacing the existing leased T-1 lines with wireless links from the hospital to the satellite buildings. Data encryption provided security for the wireless links.

We outlined procedures to follow when implementing these projects. We discussed testing methods to verify that the wireless access devices and links are working and that the range of wireless access includes all locations within the building. At the conclusion of the project, a follow-up meeting was held to gather the client's feedback on the project.

Solutions Fast Track

Introducing the Enterprise Case Study

- ☑ Hospital requires wireless access for laptops in satellite buildings.

- ☑ Hospital requires wireless access in conference room.

- ☑ Hospital requires building-to-building wireless links.

Examining Network Requirements

- ☑ The area in the satellite buildings has rooms that will cause interference to the wireless buildings.

- ☑ The area in the conference room is small.

- ☑ There is clear line-of-sight from the main hospital buildings to provide a wireless solution.

- ☑ The distance between buildings permits wireless links.

Designing a Wireless Solution

- ☑ Use two access point bridges per floor in the satellite buildings with omnidirectional antennas.

- ☑ Add an access point bridge in the conference of the main hospital building.

- ☑ Use access-point outdoor routers with directional antennas for hospital to satellite building wireless connectivity.

- ☑ Add wireless links between building pairs for redundancy.

- ☑ Use encryption for security.

Implementing and Testing the Wireless Solution

- ☑ Install, configure, and test the access points and antennas in the satellite buildings. Test that laptops can access the LAN from all locations in the floor.

- ☑ Install, configure, and test the access point in the hospital conference room.

☑ Install, configure, and test the outdoor routers and wireless links. Then install the redundant wireless links.

Lessons Learned

☑ Using multiple access point devices on a floor will provide additional access range and redundancy.

☑ Using an access point with a directional antenna in the conference room will provide wireless access for those attending meetings.

☑ Using encryption will provide data security for the wireless network.

☑ Using IEEE 802.11b outdoor routers with wireless directional antennas provides increased bandwidth to 11 Mbps between buildings.

Frequently Asked Questions

The following Frequently Asked Questions, answered by the authors of this book, are designed to both measure your understanding of the concepts presented in this chapter and to assist you with real-life implementation of these concepts. To have your questions about this chapter answered by the author, browse to **www.syngress.com/solutions** and click on the **"Ask the Author"** form.

Q: We have concerns about the security of our data—how is our data protected?

A: Wireless products come with varying levels of encryption methods to protect data. Some of the data encryption methods used are DES, 64- bit WEP, and 128- bit RC4. Also, MAC address-based access control table schemes are used.

Q: What are the ranges of wireless devices outdoors?

A: Directional antennas can provide a range of up to 16 miles.

Q: What routing or bridging functionality is available on access point devices?

A: Access point wireless devices can act as bridges or routers. They can be config- ured with static routers or with simple Routing Information Protocol (RIP). They can also be configured to filter based on a MAC address when acting as a bridge.

Q: How will wireless laptops acquire an IP address and other IP-related information?

A: When using wireless bridges, you still use your existing Dynamic Host Configuration Protocol (DHCP) servers on the network to acquire an IP address, subnet masks, default gateways, Domain Name System (DNS) server, and other IP information regularly configured via DHCP. No special changes are required to access the DHCP server.

Chapter 10

Designing a Wireless Industrial Network: Retail Case Study

Solutions in this chapter:

- Introducing the Industrial Case Study
- Designing and Implementing the Wireless Network
- Planning the Equipment Placement
- Lessons Learned

☑ Summary

☑ Solutions Fast Track

☑ Frequently Asked Questions

Introduction

Experts in the industrial environment acknowledge the growing need for wireless technology. The emerging wireless handheld devices dramatically expand mobility when applied to standard industrial activities like inventory and stock management. The increased productivity and cost savings far outweigh the cost of investing in new wireless technology. This chapter describes how to implement a wireless network in an industrial environment. Although there are various types of industrial applications, we will focus on applying wireless technology to a retail store environment. Retail stores implement wireless technology for a number of purposes, including helping their employees to track inventory using a mobile system, and allowing customers to self-scan purchases and check the price of items.

Although it may be easier to think of applying wireless technology to large superstores, the mobility that wireless provides offers a big advantage for smaller stores. Smaller stores cannot support the number of employees or merchandise that large superstores do, so if they implement wireless technology they are able to streamline the staff dramatically. By adding mobile devices, a store owner can provide customers with the ability to answer their own questions about pricing and inventory. Customers who are more self-reliant do not require as much help from staff. Similarly, employees with self-reliant customers do not require the same amount of support from their management.

Through the case study presented in this chapter, you will learn how a consulting company can apply the design principles described in previous chapters. The flow through the discovery and planning phases show typical real-world issues and events. The planning phase contains the details you must be aware of when implementing a similar type of wireless network. The implementation section of this chapter walks you through the process of integrating the existing wired network with the proposed wireless network.

Applying Wireless Technology in an Industrial Network

In the past two years, companies like 3Com Corporation have designed wireless cellular digital packet data (CDPD) networks for consumer applications on popular personal digital assistants (PDAs). More recently, by coupling wireless 802.11b technology with their IPAQ PDA, the Compaq Computer Company is successfully implementing wireless technology in the industrial setting. With data rates that support up to 11 Mbps, companies are finding useful applications for everything from network troubleshooting for corporate LANs to inventory control directly from these mobile devices. Transmeta's TM3200 chipset provides more effective processing power. As this power is incorporated with the operating system efficiency of Linux in handheld technologies, an explosion of new and enhanced applications will find their way to these powerful devices.

Although size is an issue with mobile units, companies like Symbol Technologies are finding unique ways to shrink wireless devices to allow customers and employees to perform simple retail functions while roaming through a store. Beyond these immediate examples, several key benefits are inherent when wireless technology is incorporated into business processes.

The retail side of this industry is rapidly warming up to wireless technology. Recently, large department chains like Sears, Roebuck and Co. and Wal-Mart implemented handheld devices for employees. These devices enable the employee to check inventory quickly, make price changes, enable merchandise pickup, and maintain adequate stock. Customers benefit when companies like these use handheld devices to prescan items prior to checkout.

Introducing the Industrial Case Study

In this case study, Bob Tucker, the owner of a large retail sporting goods store called Pro Sports, is interested in applying wireless technology to make his network more efficient and to increase customer service. His current sales figures are looking solid, but in his market environment, competitors could soon be moving in down the street. Future competition will drive prices, but it will more clearly drive service. Bob has kept up with wireless trends in the retail market, including the wireless checkout bays used in a few superstores in his area. It appears to him that customers are eager to use new technology.

He also knows that he needs to increase employee productivity and customer response. After analyzing current growth and predicting future sales trends, Bob believes he must either increase his full-time sales staff by three employees or implement technology that will likewise expand sales efficiency and customer response. This choice makes the incorporation of wireless technology a value proposition to weigh against future plans. One of the main reasons to implement wireless technology is to provide better customer service.

Assessing the Opportunity

Bob Tucker evaluates Pro Sports' needs and develops a list of the benefits he wants to add for employees and customers. For the employees, he seeks to automate in-store inventory. Currently, employees manually track the inventory during off-hours.

During regular hours, office personnel enter the inventory lists using the PCs in the company office. He figures that he can save money when employees take inventory via handheld devices. This eliminates the reentry performed in the office, and the employees can take the inventory during normal working hours.

Bob wants his shipping clerk to place items in inventory as they are unloaded in the docking bay. He projects that the handheld devices will enable the shipping/receiving clerk to add to the inventory real-time as merchandise comes off the truck. This activity will eliminate the extra effort it takes to provide the paperwork to the office for manual entry. This automation will also enable other employees to check stock accurately and quickly for items that just arrived. Another advantage to adding this capability to the shipping/receiving area is that wireless technology will enable the shipping/receiving clerk to access the current wired network. The PC used in the shipping/receiving department currently is not connected to the office PCs. It is a stand-alone PC with a separate software package that is not tied to the company accounting system.

Handheld devices will enable the employees to respond to customer pages. These devices will make it easy to assess the customers who need assistance and respond to them quickly. When a customer requests specific information, the sales associates with that particular expertise can respond. Since employee incentives are based on commission and customer satisfaction, handheld devices will become sales associates' pagers. What better chance of earning a commission can Pro Sports offer to the employee than answering a customer's page?

Bob figures that the customers will benefit from the wireless technology by enabling them to check for stock and prices. For example, as the sports seasons change, the shoe department is often one of the busiest departments. By giving customers handheld devices and allowing them to scan the bar codes of the display shoes, the customers can check to see if the inventory contains shoes of that type in their size. Although Bob's office team works hard to print price tags for incoming items, human errors occur and sometimes items show up without price tags. The handheld device enables the customer to scan an item's bar code for pricing if a price tag is not available. This device will also provide a map to help the customer locate items within the store.

Handheld devices can eliminate the customer's wait in long lines. Customers can scan their items and present the device to the cashier at the register. The cashier downloads the information from the handheld device. This step is particularly useful on weekends and holidays when the store is very busy. Bob figures that this feature may eliminate his need to hire extra holiday seasonal help.

After careful consideration, Pro Sports contacts your wireless networking firm to create a design to see if these goals can be met by implementing wireless technology in the store. The results of the wireless implementation in this store might open up additional opportunities throughout the chain. The opportunity for future sales and support makes it clear that your planning must be thorough, your design must be efficient, and your hardware selections must be cost-effective.

Defining the Scope of the Case Study

This is the first implementation of wireless technology in the individually owned Pro Sports chain, so the current intention for this network is limited to the single store and does not include network access to other stores at this time. However, the results could lead to adding wireless networks to other stores within the chain. The existing computer network is in place and running fine. There is no need to modify the existing computer network other than to integrate the wireless system with the existing network.

Pro Sports is located in a spacious two-story building. Both floors contain merchandise that is organized to attract customers and lead them through the store. The attached warehouse acts as a receiving dock for merchandise. Employees use the warehouse for inventory overflow and office activities, like general company accounting (accounts payable, accounts receivable, and payroll). For convenience, the company offices are located in the warehouse.

The existing wired network consists of an Ethernet local area network (LAN) that connects the registers to the computer system in the company offices. The company offices consist of several desktop computers, three network servers, an Ethernet switch, and a router for wide area network (WAN) and Internet access. The wireless technology that will be implemented includes handheld scanning devices and a wireless card for an existing PC.

We will not address data security in this case study. No confidential data will be transmitted using the handheld devices. The handheld devices have the capability to swipe credit cards for payment, but Bob Tucker has stated that he does not want any credit card transactions transmitted over the wireless network, not wanting to risk the possible interception of personal customer data.

Reviewing the Current Situation

To make sure that your team understands the situation, Bob outlines his need to add wireless technology to Pro Sports. To recap, he needs to tie the shipping/receiving PC into the existing network and enable instant stocking by the shipping/receiving clerk through the use of a handheld device. The system must provide customers more autonomy by enabling price checks, inventory checks, a virtual shopping cart, an online store directory, and customer assistance paging. These features must be implemented on handheld devices. The intended benefit is to save time and money by making the employees mobile and more responsive.

Designing and Implementing the Wireless Network

The approach is straightforward—you must determine how to address the customer's needs and make sure they are well defined; the owner and his management team must verify information about the employees and the customers for you. After the approach is determined, you'll begin the planning by defining the network elements and their placement, and gathering details about the physical space and the intended use.

After ample design time, your team will purchase the hardware elements, and then implement the design by installing and configuring the hardware elements and making the necessary software changes. During implementation, you'll have to test every aspect of the system, including the range of the handheld devices and the ability to check bar codes on the loading dock. At the end of the implementation phase, you should be able to verify that the results fulfill the needs of Pro Sports.

Creating the High-Level Design

Your team considers setting three subdomains to make it easier to divide the work and find where to place the access points. They will also make it easier for you to categorize and track progress as you set up the network. These subdomains include the first floor, the warehouse, and the second floor. Two of the subdomains are divided because of the physical boundaries between floors; it is easier to plan the integration of the wireless elements and the existing network elements by floor. The functional boundary of the warehouse naturally makes it a separate subdomain. The warehouse does not need to address any customers; only a few employees work in or around the warehouse.

It is determined that handheld devices will be used in each of the subdomains. Employees and customers will use handheld devices in the first floor and the second floor. Only a few of the employees will use handheld devices in the warehouse. These employees perform specialized tasks, like shipping/receiving or accounting. The team identifies the additional need for a wireless-enabled PC in the shipping/receiving area of the warehouse.

At this point, the owner wants to make an investment of 100 handheld devices. He feels that the majority of these devices should be available for the customer. He does not want so many that they end up hanging on the wall, but he does not want the customers to have to wait for the use of a handheld device. As a result, Bob chose a number that he felt would be balanced between the two situations. You will have to determine with his help the total number of employees that will use hand-held devices and the division of the work force per floor. His responses will help in determining where the access points will be located later on in the design.

Creating a Detailed Design

Your consulting company invests some time into the planning and design of the wireless network for Pro Sports, addressing the following tasks:

- Obtain a physical map to chart all aspects of the building, including electrical outlets, Ethernet cabling, and existing network elements. Since the new wireless network will have to interface with the existing one, knowing the details of the current network will help you make decisions.

- Talk to the owner about expected user density. How many customers does he expect to have on either of the floors at one time? How does he assign employee activities? What is the maximum number of employees scheduled to work on each floor? The answers to these questions help determine the number of access points required for efficient transmission, as well as where to put the access points.

- Identify any constraints that may limit the design of the wireless network. When you identify constraints early, you have more time to work around the issues. Constraints can be physical, such as no access to electrical outlets. The consumer can also mandate constraints.

- Conduct a walk-through to verify information on the physical map. This helps you account for any deviance from the physical map to the existing structure. A deviance can occur when store improvements are not added to existing documentation. Walk-throughs also provide you with additional information. For example, if there was no access to an electrical outlet but a light fixture was located nearby, you could assume that an electrical connection can be established close to the light fixture.

- Identify any potential radio frequency (RF) interface sources. Any electrical appliances using the 2.4GHz range can affect the reliability of the wireless network, such as microwaves and 2.4GHz cordless phones.

- Determine the size of the store and the radius of RF transmission. Apply the facts regarding the size of the store and the expected user density to determine the required range of RF transmission. While planning this radius, make sure you record any overflow coverage.

- Plan the access point locations to take advantage of transmission coverage. Make sure one or two radios are added to the access point as needed for transmission. Extend the radio antennas as needed for coverage.

- Determine Internet Protocol (IP) addresses. You can identify network elements (wired and wireless) by IP addresses. The IP addresses must be unique within the network. The Dynamic Host Configuration Protocol (DHCP) server enables you to set the IP range and monitor use of the addresses.

- Define the process to integrate the new wireless infrastructure into the existing computer network. Certain capabilities, like IP addressing and tracking are available in the existing network. The new wireless system can rely on the same DHCP server that controls the existing IP range.

All of these points must be addressed as you progress through the network development phases of planning and design.

Obtaining a Physical Map

The physical map contains information about the placement of the different areas of the store, information about the current wired network, and other physical characteristics, like access to electricity. These particulars provide the physical details used when combining the wireless elements to the wired elements. For example, if there is no access to electricity, a network element cannot be plugged in. An additional physical requirement is that access points must connect to the Ethernet cable of the existing network.

The 10 old steel-framed Pro Sports building projects a spacious feeling with 20-foot high ceilings. The 18,000 square foot retail store is composed of a 10,000 square foot first story, a 4000 square foot warehouse expansion at the back of the first story, and an 8000 square foot second story. The load-bearing first floor contains four columns that dissect the room. The second floor does not contain any columns. The drop ceilings for each floor allow for four feet between the second-story floor and the first-story ceiling, and the roof and the second-story ceiling. The drop ceilings can provide enough room to accommodate the weight and the space requirements of the access points.

Figure 10.1 illustrates the layout of the departments on the first floor of Pro Sports. The first floor contains various clothing departments, a shoe department, a baseball/soccer department, a golf department, a seasonal department, and a sunglasses department. Generally, the most active of all the first-floor departments is the shoe department, which contains tennis shoes, cleats, boots, and specialty sports shoes for everyone in the family—note that the merchandise planners placed the most active department in the back of the store. The planners implemented this store design to influence customers to buy other items as they walk through the store.

Figure 10.1 Layout of the First Floor

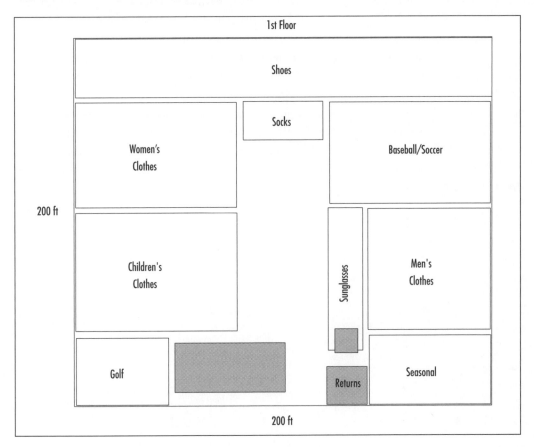

The owner has placed ten checkout registers near the outside doors on the first floor. During the weekdays, up to seven of the ten are available for checkout. On weekends and during holidays, all ten checkout registers are open for business. As a rule, Bob Tucker does not like to see more than two people waiting per checkout line. There is one register at the return counter to enable the employee to process customer requests for exchanges and returns, and to return items to inventory. One register is available in the sunglasses department. All sunglass purchases must be made using this register; this is a physical security implementation due to the ever-increasing prices in the sunglasses department and the portability of the merchandise.

The first floor spans 200 feet by 200 feet. The main entrance to Pro Sports is on the north side of the building. The entrance implements glass panes to let natural light filter into the store. An escalator, which enables customers to move to the second story, divides the store horizontally and vertically. An additional escalator on the reverse side enables customers to go to the first floor to check out at the register. The 40,000 square foot measurement does not include the warehouse on the south side of the first floor; it will be addressed as a separate subdomain.

The warehouse contains a shipping/receiving area for processing items coming into the store and items being shipped from the store. Trucks haul merchandise to the loading dock. The shipping/receiving clerk verifies the receipt of the items and stores them in the warehouse until an employee can stock the merchandise. The warehouse also contains the company office, where the administrators run the store's accounting software and track employee database information. As Figure 10.2 shows, the warehouse also contains the computer closet, which holds most of the existing network equipment.

Figure 10.2 Layout of the Warehouse

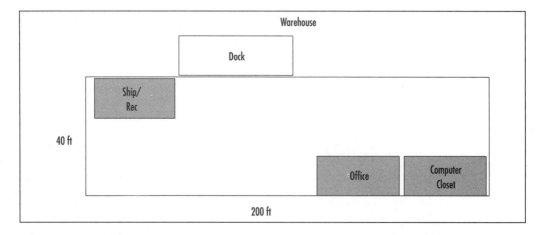

The warehouse extends 40 feet in length parallel to the south side of the first floor. Loading docks extend beyond the warehouse for merchandise that is trucked to the store. Trucks arrive at various times of the day and the shipping/receiving clerk tends to each shipment. Because the warehouse stores merchandise until an employee has time to stock it, much of the warehouse appears to be pallets that are stacked high with boxes.

The second floor contains the largest items sold by the store. Pro Sports sells merchandise for water and snow sports, camping, fishing, and hunting (the department layout is shown in Figure 10.3). The store design includes many demonstration displays on the second floor, from assembled tents to hanging kayaks. Employees carry large purchases downstairs using a freight elevator on the northeast corner of the first and second floors. Many of these departments require salespeople who are very knowledgeable about the subject to be constantly available to customers to answer questions.

Figure 10.3 Layout of the Second Floor

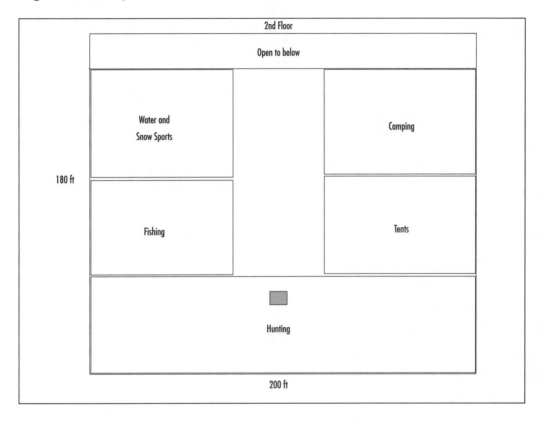

The 180 by 200 square foot second floor is open so that customers can view the last 20 feet of the first story. Future plans include the placement of a children's play area in the first-floor area that can be viewed from the second floor. A floor-to-ceiling fence surrounds this overlook. Electrical outlets exist every 20 feet across the east and west walls. Droplights hang from the ceiling at 20-foot intervals. A single register is available on the south side of the second floor in the hunting department. This register is used to license firearms, process security clearances, and purchase firearms.

Figure 10.4 shows the current wired network for the first floor and the warehouse. This network contains a server farm for the existing LAN. The server farm is located in the computer closet. There is also a router in the closet. As mentioned earlier, the router provides connectivity to the Internet as well as other Pro Sports stores. The computer closet is a basic wiring closet with a DHCP server to handle IP addressing for the store PCs and registers. Other servers in the server farm address the database and processing needs of the existing network. An Ethernet switch, located in the middle of the false ceiling in the first floor, connects the cash registers in the front of the store to the server farm.

Figure 10.4 Existing Network for the First Floor and Warehouse

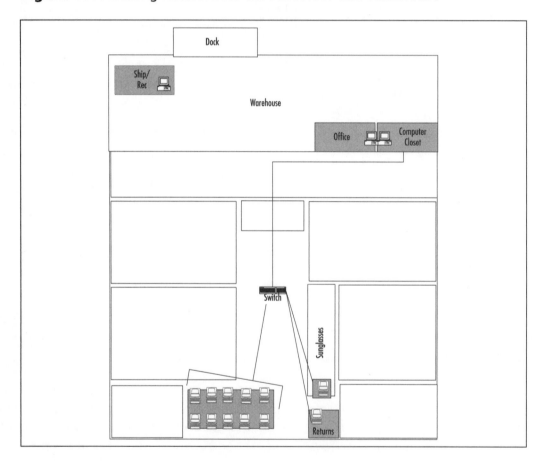

The Ethernet switch is located in the middle of the first-floor false ceiling for a single purpose. It can reach the only register found on the second story. The Ethernet switch registers connections shown in Figure 10.5. The Ethernet switch is connected to the current network via Ethernet. You decide that an access point should be placed in the proximity of the Ethernet switch so that the access point can be connected to the existing network.

Figure 10.5 Existing Network for the Second Floor

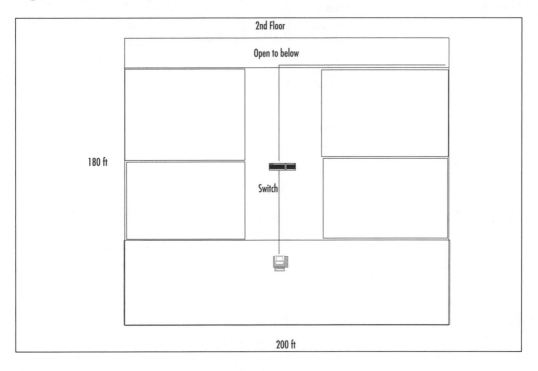

Determining User Density

Next you need to talk to Bob Tucker about user density; that is, the maximum number of customers and employees that could co-exist on the first floor, second floor, and warehouse subdomain of the store. Bob reviews current employee placement records, takes averages, and applies a slight increase (+ or − 15%) to answer questions about the number of people per subdomain. The user density helps your team determine the number of users that need to be addressed via access points.

Bob explains that he expects to provide one handheld device to every employee in the store. During store hours, management assigns employees to cover static positions in a particular department or dynamic positions within the departments of a floor for employees to roam between departments for additional coverage. These employee positions are shown in Table 10.1.

Table 10.1 Employee User Density

Floor	Department	Number
First floor	Shoes	2
	Seasonal	1
	Men's Outdoor Clothing	1
	Ladies Apparel	1
	Children's Apparel	1
	Sunglasses	1
	Roaming throughout floor	3
Warehouse	Shipping	1
	Office	1
Second floor	Camping	1
	Hunting	1
	Roaming throughout floor	3
Roaming Management	Store Manager	1
	Floor Managers	2

Two handheld devices will exist within the warehouse. The shipping/receiving clerk needs a handheld device to check stock into inventory. The office needs a handheld device to track employee time. In addition to employee positions, three managers walk the floors to ensure quality customer service. These managers are not assigned to stay on a particular floor, but may all be on the first floor, the second floor, or dispersed throughout the floors.

Identifying Constraints

All constraints must be identified to make sure that the plan can be as foolproof as possible. When you identify constraints early in the planning stage, you have more time to work around your findings; also, constraints identified during implementation clearly add to the schedule and scope of the project.

The only physical constraints are that the owner of Pro Sports does not want the access points viewable by his customers. It's a matter of aesthetics, Bob explains. He wants customers to concentrate on merchandise, not network equipment. One of the benefits to wireless technology is that the hardware is low profile and easy to conceal. Knowing this constraint, you propose that the access must be concealed in the false ceiling if at all possible.

Conducting the Walk-Through

After gathering building facts such as square feet per floor from the physical map, you conduct a walk-through to identify construction materials between ceiling and walls. Walk-throughs identify elements that do not exist on the physical map, such as improvements beyond the original design. They also help to locate additional resources, such as AC availability through light fixtures and other wired elements. This walkthrough identifies no issues, as you find the steel construction with wood floorboards and drop ceilings.

Identifying RF Interface Sources

Your team identifies potential RF interface sources. They check for these sources because certain electrical appliances can provide interference to wireless transmissions in the form of static on the receiver side. They look for cordless 2.4GHz phones and microwaves. Pro Sports does not use cordless phones, but the break room in the warehouse contains a microwave that runs at 2.4GHz. You can identify the microwave as a potential risk but you will not know the impact of the interference until the wireless network is tested during implementation.

Plan the RF Pattern for the Network

After executing the walkthrough, your team plans the RF pattern for the network. You determine the size of the store and radius of antenna transmission. Since the handheld devices will be roaming throughout the three subdomains, you overlap the RF patterns to create accessibility to more radios. Overlapping the RF patterns makes the signal stronger and transition between access points less noticeable.

The proposed design covers a dense population of handheld users in a small area where there are few planes for interference sources. The RF pattern is designed so that any spot in the store is covered by at least three radios, with some areas being covered by five radios. This extra coverage ensures that every user is provided coverage that is transparent as the user roams throughout the building.

You note that with an 802.11b extender antenna with a 100 foot by 100 foot footprint, distance is not an issue, but density may be an issue if employees and customers use the 100 handheld devices at the same time. It is assumed that if the 100 handheld devices are used at once, they will be divided within the floors. So, you design the first floor for a density of 13 employees and 80 shoppers. The density for the second floor can be set at eight employees and 50 shoppers. You count the density for the warehouse to be five employees and managers.

Your team follows the general application rules. Current equipment guidelines state no more than 30 users per radio. You apply the density to the amount of users that one radio can handle. From the outset, with the projected densities per floor, it appears that there will be four access points throughout Pro Sports.

Planning the Equipment Placement

Before you can determine the placement of the equipment, your team checks for Ethernet connections, electricity availability, and physical access. The access points must connect to the existing network using a RJ-45 wire. An electrical outlet must be available for the access point. Physical access can be an issue if wires cannot be concealed or wire length exceeds CAT5 limitations.

Ethernet connections are available in the center of the drop ceiling for the second floor. This is the location of the existing wired switch. Ethernet connections are also available when strung to the computer closet in the warehouse. The amount of wire needed does not appear to be an extraordinary amount.

Electrical outlets exist on the east and west sides of the building on both floors. Lights hang from the false ceiling on the first floor and the ceiling on the second floor. You arrange to have an electrician install an electrical outlet six feet away from the light in the center of the false ceiling.

The building is steel construction and there are no columns on the second floor. Measurement of the length comes to 20 feet of slack in the computer closet, 20 feet up the wall to the ceiling, and 20 feet up to the second floor. A diagonal to the center of the second-floor ceiling requires 70 feet. This makes a total of 130 feet of cable (this amount of cable is reasonable because it is under the CAT5 cable length limitation of 100 meters).

Determining Where to Place the Access Points

Handheld devices search for and use the access point with the strongest signal. Therefore it is important that the access points be placed where the signal will optimally cover the area. Since the handheld devices will be roaming throughout the three subdomains, you overlap the access point ranges to create a stronger signal. Overlapping makes the transition between access points invisible to the mobile consumer or employee carrying the handheld device.

Your team decides on the best location for the access points. They take into consideration that the location must be close to the Ethernet switch in the center of the first-floor ceiling. They place the access points in a centralized location because the dimensions of the store are small enough that they do not have to worry about coverage within the building. They will install the equipment in fairly close proximity at the center of the building, to make it easier and more convenient to service the equipment.

To address the second-floor density of 58, one access point will be placed in the center of the ceiling. The access point contains two radios. To span the width of the floor, the antennas for each radio are extended six feet to the east and west of the access point, as shown in Figure 10.6. The access point diagrams shown in Figures 10.6, 10.7, and 10.8 represent the access point as a diamond, the radios as ovals, and a small antenna tower as an extended radio antenna. The RF patterns for the second-floor overlap, as shown in Figure 10.6. This overlap provides double coverage on the second floor.

Figure 10.6 Access Points on the Second Floor

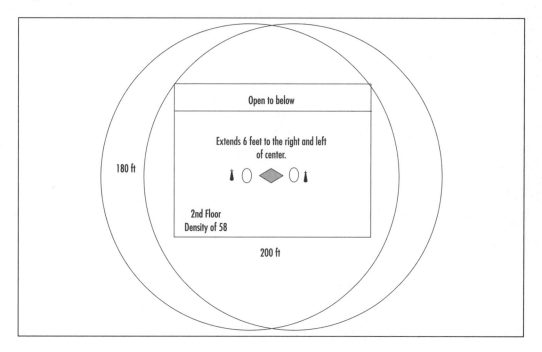

To address the first-floor density of 93, two access points are used. One access point is located three feet west of the center of the false ceiling on the first floor. This access point contains one radio. The second access point is located ten feet diagonally to the northwest (front) corner of the building. This access point contains two radios. To make sure that connection is made evenly throughout the store, the antenna on one of the radios is extended toward the middle of the store. The warehouse needs a single access point located near the computer closet. This access point contains one radio, as shown in Figure 10.7. This access point provides an overflow into the southwest corner of the store.

Note how the RF patterns overlap on the first floor and the warehouse. The overlap ensures that the large amount of handheld users will have adequate coverage. The RF pattern for the warehouse extends outside of the physical building. This pattern allows for coverage at the dock so the shipping/receiving clerk can check merchandise into inventory. It also gives the clerk space so that he can actually go in the trucks while he checks the merchandise.

Figure 10.7 Access Points on the First Floor and Warehouse

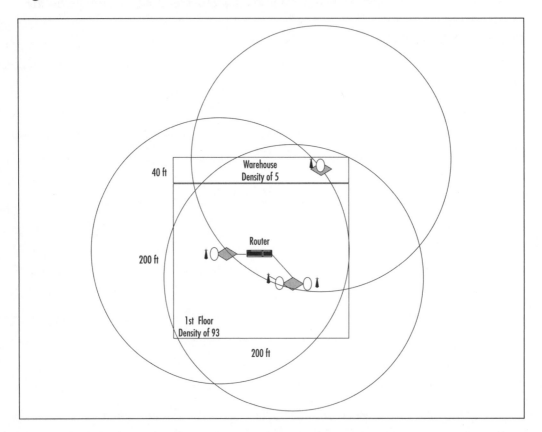

The access points provide plenty of coverage for the predicted user density. Remember that these figures represent only a two-dimensional representation of the pattern—the RF pattern actually covers a three-dimensional range. Coverage for both floors is represented more accurately in Figure 10.8. This figure shows the RF pattern for the first floor overlaid with the RF pattern of the second floor. The RF pattern shows that almost all areas of the store are covered by at least three RF patterns. In fact, most areas in the store are covered by five RF patterns. This extra coverage ensures coverage for the estimated user density.

You have determined a total of four access points and six radios. To recap, you plan one access point and two radios on the second floor, one access point and one radio in the warehouse, and two access points and three radios on the first floor. This plan allows for the future expansion of two radios.

Figure 10.8 The RF Pattern Overlay for Both Floors

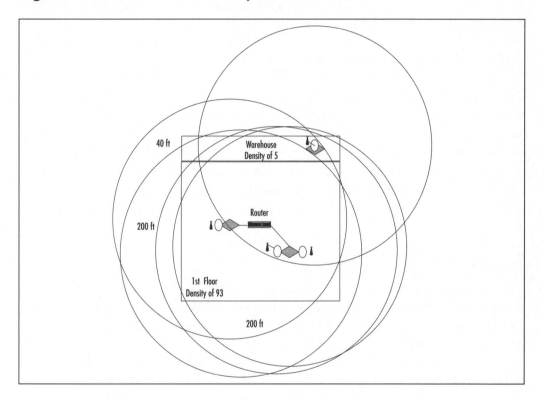

Determining the RF Channel Optimization

Since the radios in the access points will be very close in proximity to each other, it is very important that they all operate on different 802.11b frequency channels. The 802.11b specification provides 11 different frequency channels. It is also important that the channels that are chosen be as separate as possible from each other. In other words, you cannot set one radio to channel 1 and the next one to channel 2. Most access point vendors recommend a three-channel spacing between usable channels; however in certain cases it is possible to push that limit to a two-channel spacing. A three-channel spacing will allow for three usable channels. Since you have split the

RF spaces into three separate subdomains, you will not have a problem with channel overlap.

There are a total of two access points and three radios on the first floor. The access point with a single radio will be configured to operate on channel 3. The other access point will have one radio on channel 7 and the other on channel 11. When your team has determined the operational frequency channels for the first subdomain, they will reuse the same settings for the other subdomains. For the second floor, they will set one radio to channel 3 and the other to channel 7. In the warehouse, they will set the access point to channel 11. If Bob Tucker decides to add additional capacity in the future, the RF channel pattern will most likely need to be modified. Configuring and modifying channel patterns are simple software changes.

With this configuration, a shopper will connect to the radio with the strongest signal. Since the flooring between the first floor and the second floor, as well as the wall between the store and the warehouse, is made of concrete, the radios operating on the same channel will not be a problem (the concrete will absorb most of the RF energy from one subdomain to another, and the stronger access point will always be in the subdomain of the user). If all of the radios were set to operate on the same channel there could very possibly be any number of unexplainable RF problems. It is always a good idea to ensure that access point radios that overlap RF patterns be set to different channels.

Identifying IP Addresses

The existing wired network has a DHCP server in the server room of the warehouse. Pro Sports currently is using typical Class C IP addresses in a range from 192.168.1.0 to 192.168.1.255. Of these 256 addresses, they currently are using 20. That leaves 236 available to use in the same range.

You plan to use the same address range for the handheld units through DHCP, reserving addresses from 192.168.1.100 to 192.168.1.225 for the full 100 handheld devices. This schema leaves 25 additional IP addresses for future growth.

Implementing the Wireless Network

The design is in place and your team begins to implement the design. Begin by identifying the hardware required for the installation. In this case, hardware must be provided for the access points, the handheld units, and the wireless card required in the shipping/receiving PC. After the hardware is selected, begin installing the wireless network.

Selecting the Hardware

As stated, the wireless hardware elements required to connect the wireless aspects of the network to the existing network include access points, handheld devices, and a wireless PC card. You review the hardware elements used in previous commitments against new wireless network element technology, also taking into account pricing and availability.

The first element you select is the access point. You select this element first because the other wireless components must be compatible to it. The access point acts as a wireless hub, receiving and transmitting information over a radio frequency of 2.4 GHz. Requirements for the access point include 802.11b compliance and a throughput of up to 11 Mbps. After weighing many of the available access points, you select the Agere Orinoco AP-1000 access point, because it is expandable. The AP-1000 provides two radio slots. Multiple radio slots enable you to load balance access points when they are heavily used. Each of the radios can operate on a different frequency channel. Also, if only one radio is used, the network becomes scalable. This access point is easy to configure using Windows-compatible software and provides an integrated Ethernet interface. The AP-1000 can perform many functions, serving as a router, bridge, or DHCP server.

The compatible Agere Orinoco PC Memory Card International Association (PCMCIA) cards serve as the radios in the AP-1000 access point. Besides the six radios that are chosen for the access points, you purchase one hundred more to go in the handheld devices.

An Orinoco Range Extender antenna is purchased for each radio attached to an access point. The Range Extender is compatible with the AP-1000 and the Peripheral Component Interconnect (PCI) cards, and is a 5-dBi indoor omnidirectional antenna. These antennas can boost coverage up to 50 percent, based on the physical environment.

The manufacturer recommends that at least one of the two PC cards should be equipped with a range extender to create a distance of at least one meter between the antennas of the two PC cards. You decide that all of the radios should have the Range Extender. Orinoco also recommends that you set each of the two PC cards to a different frequency channel and to optimize capacity and minimize channel crossover, and suggests that you separate the two channels as far as possible. The shipping/receiving PC requires an Orinoco PCI/MCA (microchannel architecture) card to interface with the wireless and existing networks. These cards are designed to interface with the AP-1000. The card fits in the PC casing and boasts sufficient range and stability, and will transmit data over a radio frequency of 2.4 GHz.

Since the shipping/receiving PC runs stand-alone applications, the application must be integrated with the standard networked PCs used in the office. You evaluate the changes (and your software consultants join the process) and revamp the networked software with the shipping/receiving application.

Price checks and inventory control require a means for efficient scanning. You review a number of handheld devices that have 802.11b LAN access with scanning capabilities. It is important to note that these are two very different functions. 802.11b allows access to the existing LAN and is the Institute of Electrical and Electronics Engineers (IEEE) standard for wireless. These scanning devices must interface with existing peripherals, including the registers. Because employees must carry a handheld device throughout their shift, the device must be lightweight. It must also have a large viewing screen for ease of use by the customer and employee. Bob Tucker has also asked that the cost of the handheld devices be within an estimated cost range. Handheld devices that did not meet these needs were not considered.

The SPT1700 model from Symbol Technologies met all the requirements, including the scanning capability and 802.11b; it is lightweight and easy to use. The devices implement a Web browser interface for accessing the in-store network. A great feature of the SPT 1700 is that the IP address is stored in the handheld device. Perhaps the most important aspect, however, is the cost—it fits the projected budget supplied in the consultant's equipment proposal.

When the owner of Pro Sports is presented with the pricing scheme for this model, he asks about security of the handheld devices. You assure him that tags are embedded into the devices and that the outside doors scan for these tags just like the ink tags placed on large ticket inventory items within the store.

Installing the Wireless Components

At this point in the process, you and your team must install the wireless network elements that were chosen. The installation includes adding the PC card and testing its functionality, setting up the access points, configuring the access points, setting up the IP address range, and testing the handheld devices. Testing, performed after each step of the implementation, ensures proper communication with the existing network.

Create an installation checklist and verify the steps on the list. The checklist contains the following high-level actions, which are described in detail in the following sections:

- Set up the IP information

- Install the access points

- Install the AP Manager software

- Test the wireless network

- Review the client's objectives

Setting Up IP Information

As the first step in the implementation, set up the IP addresses by adding the media access control (MAC) addresses of the access points to the IP configuration table in the existing DHCP server, which is located in the server closet on the first floor. Reserve the IP address range of 192.168.1.100 to 192.168.1.225 for the 100 hand-held devices.

Installing the Access Points

After the IP information is provided, mount the access points as shown previously in Figure 10.6 (second floor) and Figure 10.7 (first floor and warehouse). The following list summarizes the placement of the access points and the placement of radios (PC cards):

- **Warehouse** This subdomain contains one access point located in the ceiling above the computer closet. Mount this access point four feet from the southwest corner of the warehouse. Insert one PC card into this access point.

- **First Floor** This subdomain contains two access points. Mount the first access point in the drop ceiling four feet west of the center of the room. Insert one PC card in this access point. Mount the second access point in the drop ceiling, ten feet from the center of the room diagonally across to the southeast corner. Insert two PC cards and extend the antenna three feet towards the center of the ceiling.

- **Second Floor** This subdomain contains one access point, mounted in the center of the drop ceiling. Insert two PC cards and extend antennas from both cards six feet east and west to the outer walls.

Your team performs the following steps for each of these access points:

1. Mount the power supply in the desired location.

2. Mount the processor module.

3. Connect the network interfaces by inserting the PC cards into the processor module.

4. Connect the Ethernet cable to the 10/100 Base-T Ethernet interface on the access point.

5. Mount the cover plate.

6. Power up the unit.

7. Verify that the LCD lights show the availability of the unit.

Install the AP Manager Software

After the access points are installed, install the AP Manager software on a Windows NT server in the server closet. This server has a 486 processor with 32 MB of RAM and two gigabytes of hard disk space. The consultants compare the specifications of this PC with the required specifications to make sure that they can run the AP Manager software on the PC.

This software establishes the connection of the AP-1000s. Since there are multiple AP-1000s, the consultants configure the other AP-1000s to match the values for the first access point. Make the following setting changes for each access point:

1. Set the PC card settings to "Access Point."

2. Set the network name to the name of the existing network.

3. Verify the IP addresses the consultants provided to the DHCP server that were automatically assigned.

4. Continue steps 1 through 3 to configure each access point.

Installing the PC Card in Shipping/Receiving

Add a Lucent PCI-to-PCMCIA card to the shipping/receiving PC to enable communication to the network. To test this card, you deploy a handheld device. When the hardware is deployed you make minor changes to the shipping/receiving software so that the scan directly feeds to the wired accounting system. Test this functionality and make adjustments as needed.

Testing the Wireless Network

After the configuration is performed, test the links to make sure they are active on the network. At this point the links test correctly. However, in the middle of the testing process, your team learns that Bob Tucker just received funding for an extension to the warehouse. Bob provides you with the physical layout of the extension and it appears as if the wireless design will cover the extension without a problem.

Just before you install the access point in the warehouse, you talk to the heating and air conditioning contractor who will work on the warehouse extension. When your team had performed the walk-through, you made sure there were no potential interference or physical placement issues with the ventilation system. However, the contractor explains that the ventilation duct close to the hub will now need to be split to flow properly into the extension. Rather than install the access point now and move it in a few months, you move the warehouse access point ten feet to the west.

During the RF pattern discovery, another microwave was found. After the network is installed, you run the microwave at various power levels while using handheld devices. The test proves that the microwave does not interfere with network communications.

You and your team thoroughly test the handheld devices using the wireless access points to ensure connectivity. You test the devices in major area of the store as well as areas with less regular traffic. The devices prove to be functional and responsive in testing. Access is also addressed, and you restrict access accordingly based on IP addresses.

Reviewing the Client's Objectives

After thoroughly testing the wireless portion of the network and testing the interaction between the wired and wireless aspects of the network, you can take the owner of the store on a tour. You show Bob Tucker how the shipping/receiving clerk can enter inventory at the dock. As you do this, a truck rolls up to the dock. You follow the clerk into the truck and watch as the clerk records the merchandise. At the shipping/receiving desk, the merchandise information is downloaded to the wireless PC, which in turn adds the information to the store database and accounting system.

On the first floor, you show Bob how to use the employee handheld devices to scan items for pricing and inventory. He takes a consumer handheld device, goes to the shoe department, and looks up the price and inventory for several types of shoes. He takes both handheld devices to the second floor and performs price checks; he successfully pages an employee for the hunting department; the store map also works. He takes the employee's handset and scans items for a customer. When that information is successfully downloaded to a register, Bob is satisfied that all his objectives were met.

Lessons Learned

After the job is finished, your team meets to perform a *post mortem* of the installation. In this meeting, you can identify major lessons to apply to future jobs. The most important lesson is to adequately evaluate software development. The accounting software was proprietary software. It required changes from the software vendor, the Accounting/Informational Technology expert, and your team of software developers. It was obvious that you should have included the software team much sooner in the design process.

The warehouse extension was not planned at the outset of the wireless network planning stage. The owner did not get funding until the wireless network was in the implementation stage, but you had not known the changes were even imminent, so the possible ductwork changes had not been factored in when you had evaluated impacts in the ceiling. Fortunately, you found out about the changes in time to move the access point in the warehouse before the work on the ventilation ducts began.

Summary

To summarize, combining wireless technology with an existing wired network empowers industry owners, managers, employees, and customers with scalability, flexibility, and mobility. In this case study, the owner of a retail store called Pro Sports defines the updates he wants to provide his employees and customers. He is able to tie his shipping/receiving PC into the existing network and enables instant stocking. The system gives customers the ability to check prices, inventory, stock a virtual shopping cart, and find items using the online store directory. The same mobility is provided for customers as it is for the employees. By checking prices, inventory, and store maps, customers can be more self-reliant and more efficient.

The owner successfully assessed the opportunities that wireless technology could provide to his store. By taking time to develop his goals, he was able to present you, the consultant, with a vivid picture of the expected results. This eliminated what is sometimes the hardest part of a project—getting the client to provide a set of goals. Although it is tough getting this information, it's the only way to measure whether the job is completed satisfactorily.

Your team walks through the planning stages quite efficiently, obtaining the background information, including the physical map, talking to the client about the expected user density, and recording any constraints. They perform a thorough walk-through of the building to look at building materials, access in the ceiling, and the current network elements. Potential radio frequency (RF) interface sources are found to be negligible. The team determines the location of the access points and tracks the RF patterns to make sure that adequate coverage is provided. IP ranges are established.

When the planning stage is complete, implementation begins with a selection of hardware. Most of the equipment is manufactured by the same company (in this case, Orinoco) to ensure compatibility. After the hardware is purchased, it is installed and configured, and you test the component functionality and review all new features and functionality with the client.

The resulting network proved to meet all of the client's requirements. It has added extensibility. When there is a need for more handheld devices for customers or employees, the owner can purchase more of the SPT1700 devices and PCI cards. When the serial numbers for the new hardware are configured in the access point software, the devices are ready to use.

Solutions Fast Track

Introducing the Industrial Case Study

☑ Wireless technology addresses the emerging mobility needs in the industrial setting. Recent coupling of 802.11b technology with handheld devices promotes widespread uses, from mobile inventory to network administration, to increase employee productivity and customer service.

☑ In the case study, the store owner wants to make his existing wired network more efficient and address customer needs. Handheld devices must be implemented to provide mobility.

☑ By streamlining the network, the store owner provides employees and customers easy access to store data, such as pricing and inventory.

Designing and Implementing the Wireless Network

☑ The network consultants approach the design by categorizing the physical store into three subdomains: the first floor, the warehouse, and the second floor.

☑ The consultants obtained a physical map and reviewed the existing network.

☑ The store owner provided estimates of the maximum number of customers and employees on each subdomain.

☑ The store owner also provided the constraint that all network elements must be hidden for aesthetics.

☑ Planning for the RF patterns took place. The consultants planned the placement of the network elements. IP addresses were established.

Planning the Equipment Placement

☑ The following hardware was selected: the Orinoco AP-1000 access point, the Orinoco PCI card, the Orinoco Range Extender, the Orinoco PCI/MCA card, and the SPT1700 handheld device.

☑ The consultants set up the IP addresses, installed the access points, and installed the related software. They installed the radios in the access points and handheld devices and installed the PCI/MCA card in the shipping/receiving PC. All of the hardware and software underwent testing to ensure functionality.

Lessons Learned

☑ You learned how a consulting company can apply the design principles described in previous chapters.

☑ The planning phase contains the details you must be aware of when implementing a similar type of wireless network.

☑ The implementation section of this chapter walks you through the process of integrating the existing wired network with the proposed wireless network.

☑ The most important lesson is to adequately evaluate software development.

Frequently Asked Questions

The following Frequently Asked Questions, answered by the authors of this book, are designed to both measure your understanding of the concepts presented in this chapter and to assist you with real-life implementation of these concepts. To have your questions about this chapter answered by the author, browse to **www.syngress.com/solutions** and click on the **"Ask the Author"** form.

Q: What are my choices in alternative handheld devices for a retail application?

A: There are many handheld devices on the market. Any industrial handheld device that is capable of scanning and is 802.11b compatible would work—the greater decision is in the pricing of the device.

Q: How can I make sure that the handheld devices do not leave the store?

A: There are various ways to add security to the physical device itself. One method is to implant a chip inside the device. Alternatively, you could add a magnetic bar code on the bottom of the cover. Either of these methods requires you to add the code information to a sensing mechanism at the exit, which will activate an alarm when the handheld device nears the door. Similar security is often used in retail stores that attach sensor tags to merchandise.

Q: Can wireless technology actually save you money?

A: The flexibility of a wireless network can save you money. This flexibility enables you to quickly add networked devices and peripherals, temporary networks, or make changes within the company. When your needs change, modification costs are low. You save a tremendous amount of money by not paying utility companies for leased lines, construction workers for trenches and holes, or linesman to string cable. With wireless, you do not have to worry if a cable is cut or goes bad. Wireless technology is so flexible that you can quickly and easily network hard-to-reach areas like a connection between buildings.

You can count other savings in personnel. The network is efficient, extensible, and static. There is less need for senior IT personnel. The software (in the case study, the AP Manager tool) often runs on Windows 98, Windows 2000, or Windows NT. As compared to network administration in fixed network elements, wireless software is explanatory and user-friendly. For example, you do not need an IT manager to set up the security for accounts. Security management is more manageable because of the application of per-user, per-session keys. These keys make it easy to create and maintain security.

Q: How can the Pro Sports store in the case study increase the number of supported units?

A: The design has made the wireless network extensible. Using the current access points, you can add an additional radio to the warehouse access point. You can add another radio to one of the first-floor access points. These additions will add two more overlays of RF patterns. Since each radio can cover 50 units, you could add up to 100 more handheld devices or wireless PCs. These points assume that you are not going to buy additional hardware. You can also connect additional AP-1000 access points to the network to extend it further.

Q: What additional wireless technologies and improvements could be applied to the retail market?

A: Traditionally static products like printers, weight scales, and time clocks can be integrated with new technology. Symbol Technologies is a leader in this industry. Specialty products like IP video cameras also adapt well to wireless implementation. Other improvements benefit retail management. For example, consider that a store manager can access sales reports, current transactions, inventory, and employee scheduling from anywhere in the store. This effort brings new meaning to the management style "management by walking around."

Chapter 11

Designing a Wireless Home Network: Home Office Case Study

Solutions in this chapter:

- **Introducing the Wireless Home Network Case Study**

- **Designing the Wireless Home Network**

- **Implementing the Wireless Home Network**

- **Designing a Wireless Home Network for Data, Voice, and Beyond**

- **Lessons Learned**

☑ Summary

☑ Solutions Fast Track

☑ Frequently Asked Questions

Introduction

One of the most exciting applications for wireless technologies is the wireless home network. Home networks allow you to network PCs and other devices for peripheral and file sharing, online gaming, and shared Internet access. As new Internet-ready devices flood the marketplace and a whole new range of household, business, and entertainment services become available with expanded broadband access, a home network will become a must for many households. With a wireless home network, you will be free from the need to install wired connections where fixed Internet-ready devices are desired. You will also be able to control those devices as you move in and around your house.

The business-related advantages of a network are widely recognized, but most home PC users have not yet recognized the advantages a home network can provide. Online gamers have long been using networked PCs in the home to play multiplayer games. Small home office users, along with some other multi-PC families use home networks for peripheral, file, and Internet sharing. But for the most part, the possibilities of a home network, particularly a wireless home network, have yet to be tapped.

This chapter and its case study explore the possibilities of a wireless home network, both today and in the near future. It explains the potential benefits and the options available for the type of home network that will meet your needs and your budget.

Advantages of a Home Network

Already, the popularity of online music services has begun to demonstrate the potentials and the pitfalls of electronically distributed entertainment. As broadband access has expanded, we've also begun to see video-on-demand services appear on Internet sites. What's more, a proliferation of Internet-ready entertainment devices is hitting the marketplace. Set top boxes are currently available for using Web services from your analog television. Network-ready MP3 players are available for your home music systems. Moreover, the market will soon be flooded with a range of telephony products (fixed and mobile) that support both voice and Internet services. With technologies and services available today, you can control and distribute entertainment services throughout your home.

In the area of household automation, appliance makers are building or considering network-ready appliances of nearly every kind. Home network and Internet services are envisioned for virtually every type of kitchen appliance, as well as heating, cooling, and lighting systems. Services range from remote control and maintenance of your stove, washing machine, furnace, or coffee pot, to enhanced services, such as automated grocery lists generated from your refrigerator, or home security systems that will alert you remotely when they are activated. Appliance manufacturers envision a 21st century kitchen where many of today's routine household tasks can either be automated or remotely controlled, either in your home or out of it, from mobile wireless devices.

For the home office user, broadband services can now offer complete integration into corporate networks. Large business users can operate from the home with much the same security and access to network resources as those at work. Small business users are better able to host their own Web sites and will see a whole new array of small-business services.

Telecommunication and cable companies are integrating expanded home services into their broadband portfolios. Many complex services will require more than broadband Internet access, especially when voice and data integration are required or when access is required away from your home. Remote, mobile, and integrated access, whether to outside services when you are at home or to home devices when you are away, will be the next great achievement of the broadband industry.

> **NOTE**
>
> As service logic is developed in our telecommunication networks, services such as enhanced home security systems and automated shopping will begin to emerge.

Enhanced security systems will allow you to monitor your home, whether you are in it or away. With cameras mounted in and around your house, your home security system could feed full-motion video of what is currently happening in your home to your mobile wireless device. When notified that the alarm has sounded, you would be able to alert your security service or notify them of a false alarm. With connections to the lighting and entertainment subsystems of your network, you could make your lights flash to alert your neighbors to a burglary, or control lights and entertainment devices as if you were home, to prevent one.

Appliance manufacturers and grocers envision a day in the very near future where you'll use scanners on your refrigerator, your cabinets, and possibly even your trashcan to generate grocery lists. Using the refrigerator's video screen, you'll be able to edit your grocery list if necessary and send it to the supermarket for home delivery. Alternatively, you could use the screen to browse the Internet for recipes or store them for later retrieval.

Advantages of a Wireless Home Network

Although home networks can be created with wired technologies, wireless technologies offer far greater convenience and mobility than the wired options. Wireless networks are more convenient because they don't require the installation of new wires or new network access points where broadband services are desired. Even though new technologies for providing broadband data access over existing home wiring (telephone and even power lines) are becoming available, the convenience of wireless cannot be matched. Even if every power and telephone outlet in your home could become a potential broadband data port, wireless still offers the convenience of locating your Internet-ready devices in places where the physical outlets do not exist.

Even more advantageous, wireless networks allow you to use Internet-ready devices while mobile. Whether you want to move your laptop to your living-room couch, to your bed, or to your deck at the back of the house, a wireless network will let you move without the need to "plug in" to a new connection. Even better, you can stay connected while you are moving. This becomes particularly useful when using devices such as personal digital assistants (PDAs) or cell phones. With a wireless home network, you could have the power to control lighting, music, or other services while moving about your house, all from your hand-held control center.

Introducing the Wireless Home Network Case Study

The following case study illustrates the design of a simple home network intended for a home-office user. The user is interested in high-speed data services only and needs to build the network with technology available today. She has no immediate plans for expanding the network beyond her current home-office needs. This section will describe the user's current situation, a statement of her problem, her proposed solution, and how she implements her solution. It will also describe the lessons she learned during the process.

Assessing the Opportunity

Under doctor's orders for more bed rest, Jan received authorization from her employer to work from home during and immediately after her pregnancy.

However, to do her job effectively, she routinely needs to retrieve large files from the corporate local area network (LAN), modify them, and return them to another location on the LAN. She also occasionally likes to print something for convenience or record keeping, and she needs convenient access to a telephone. Jan has received instructions from her company's Information Technology (IT) staff regarding how to connect to the corporate LAN.

Jan currently has broadband access to her home. However, the only access is to a PC in a family room in the home's finished basement. Jan's family uses this PC for Internet access, online gaming, and as a resource for school projects. The PC is connected to a color printer.

Jan wants to create a home office in an unused upstairs bedroom. During the later stages of her pregnancy, she wants the convenience of working from more comfortable locations, such as her couch or bed. She already has a cordless phone and is planning to purchase a laptop PC and perhaps a second printer. However, a quick call to her broadband provider has caused her to question the financial feasibility of running new wiring for broadband access in other locations of her home. Perplexed with her problem, Jan talks to some of her coworkers, and one of them mentions wireless. Jan does a little investigation of the wireless LAN products available and decides many of the products are within her budget.

Defining the Scope of the Case Study

The scope of Jan's solution will be limited by the fact that she already has broadband access installed in her home. She also has instructions for connecting to the corporate LAN. However, she has not checked to see if the wireless home network will affect these instructions.

Jan's challenge is that she needs reliable high-speed access to the corporate LAN from the new home office and other convenient locations in the house. She wants to interfere as little as possible with use of the PC in the family room, and she needs her laptop to be inaccessible from the family PC. She would like to have printing capability in the home office. However, her solution must fit within a limited budget. Finally, all equipment that she uses for her solution must be immediately available.

Designing the Wireless Home Network

This section explains how Jan determines the need for, plans, designs, and implements a wireless home network. As a part of these processes, Jan learns more about the strengths and weaknesses of wireless networks, and about the costs and advantages of different vendor solutions. The processes she follows are:

- Determining the requirements

- Analyzing the existing environment

- Creating a preliminary design

- Developing a detailed design

- Implementing the network

Using this design methodology, Jan decides to conduct her investigation as if she is designing a network for a business, making appropriate changes as the situation warrants. Jan begins her investigation by performing the following tasks:

- Determining the functional requirements of her manager and family

- Talking to her company's IT staff

- Drawing a physical map of her home

Determining the Functional Requirements

The actual users of Jan's home network will be Jan, her husband, and their children. Since Jan's manager will be auditing her work, she also feels that her manager must give her advice regarding what is expected. Jan works with her manager and family to define their expectations of the home network.

Determining the Needs of Management

At work, Jan discovers that her manager is concerned primarily about the security of the files she will be using. Will the security of the corporate LAN be compromised by the wireless connection? Can the home network be child-proofed? Based on this conversation, Jan decides that the connection between her laptop and the corporate LAN must be secure from the family computer and safe from Internet hackers. Another concern is risk mitigation—basically, what happens if Jan's laptop goes down? What backup procedure does Jan envision? Jan believes that she will copy her work to the company network on a daily basis. This practice should limit the amount of loss to a single day.

Determining the Needs of the Family

Though intrigued by the possibilities of a wireless network, Jan's husband is concerned primarily about the impact on the family's budget, and the future value of the new equipment. Although the home network will benefit Jan's employer, the employer will not finance any of Jan's home networking needs. Since home networks, and wireless technology in particular, are considered "new technology," he reasons that the costs will be significantly higher now than they will be in the future.

Jan and her husband are so afraid that the cost will be phenomenal that she limits her desires to the basic necessities. Since Jan plans to buy a new laptop PC and another printer, they want to hold the cost of the network to a few hundred dollars. They consider running wires to her home office themselves if that would be a less expensive alternative. Her husband even suggests, somewhat jokingly, moving the printer to Jan's office during the day and back to the family room for schoolwork at night. Although moving the printer is not practical, Jan considers moving the printer permanently to the home office. The children debate this idea because they frequently need the printer to print papers and book reports for school. The children also are concerned about how a network will affect the bandwidth for their online gaming.

Talking to the IT Department

Jan calls Diane, a network engineer in the company's IT department. Diane tells her that to secure her laptop from the family PC, she must purchase a wireless access point (AP) rather than network the PC and the laptop. Without the access point, the family PC would have to act as a server to the laptop, since the wired broadband connection is near that PC. However, with the wireless access point, Jan can make either PC the server, or even purchase an access point that would perform that function. She also needs the access point if she wants to connect any other devices wirelessly, such as the printer or another PC. Of course, in doing so, Jan needs to remember that each device requires a wireless network card.

Diane regards the security risks of the wireless LAN to be acceptable as long as Jan's browser uses standard encryption technology. Since the range of home wireless LANs on the market today is about 100 meters, she does recommend that Jan not make it well known outside of the office that she's using a wireless LAN for company business. Diane also recommends that Jan purchase a home firewall to protect her from Internet hackers over her broadband connection. However, she assures Jan that the wireless network will cause no serious configuration issues in connecting to the corporate LAN. The configuration steps will be the same.

Jan also discovers that one of her coworkers has a wireless network at home, so she talks to him about his experiences. He is largely happy with his home network. His brother connected it for him, so he can't say much about network design or the advantages of various vendor solutions. However, one problem he's had is that the network seems to cause a "popping and cracking" noise in his cordless phone. He has noticed that the noise is more serious when he is transferring data.

Creating a Site Survey of the Home

In preparing to conduct her site survey, Jan decides she needs to consider the following factors:

- Whether any locations where she wants to use the laptop will be more than 100 meters (over 300 feet) away from the access point.

- Whether any potential sources of interference will cause any problems with the network. The attention Jan's coworker brought to this issue made her realize that she needs to learn more about interference issues. Her cordless telephone will be important for her work activities.

Since the access point must be located near the Broadband connection, Jan decides to measure her house and create a diagram showing all the relevant distances. She also decides to note any sources of interference. With a little investigation, she finds that many cordless telephones do in fact experience interference from the current generation of wireless LANs. Even more, she finds that radio frequency (RF) leakage from microwave ovens also can cause wireless LANs to experience a loss in the data rate.

Assessing the Functional Requirements

Based on her preliminary investigation, Jan comes up with the following list of design considerations:

- She needs to purchase a wireless access point.

- She needs to purchase wireless network cards for any devices she may want to connect wirelessly.

- She should purchase a home firewall to protect from Internet hackers.

- The location of all wireless devices must remain within 100 meters of the access point for connectivity.

- She must consider sources of interference and their locations.

- The printer in the family room is used considerably.

- The wireless network will not cause any problems in configuring access to her corporate LAN.

Jan also completes a site map of her house showing the approximate location of the wireless accent point, all relevant dimensions of her house, and the types and locations of any interference sources. Her diagram is shown in Figure 11.1.

Figure 11.1 Jan's Site Map

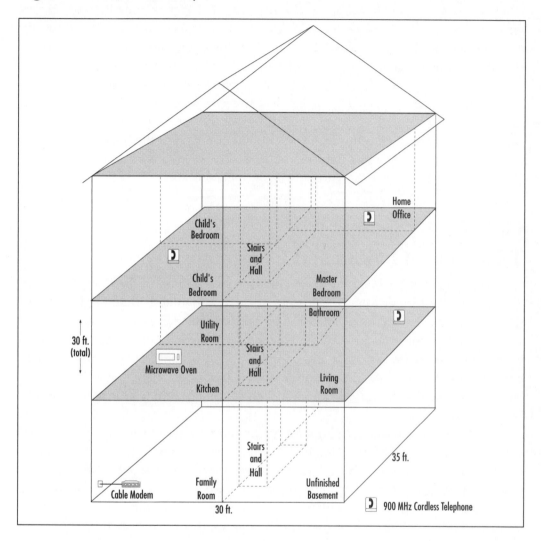

Analyzing the Existing Environment

The next step in Jan's design methodology is to analyze her existing environment. Her analysis includes the following processes:

- Identifying current technology options and constraints
- Investigating the costs
- Weighing the costs and benefits

Jan decides that her current applications consist mostly of entertainment and school content. She and her husband also occasionally use the family PC and printer for work. The two children both use the PC and printer for homework. By observing the computer use, Jan has discovered that the kids use the printer more than she had thought, so moving the printer from the family room is probably not a viable option. Both also use the Internet for various activities, including online games, so Jan is concerned about how that might affect her bandwidth while she is working.

Jan's existing network is simple. She has Broadband Internet access available in the family room only. The Broadband service is delivered into the house from a cable modem. From the cable modem, the service is wired to the family PC via a Category 5 Ethernet cable. The printer is a peripheral of the PC via the PC's serial port.

Identifying Current Technology Options and Constraints

Although Jan has already done some preliminary investigation of the technology, that investigation has led her to realize that she needs to know more. By using key words such as *wireless LAN* and *home networking* on her Internet searches, Jan is able to learn a considerable amount about both the benefits and drawbacks of wireless home networks. She also discovers that she can consider using existing telephone wires in her home to wire her network. Although she doesn't have a telephone outlet in her home office, she decides to add existing telephone outlets throughout her house to her site map.

Jan learns that she has several options in configuring her wireless network. Some wireless access points can be configured as routers or Dynamic Host Configuration Protocol (DHCP) servers, whereas others simply bridge Internet traffic from the modem to a single device. She also learns that a home firewall can act as a server or router, making all other devices (including the wireless access point) clients of the firewall. Finally, she finds at least one vendor solution that serves as both firewall and wireless access point.

Jan also learns more about interference issues. She discovers that the current generation of home networks operates on the IEEE 802.11b specification. Thus, all current devices will suffer from interference from certain cordless phones. However, she discovers that the interference is limited to phones operating on the 2.4 GHz band. Phones using the 900 MHz band won't suffer from the same interference.

In addition to the interference from microwaves and 2.4 GHz phones, Jan learns that glass objects, particularly windows and mirrors, can reflect the wireless signal occasionally, causing some minor interference issues (from multiple reflections). She also discovers that dense material, such as concrete and metals, can block the signal.

In the area of security, Jan confirms Diane's risk assessment. Although the wireless signal can be intercepted anywhere within 100 meters of the wireless access point, encryption on her browser offers the same security as crossing the Internet.

Investigating Costs

While investigating her technology options, Jan has been noting the costs of various components. She now makes a more thorough cost investigation by documenting the expense for purchasing and installing any components she thinks she might use in her design. The cost factors Jan considers for three different network types are:

- **Completely Wired Solution** Jan uses the installation costs quoted by her Broadband provider for extending wired access, which are relatively high. She also investigates the cost of having her husband run the wire instead. Although the costs of this are low, the level of effort required is very high.

- **Completely Wireless Solution** In a totally wireless solution, Jan will need not only a wireless access point and wireless network cards for every device, but she will need to buy two new printers, since her current printer won't support the wireless card. The cost of the network-compatible printers and network cards make this solution even more expensive than paying for wire installation.

- **Hybrid Wired/Wireless Solution** This solution seems to offer the most cost-effective approach. By maintaining her current wired connection to the family PC, Jan can achieve her primary requirements with the purchase of only two new components: a wireless access point and a wireless network card for the laptop PC. However, if she wants to connect her home-office printer wirelessly, the cost will be relatively high.

Weighing Costs and Benefits

Given her investigation of costs, Jan is now prepared to weigh the costs and benefits of various designs for her network and review them with the only other decision maker: her husband. Although a wired solution would be inexpensive if they do the wiring themselves, they decide that it's probably beyond their expertise to install the wiring in an inconspicuous fashion. A wired solution would also offer Jan less convenience and no mobility.

A completely wireless solution, though offering the maximum in mobility and convenience, is far beyond their budget for the project, mostly due to the cost of the wireless printers. This solution also goes well beyond the family's network needs. There is not really any reason to make the family PC and printer wireless components. They serve their functions well where they are.

They agree that the best choice is probably a hybrid wired/wireless network. The wireless network can be purchased inexpensively, and it offers the convenience and mobility that Jan considers the most important of her requirements. Adding a wireless printer in the home office will probably be outside of their budget, but Jan decides convenient printing is a less important requirement. She figures she'll still have access to the printer in the family room, or she can buy a standard printer for the home office and connect it to the laptop when she needs it.

Assessing the Existing Environment

Jan comes up with the following list of additional considerations and conclusions:

- She has determined her current applications and network design.

- She learned that she has some options for which device to use as her server.

- She has discovered that interference can also be caused by windows, mirrors, and dense metal or concrete objects.

- She has discovered that the wireless network's interference with cordless phones is limited to those operating in the 2.4 GHz band. There is no issue with phones that operate in the 900 MHz band.

- Through her cost/benefit analysis, she has decided that the most feasible design is probably a hybrid wired/wireless solution.

- She is still uncertain about exactly how she will solve her printing problem.

Jan also updates her site map to identify her existing network, the location on the network where particular applications are used, and additional sources for potential interference. Jan's updated site map is shown in Figure 11.2.

Figure 11.2 Jan's Updated Site Map

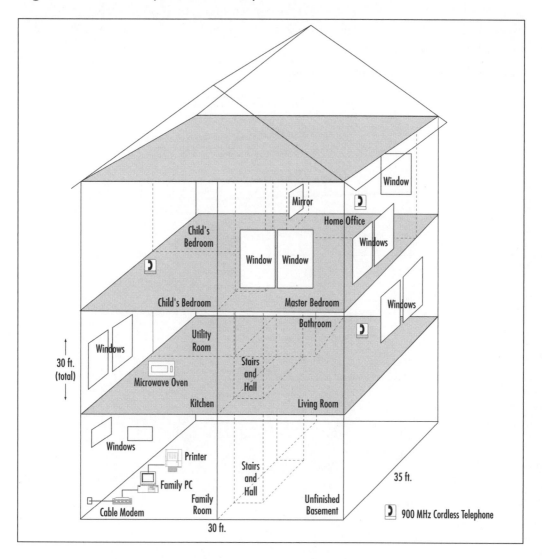

Developing a Preliminary Design

In this section, Jan plans her preliminary design and chooses the vendor solutions. Jan begins designing her network by drawing it out on her site map. She assumes for now that she will buy a combined firewall and wireless access point. Her initial network design is shown in Figure 11.3. Even though her network is very simple, Jan quickly realizes the benefit of drawing it out. First, she sees that her wireless access point/firewall must serve as a DHCP server since multiple PCs will connect as clients to it. Secondly, the wireless access point/firewall must also have an Ethernet port for her family PC.

Figure 11.3 Jan's Preliminary Design

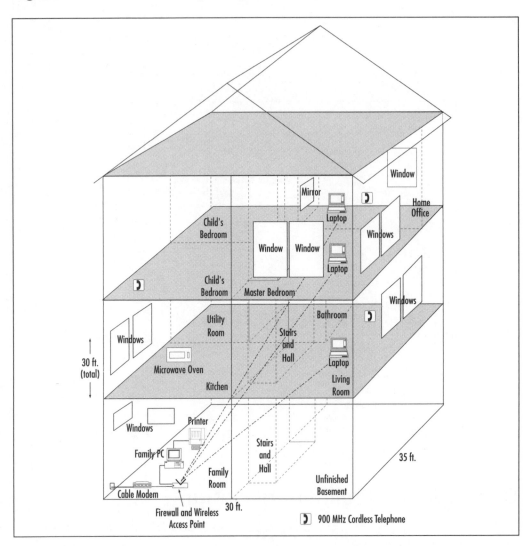

In reviewing her site map, Jan decides that the only serious interference source (the microwave oven) is far enough away from her access point and the places she will be using the laptop that it won't be a significant problem. However, the telephone connection near the family PC causes her to recollect that on one occasion in the past, her Broadband connection went down for several days. If this were to happen again, the only way she could work at all would be to dial up to the corporate LAN using a 56 K modem. She decides she should consider support for dial-up access as an additional backup requirement for her network.

Choosing Vendor Solutions

Finally, Jan considers her options with various vendor solutions. Based on her previous investigation and analysis, her requirements for the wireless access point are as follows:

- Support for DHCP
- Support of both wireless and Ethernet
- Firewall protection
- v90 modem support

Based on product reviews, prices, and product features, Jan narrows her options to the following two solutions:

1. Linksys Wireless Access Point and Home Firewall. This product supports DHCP, provides firewall protection, and has multiple Ethernet ports. However, it does not provide a modem port.

2. Agere Systems Orinoco RG1000 Wireless Access Point. This product supports DHCP and does provide a modem port. It doesn't provide firewall protection for the wired computer and doesn't have any Ethernet ports. However, Jan can purchase a separate home firewall from Linksys, which will provide both the firewall function and the Ethernet port.

Although the first solution will be somewhat less expensive and will be contained in a single box, Jan decides that she is concerned enough about losing her Broadband connection that the analog modem support is worth paying a little more. She therefore chooses the second solution.

The RG1000 requires that the wireless network card for her laptop support 64-bit encryption. She is also concerned that with a new technology, she should use the same vendor as she does for the access point. She therefore decides to purchase her wireless network card from Agere Systems as well.

Developing a Detailed Design

Jan purchases her products and makes her final considerations. She updates her site map to show the final components, and she considers her configuration options. For configuration, her primary consideration is which devices should implement DHCP.

Since the firewall will have multiple clients, she decides it should implement DHCP. The access point, on the other hand, can serve simply as a bridge between the laptop and the firewall. She decides to disable DHCP on it. Jan's detailed design is shown in Figure 11.4.

Figure 11.4 Jan's Detailed Design

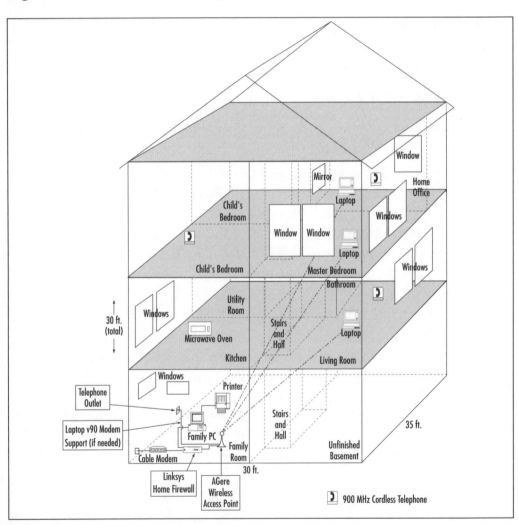

Implementing the Wireless Home Network

This section describes, at a high level, how Jan builds her home network. Jan approaches the implementation by:

- Assembling the network components
- Determining Broadband configuration
- Installing the hardware
- Installing and configuring the software
- Testing the network

Assembling the Network Components

Having planned and designed her network, Jan purchases the following components:

- One Agere Systems Orinoco RG1000 Wireless Gateway
- One Linksys BEFSR41 4-Port, 10/100Mbps Home Firewall
- One Agere Systems Orinoco Silver PCMCIA Wireless Network Card for her laptop PC
- One Dell laptop PC with open PCMCIA slot with Windows 2000 installed and an open parallel port
- Two short Category 5 Ethernet cables

The other components of Jan's network that she already owns are:

- One fully equipped Gateway PC with Ethernet network card and Windows 98 installed
- One Hewlett-Packard color printer with parallel port and cable

Jan assembles all of the components in her basement family room since all of her network installation and configuration can be done from there.

Determining Broadband Configuration

Jan begins by reading the instructions for all the components of her network. She discovers that before installing her network, she needs to know whether her existing PC is given a static IP (Internet Protocol) address or whether her Broadband provider supplies her a dynamic address from their DHCP server. Whichever the case, Jan will need to set her firewall to the same setting. To determine her Broadband settings, Jan completes the following procedure:

1. From the Windows Start menu, she selects **Settings | Control Panels**.

2. In the Control Panel window, she selects the **Network** icon.

3. In the Network Properties window (**Configuration** tab), she selects **TCP/IP** and then the **Properties** button.

4. In the TCP/IP Properties window, shown in Figure 11.5, Jan sees that her IP address is dynamically assigned to her PC (**Obtain an IP address automatically** is checked). Thus, she knows to configure her firewall in the same fashion.

5. Jan closes the TCP/IP Properties window and all other windows without making any changes.

Figure 11.5 TCP/IP Properties Window with Dynamic IP Address

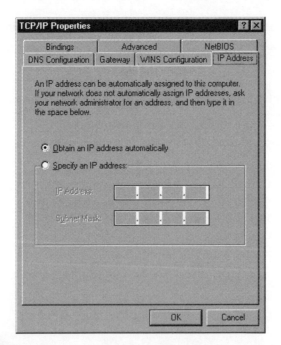

Had the other option (**Specify an IP address**) been selected, Jan would have needed to take note of her IP Address and Subnet Mask information from the IP Address tab of the TCP/IP Properties window. She would have also needed her Gateway and WINS Configuration information from the TCP/IP Properties window. However, most Broadband service providers set up services using a DHCP server, as Jan's did.

Installing the Hardware

Jan decides to install the hardware. With the exception of the software for the wireless network card, she installs all of the network's hardware before doing any software installation or configuration. She uses the following procedure:

1. Jan powers on the laptop and inserts the wireless network card in the PCMCIA slot. Windows recognizes the hardware and offers to configure it for her. She declines and instead uses the software accompanying her network card.

2. She shuts down and disconnects electrical power from all other network components except the cable modem. Some cable and DSL service providers recommend that you do not disconnect the power supply from their network devices. They may be grounded against electrical storms through the power line.

3. She disconnects the family PC from the cable modem (at the cable modem), leaving the Ethernet cable attached to the PC.

4. Using one of the new Category 5 cables, she connects the cable modem to the **In** Ethernet port on the home firewall.

5. Using the other new Category 5 cable, she connects an **Out** Ethernet port on the firewall to the **In** port on the wireless access point.

6. She connects the Ethernet cable from the family PC to another **Out** port on the home firewall.

7. She connects (or reconnects) all wired network components to electrical power.

Installing and Configuring the Software

Jan now begins to install software and configure her network. To install and configure the software for the firewall and wireless access point, she needs to use a computer that is directly attached to them. She therefore uses the family PC to configure the firewall and the laptop to configure the wireless access point. Jan follows the procedures outlined in the following section.

Installing and Configuring the Software for the Home Firewall

The Linksys instructions indicate their software is configured directly to the firewall through a Web interface. Following the instructions, Jan performs the following steps (note that her installation and configuration are particular to her situation):

1. She turns on both the home firewall and the family PC.

2. She opens a Web browser and enters the default IP address into the browser of **http://192.168.1.1**.

3. She enters the default user and password (no user and admin).

4. Once she has logged into the firewall, she sees the setup page shown in Figure 11.6.

Figure 11.6 Jan's Completed Setup Page

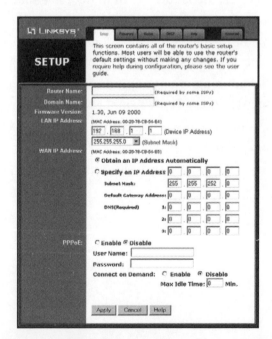

5. Jan follows the instructions provided by Linksys to set up features of her firewall. She accepts the defaults on most features, such as whether to use Network Address Translation (NAT). Typically, the defaults are the most desirable configuration.

The most important configuration features for Jan to consider are how the firewall communicates with her Broadband provider and how her network devices communicate with her firewall. Since she checked the configuration information on her PC earlier, Jan knows her Broadband provider supplies a dynamic (DHCP) address rather than a static IP address. She therefore needs to instruct her firewall to **Obtain an IP address automatically**. Jan also knows she wants to set up her firewall as a DHCP server to the rest of her network. She therefore wants to **Enable** DHCP on the firewall. These are typically the default settings for these two features, since most users will want this configuration.

Jan's completed Setup page for the Linksys configuration software is shown in Figure 11.6. Note that she has selected **Obtain an IP Address Automatically**. She has also left the settings for the LAN IP address and Subnet Mask at the defaults supplied by the software. These two addresses are supplied by default because Jan "enabled" DHCP on an earlier page. They define how the firewall will communicate with its DHCP clients.

Installing and Configuring the Software for the Wireless Access Point

As with the Linksys firewall, the Agere Orinoco instructions indicate their software must be installed on a PC to configure it. In this case, Jan installs the software for the wireless access point on the laptop since it has a connection (a wireless one) to the access point.

NOTE

Because Jan purchased her wireless network card from the same vendor as the wireless access point, her laptop was configured to communicate with the access point during the network card installation. Had she used a different vendor, she would have needed to set the Subnet Mask in the laptop's TCP/IP Properties to the correct Subnet Mask for the wireless access point.

Jan follows this procedure to configure the wireless access point:

1.   She turns on both the wireless access point and the laptop PC.

2.   When her Windows desktop has appeared on the PC, Jan inserts the Orinoco CD and installs the software.

3.   She selects the Custom installation because she does not want to enable DHCP on her wireless access point. There is no reason to set up DHCP since she has only one wireless device. If she were creating a complex network, Jan might want to create a wireless subsystem (most likely for security reasons). In that case, she would enable DHCP.

The key window in the custom installation is Network Topology. The three tabs of this window define how the wireless access point communicates with the firewall and the laptop as follows:

■   On the DHCP Server tab, Jan leaves the boxes unchecked because she does not want the access point to act as a server (see Figure 11.7).

Figure 11.7 Network Topology DHCP Server Tab

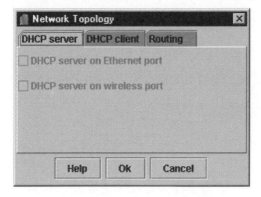

■   On the DHCP Client tab, she identifies the access point as a DHCP client of the firewall attached to its Ethernet port (this performs the same function as selecting Obtain an Address Automatically in Windows and the Linksys software). This is shown in Figure 11.8.

Figure 11.8 Network Topology DHCP Client Tab

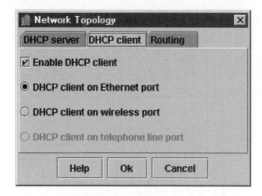

- On the Routing tab, she identifies the access point as a bridge, as shown in Figure 11.9. Notice she does not enable NAT because it was already enabled.

Figure 11.9 Network Topology Routing Tab

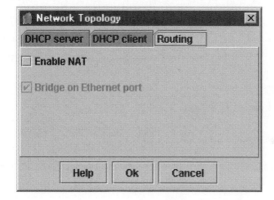

Testing the Network

After completing her installation and configuration, Jan tests everything to make sure she has all the functionality she was expecting. She verifies that both her family PC and her laptop have Internet access. After configuring her corporate LAN access, she verifies that it is functional. Access from her laptop doesn't appear to be affected by where she is in the home, except she thinks it may be slower when she's in the kitchen with the microwave oven in operation.

Jan realizes at this point that she isn't able to access her family PC from the laptop. Thus, she can't reach the family printer either. When she is unable to understand why, she calls her IT contact Diane. Diane explains that for a peer-to-peer session on her LAN, she would need to configure the NetBEUI protocol in Windows (which creates the Network Neighborhood). However, Diane also explains that this would expose her laptop to the family PC. She could password-protect her laptop, but it would still be an unacceptable security risk for the company's information. Jan therefore decides to just buy a printer for the home office and connect it to the laptop when she needs it.

Designing a Wireless Home Network for Data, Voice, and Beyond

Another colleague and friend of Jan's, called Dennis, is very excited by the convenience and mobility offered by Jan's wireless network. Dennis is an audio and video enthusiast and is interested in using a home network to create an audio/video server on his home PC. He begins reading about home networks in general, and wireless home networks in particular. He learns that he easily could build his audio/video server with an existing IEEE 802.11b-based home LAN, a network-ready MP3 player for his home sound system, and a network-ready set top box connected to his existing TV. However, Dennis also discovers that there may be good reason for him to wait just a little while.

Current State of the Home Wireless Marketplace

First, Dennis learns that there are currently three wireless standards competing for the wireless home-network space: IEEE 802.11b, Bluetooth, and HomeRF. Unfortunately, the technologies are, for the most part, incompatible, and it is still unclear which will eventually emerge as the technology (or technologies) of choice. However, the capabilities of each are beginning to suggest some trends.

Products based on the IEEE 802.11b standard have been available for some time, particularly for business applications. The major drawbacks of 802.11b products are their interference with 2.4 GHz phones and the fact that they support data only (no native voice integration). New products based on the IEEE 802.11a standard will be emerging in the near future. These products will support an even higher bandwidth and will not interfere with the cordless phones. Major players in the industry, such as Intel and Microsoft, are currently moving toward adoption of the 802.11a standards. However, the lack of an integrated voice signal in these standards severely restricts their applications.

Bluetooth is another standard that is likely to find a place in the home network marketplace. Bluetooth provides for voice and data integration. However, it currently operates on Class 2 devices, and will therefore be limited to bandwidths under 1 Mbps. Bluetooth devices will most likely be limited to voice and command-and-control services. However, its strength is in merging the home and public network spaces. Bluetooth devices are a likely solution for control of home devices when at work or in a public space such as an airport or retail establishment with a wireless public network.

Probably the most exciting of today's home wireless technologies are based on the HomeRF 2.0 standard. HomeRF 2.0 delivers up to 10 Mbps of bandwidth for data. But even better, it provides a fully integrated 2.4 GHz voice signal with up to 8 high-quality 2.4 GHz voice channels and all the Custom Local Area Signaling Service (CLASS) calling features like call waiting and caller ID. The HomeRF standard also uses a *frequency hopping* technology that avoids interference with existing 2.4 GHz devices. It will also likely provide greater security from someone intercepting your RF signal.

The key advantages of the HomeRF 2.0 standard are that it integrates the voice and data channels over the same wireless transport protocol, handles multimedia streams effectively, and supports synchronous full-duplex voice traffic. Because the voice and data signals are integrated, products using the HomeRF standard should find voice recognition and automation applications easier to develop and support.

Products based on the HomeRF 2.0 standard will likely be emerging in the second half of 2001. Siemens has been working closely with Proxim (the HomeRF 2.0 chipset manufacturer) to integrate HomeRF 2.0 and the Digital Enhanced Cordless Telephone (DECT) specification natively. It is expected that Siemens will leverage these integrated capabilities to support new and innovative products.

Designing & Planning...

Home Networking Technologies

Although wireless offers the greatest convenience and mobility for home networking products, it is certainly not the only solution for building a home network. Products using Home Phoneline Networking Alliance (HPNA) standards are currently on the market, which allow you to use existing Category 3 telephone lines in your home to deliver your existing voice signal and up to 10 Mbps of data simultaneously. Similarly, power-line technologies are emerging that will carry even larger data band data simultaneously. Similarly, power-line technologies are emerging that will carry even larger data bandwidths over your electrical power lines (simultaneously with the electrical power).

Most likely, all of these technologies eventually will be used in the home network. Fixed devices with ready access to an electrical plug-in may use power-line technology, whereas mobile devices or those you move frequently may use wireless. The access method you'll want to use for any given device will probably be determined by the network access points available in the locations where you expect the device to reside.

A key question in all of this is where the network hub will reside. Most likely, you will want to have control of many of your devices from a single mobile device such as a cell phone or PDA. However, for security reasons, you will also want to have network subsystems (requiring DHCP servers) for general categories of devices (for example, heating and cooling, lighting, kitchen appliances, and entertainment devices).

Two major players are emerging in the command and control arena: Microsoft's Universal Plug and Play (UPnP) and Sun Microsystems' Genie. Not surprisingly, Microsoft's approach is "PC-centric," meaning a PC will serve as the central hub and quite likely as the servers for the various subsystems. Sun's approach, on the other hand, is device-centric, meaning that a wide array of devices could serve these functions. Which solution will win the battle and which device will be the central hub remains to be seen, but the solution should certainly become apparent in just a few years.

A Proposed Solution for the Future

Jan's colleague Dennis was initially planning to use a wireless home network just to build an audio/video server. However, after learning more about the home networking marketplace and its future, he decides he would rather choose a solution that will be expandable to meet his future home-networking needs. He therefore decides to apply Jan's design methodology to his own situation with an eye to the future. Although the technology is not yet available for much of the home automation possibilities, Dennis decides to include these in his preliminary design to better decide which products he will eventually purchase.

Dennis completes the same investigation, analysis, and design process as Jan did. The preliminary design that Dennis develops is shown in Figure 11.10.

Figure 11.10 Dennis's Preliminary Design

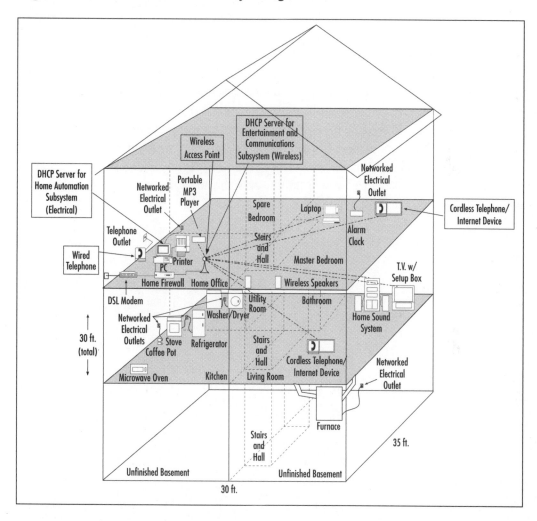

Lessons Learned

Both Jan and Dennis have learned a number of lessons while planning, designing, and building their home networks. First, Jan has learned that no matter how much she investigates the technology and plans her network, there is always something new to learn and there will always be missteps. Even though a simple home network can be built without a lot of difficulty, there is always another technology to consider.

Jan has learned that the processes outlined in her methodology are not as neat and clean as they may first appear. As with any technology, there are drawbacks (such as interference and technology incompatibilities) that must be considered and a complex array of choices to be made.

Both Jan and Dennis have learned that there are considerable risks in purchasing any home wireless technology today, given that it is likely to become obsolete as new technologies, services, and applications become available. The key is to determine as much as possible your immediate and long-term needs, the costs you are willing to incur for various benefits, and the solutions that will address these issues most effectively. However, just as with other emerging technologies, the chances are that the products available today will have a very limited capacity to address the needs of future services and applications.

Summary

Along with other home networking technologies, wireless home networking is seeing an explosion of activity. Home networking applications are foreseen for nearly every electronic device in your home. The advantages of wireless are that it offers you mobility and is available in locations where a wired outlet doesn't exist. The disadvantages include some minor interference problems and a quickly evolving and competitive group of standards.

Building a simple home network is a relatively easy and inexpensive task that can be completed with technology available today. However, as with any network, you should complete a thorough process of investigation, analysis, and design before purchasing any wireless networking solutions.

Preparing for the more complex wireless home networks of the future can be a more daunting task and will require a more thorough investigation of emerging technologies.

Solutions Fast Track

Introducing the Wireless Home Network Case Study

☑ New technologies and services include online music services, video-on-demand, and Internet-ready entertainment devices. Companies are also networking kitchen appliances, heating and cooling appliances, lighting, and security functions.

☑ This case study explores the user's opportunities in designing and implementing a home network with high-speed access to the company LAN and to a printer. The user in the case study has an existing PC with broadband access that is for family use only, and is located far from the home office.

Designing the Wireless Home Network

☑ The functional requirements include the security of files passed to the company's LAN, the family budget, and the accessibility of the home office.

☑ The user identifies her options, investigates costs, and weighs the results.

☑ In the preliminary design, the user checks for interference, opts for a DHCP server, and reviews the access points within her budget.

Implementing the Wireless Home Network

☑ In the detailed design, the user assembles the network components, determines the Broadband configuration, installs the hardware, installs and configures the software, and tests the resulting network.

Designing a Robust Wireless Home Network for Data, Voice, and Beyond

☑ A user in a related case study is excited by the convenience and mobility of a wireless home network. He wants to create an audio/video server on his home PC.

☑ Currently there are three wireless standards competing for the wireless home-network space: IEEE 802.11b, Bluetooth, and HomeRF. The technologies are, for the most part, incompatible, and it is still unclear which will eventually emerge as the technology (or technologies) of choice.

☑ The major drawbacks of 802.11b products are their interference with 2.4 GHz phones and the fact that they support data only (no native voice integration). New products based on the IEEE 802.11a standard will be emerging in the near future. These products will support an even higher bandwidth and will not interfere with the cordless phones.

☑ Because the voice and data signals are integrated, products using the HomeRF standard should find voice recognition and automation applications easier to develop and support.

Lessons Learned

☑ Building a simple home network is a relatively easy and inexpensive task that can be completed with technology available today. However, as with any network, you should complete a thorough process of investigation, analysis, and design before purchasing any wireless networking solutions.

☑ The advantages of wireless are that it offers you mobility and is available in locations where a wired outlet doesn't exist.

☑ The disadvantages of wireless include some minor interference problems and a quickly evolving and competitive group of standards.

Frequently Asked Questions

The following Frequently Asked Questions, answered by the authors of this book, are designed to both measure your understanding of the concepts presented in this chapter and to assist you with real-life implementation of these concepts. To have your questions about this chapter answered by the author, browse to **www.syngress.com/solutions** and click on the **"Ask the Author"** form.

Q: What wireless devices are available today for use in the home?

A: Today's wireless devices range from PCs, printers, and scanners, to MP3 players for your home stereo system, to set top boxes for your analog television. You may also soon be seeing integrated data and telephony devices. Some of these may allow for voice command of other networked devices.

Q: How much will a wireless network cost me to build?

A: A simple wireless network can be built today for only a few hundred dollars. However, the newer the technology you use, the more likely the expense is to rise. Wireless peripherals such as printers and scanners are still expensive, and other new devices that emerge will likely be expensive at first, too. You could easily spend several thousand dollars networking just the devices available today.

Q: Is a wireless home network the right technology for me?

A: Most home networks will eventually be a patchwork of wireless and wireline technologies. However, any device that is mobile or located where no Broadband connection is readily available will require a wireless connection. Almost certainly, you will want a mobile device for command and control, which will require a wireless connection.

Q: How can I find out more about wireless home networking?

A: The Internet is probably the easiest and least expensive source. All of the major standards bodies have Web sites. Product manufacturers are, of course, advertising their existing products on the Web as well, and the major technology publications are also reviewing wireless (and wireline) home networks.

Chapter 12

Wireless Penetration Testing

Solutions in this chapter:

- WLAN Vulnerabilities, Discovery, and Encryption
- Wired Equivalent Privacy (WEP
- WiFi Protected Access (WPA/WPA2)
- Extensible Authentication Protocol (EAP)
- Virtual Private Network (VPN)
- Attacks Against WEP, WPA, LEAP and VPN
- USENET Newsgroups
- Google (Internet Search Engines)
- Wellenreiter
- Kismet
- MAC Address Spoofing
- Deauthentication with Void11
- Cracking WEP with the Aircrack Suite
- Cracking WPA with the CoWPAtty

Introduction

This chapter discusses performing wireless penetration tests using the Auditor Security Collection, which is a Live-System based on KNOPPIX containing over 300 open source security tools. It is distributed freely under the GPL 2.0 license. With no installation whatsoever, the analysis platform is started directly from the CD-ROM and is fully accessible within minutes. The Security Auditor Collection and the newly created *BackTrack* can be downloaded for free from **www.remote-exploit.org**. After reading this chapter, you will be able to identify your specific WLAN target and determine what security measures are being used. Based on that information, you will be able to assess the probability of successfully penetrating the network, and determine the correct tools and methodology for successfully compromising your target.

The Auditor Security Collection provides an incredible suite of wireless network discovery and penetration test tools. To perform successful penetration tests against wireless networks, you need to be familiar with the use of many of these tools and their specific roles in the pen testing process.

To attack your target network, you first need to find your target network. Auditor provides two tools for wireless local area network (WLAN) discovery: Kismet and Wellenreiter

After penetration testers have located the target network, many options are open to them, and Auditor provides many of the tools necessary to accomplish attacks based on these options.

Change-Mac can be used to change your client's Media Access Control (MAC) address and bypass MAC address filtering. Both Kismet and Ethereal can be used to determine the type of encryption that is being used by your target network, and can capture any clear text information that may be beneficial to you during your penetration test.

Once you have determined the type of encryption in place, several different tools provide the capability to crack different encryption mechanisms that may be in place. Void11 is used to de-authenticate clients from the target network. The Aircrack suite (Airodump, Aireplay, and Aircrack) allows you to capture traffic, reinject traffic, and crack WEP keys. CoWPAtty performs offline dictionary attacks against WPA-PSK networks.

Approach

Before beginning a penetration test against a wireless network, it is important to understand the vulnerabilities associated with WLANs. The 802.11 standard was developed as an "open" standard; in other words, when the standard was written, ease of accessibility and connection were the primary goals. Security was not a primary concern, and security mechanisms were developed almost as an afterthought. When security isn't engineered into a solution from the ground up, the security solutions have historically been less than optimal. When this happens, there is often multiple security mechanisms developed, none of which offers a robust solution. This is very much the case with wireless networks as well.

Understanding WLAN Vulnerabilities

WLAN vulnerabilities can be broken down into two basic types: vulnerabilities due to poor configuration, and vulnerabilities due to poor encryption

Configuration problems account for many of the vulnerabilities associated with WLANs. Because wireless networks are so easy to set up and deploy, they are often deployed with either no security configuration or inadequate security protections. An open WLAN, one that is in default configuration, requires no work on the part of the penetration tester. Simply configuring the WLAN adapter to associate to open networks allows access to these networks. A similar situation exists when inadequate security measures are employed. Since WLANs are often deployed because of management buy-in, the administrator simply "cloaks" the access point and/or enables MAC address filtering. Neither of these measures provides any real security, and both are easily defeated by a decent penetration tester.

When an administrator deploys the WLAN with one of the available encryption mechanisms, a penetration test can often still be successful because of inherent weaknesses with the form of encryption used. Wired Equivalent Privacy (WEP) is flawed and can be defeated in a number of ways. WiFi Protected Access (WPA) and Cisco's Lightweight Extensible Authentication Protocol (LEAP) are vulnerable to offline dictionary attacks.

Evolution of WLAN Vulnerabilities

Wireless networking has been plagued with vulnerabilities throughout its short existence. WEP was the original security standard used with wireless networks. Unfortunately, when wireless networks first started gaining popularity, researchers discovered that WEP was flawed. In their paper, "Weaknesses in the Key Scheduling Algorithm of RC4" (www.drizzle.com/~aboba/IEEE/rc4_ksaproc.pdf), Scott

Fluhrer, Itsik Mantin, and Adi Shamir detailed a way in which attackers could potentially defeat WEP because of flaws in the way WEP employed the underlying RC4 encryption algorithm.

Attacks based on this vulnerability (dubbed "FMS attacks" after the first letter of the last names of the paper's authors) started to surface shortly thereafter, and several tools were released to automate cracking WEP keys.

In response to the problems with WEP, new security solutions were developed. Cisco developed a proprietary solution, LEAP for its wireless products. WPA was also developed to be a replacement for WEP. WPA can be deployed with a pre-shared key (WPA-PSK) or with a RADIUS server (WPA-RADIUS). The initial problems with these solutions were that LEAP could only be deployed when using Cisco hardware and WPA was difficult to deploy, particularly if Windows was not the client operating system—an issue that exists to this day. Although these problems existed, for a short while it appeared that security administrators could rest easy. There were secure ways to deploy wireless networks.

Unfortunately, that was not the case. In March 2003, Joshua Wright disclosed that LEAP was vulnerable to offline dictionary attacks and shortly thereafter released a tool that automated the cracking process. WPA, it turns out, was not the solution that many hoped it would be. In November 2003, Robert Moskowitz of ISCA Labs detailed potential problems with WPA when deployed using a pre-shared key in his paper, "Weakness in Passphrase Choice in WPA Interface." This paper detailed that when using WPA-PSK with a short passphrase (less than 21 characters), WPA-PSK was vulnerable to a dictionary attack as well. In November 2004, the first tool to automate the attack against WPA-PSK was released to the public.

At this point, there were at least three security solutions available to WLAN administrators, but all were broken in one way or another. The attacks against both LEAP and WPA-PSK could be defeated by using strong passphrases and avoiding dictionary words. Additionally, WPA-RADIUS was (and is) still sound. Even the attacks against WEP weren't as bad as was initially feared. FMS attacks are based on the collection of weak initialization vectors (IVs). To collect enough weak IVs to successfully crack WEP keys required, in many cases, millions or even hundreds of millions of packets be collected. Although the vulnerability was real, practical implementation of an attack was much more difficult than many believed.

This didn't last for long. Even as the initial FMS paper was being circulated, h1kari of Dachboden labs detailed that a different attack, called "chopping" could be accomplished. Chopping eliminated the need for *weak* IVs to crack WEP, but rather required only *unique* IVs. Unique IVs could be collected much more quickly than weak IVs, and by early 2004, tools that automated the chopping process were released.

Because of the weaknesses associated with WEP, WPA, and LEAP, and the fact that automated tools have been released to help accomplish attacks against these algorithms, penetration testers now have the ability to directly attack encrypted WLANs. If WEP is used, there is a very high rate of successful penetration. If WPA or LEAP are used, the success rate is somewhat reduced. This is because of the requirement that the passphrase used with WPA-PSK or LEAP be included in the penetration tester's attack dictionary. Furthermore, there are no known attacks against WPA-RADIUS or many of the other EAP solutions that have been developed. In addition, WPA-PSK attacks are also largely ineffective against WPA2. The remainder of this chapter focuses on how a penetration tester can use these vulnerabilities and the tools to exploit them to perform a penetration test on a target's WLAN.

Core Technologies

To successfully pen test a wireless network, it is important to understand the core technologies represented in a decent tool kit. What does WLAN discovery mean and why is it important to us as pen testers? There are a number of different methods for attacking WEP encrypted networks, and why are some more effective than others? Is the dictionary attack against LEAP the same as the dictionary attack against WPA-PSK? Once a pen tester has an understanding of the technology behind the tool he is going to use, his chances of success increase significantly.

WLAN Discovery

There are two types of WLAN discovery scanners: active and passive. Active scanners rely on the SSID Broadcast Beacon to detect the existence of an access point. An access point can be "cloaked" by disabling the SSID broadcast in the beacon frame. While this does render active scanners ineffective, it doesn't stop a penetration tester, or anyone else for that matter, from discovering the WLAN. A passive scanner does not rely on the SSID Broadcast Beacon to detect that an access point exists. Rather, passive scanners require a WLAN adapter to be placed in *rfmon* (monitor) mode. This allows the card to see all of the packets being generated by any access points within range, and therefore allows access points to be discovered even if the SSID is not sent in the Broadcast Beacon.

When a passive scanner initially detects a cloaked access point, the SSID is usually not known (because it isn't included in the broadcast frame, as shown in Figures 12.1 and 12.2.

Figure 12.1 The SSID Is Broadcast (This Person Was Obviously Very Astute)

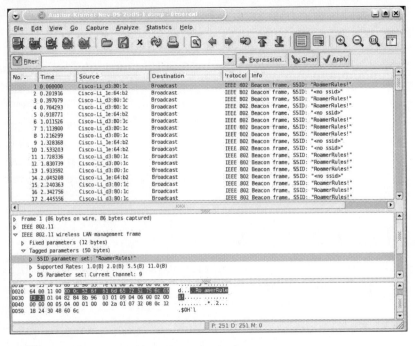

Figure 12.2 The SSID Is Not Broadcast

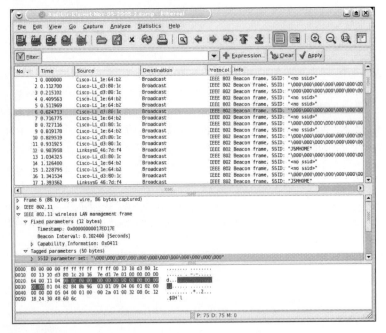

As you can see in Figure 12.2, the beacon frame is still sent, or broadcast, but the SSID is no longer included in the frame. This does not mean that you can't discover the SSID, however. When a client associates to the WLAN, even if encryption is in use, the SSID is sent from the client in clear text. Passive WLAN discovery programs can determine the SSID during this association.

Once we have identified the SSID of all wireless networks in the vicinity of our target, we can begin to hone in on our specific target.

Choosing the Right Antenna

To home in on a specific target, you need to choose the correct antenna for the job. While it is beyond the scope of this book to go into all of the possible antenna combinations, there are some basics truths to understand when choosing your antenna. If you are interested in gaining an in-depth understanding of antennas, check out the *ARRL Antenna Handbook* ISBN: 0872598047.

There are two primary types of antennas you want to be familiar with: directional and omni-directional. A directional antenna, as the name implies, can send and receive in a single direction, the direction the antenna is pointed. An omni-directional antenna, on the other hand, broadcasts and receives in all directions.

For WLAN discovery, an omni-directional antenna is usually the best initial choice, because we may not know exactly where our target is located. An omni-directional antenna provides us with data from a broader surrounding range. Note that with omni-directional antennas, bigger is not always better. The signal pattern of an omni-directional antenna resembles a donut. An antenna with a lower *gain* has a smaller circumference, but is taller. An antenna with a higher gain has a larger circumference, but is shorter. For this reason, when performing discovery in a metropolitan area, with tall buildings, an antenna with a lower gain is probably a better choice. If, however, you are performing discovery in a more open area, an antenna with a higher gain is probably the better option.

Once a potential target has been identified, switching to a directional antenna is very effective in helping to determine that the WLAN is our actual target. This is because with a directional antenna we can pinpoint the location of the WLAN and determine if it is housed in our target organization's facility. Directional antennas require line of sight, as do omni-directional antennas, and any obstructions (buildings, mountains, and so forth) reduce their effectiveness. Higher gain directional antennas are almost always a better choice.

WLAN Encryption

WLAN encryption has had a *lot* of bad press and unfortunately has fallen flat on its face many times. There are four basic types of "encryption" with which pen testers should be familiar:

- Wired Equivalent Privacy (WEP)
- WiFi Protected Access (WPA/WPA2)
- Extensible Authentication Protocol (EAP)
- Virtual private network (VPN)

Wired Equivalent Privacy (WEP)

WEP was the first encryption standard available for wireless networks. WEP can be deployed in two strengths, 64 bit and 128 bit. 64-bit WEP consists of a 40-bit secret key and a 24-bit initialization vector, and is often referred to as 40-bit WEP. 128-bit WEP similarly employs a 104-bit secret key and a 24-bit initialization vector and is often called 104-bit WEP. Association with WEP encrypted networks can be accomplished through the use of a password, an ASCII key, or a hexadecimal key. WEP's implementation of the RC4 algorithm was determined to be flawed, allowing an attacker to crack the key and compromise WEP encrypted networks.

WiFi Protected Access (WPA/WPA2)

WPA was developed to replace WEP because of the vulnerabilities associated with it. WPA can be deployed either using a pre-shared key (WPA-PSK) or in conjunction with a RADIUS server (WPA-RADIUS). WPA uses either the Temporal Key Integrity Protocol (TKIP) or the Advanced Encryption Standard (AES) for its encryption algorithm. Some vulnerabilities were discovered with certain implementations of WPA-PSK. Because of this, and to further strengthen the encryption, WPA2 was developed. The primary difference between WPA and WPA2 is that WPA2 requires the use of both TKIP and AES, where WPA allowed the user to determine which would be employed. WPA/WPA2 requires the use of an authentication piece in addition to the encryption piece. A form of the Extensible Authentication Protocol (EAP) is used for this piece. There are five different EAPs available for use with WPA/WPA2:

- EAP-TLS
- EAP-TTLS/MSCHAPv2

- EAPv0/EAP-MSCHAP2
- EAPv1/EAP-GTC
- EAP-SIM

Extensible Authentication Protocol (EAP)

EAP does not have to be used in conjunction with WPA. There are three additional types of EAP that can be deployed with wireless networks:

- EAP-MD5
- PEAP
- LEAP

EAP is not technically an encryption standard, but is included in this section because of vulnerabilities associated with LEAP, which is covered later in the chapter.

Virtual Private Network (VPN)

A VPN is a private network that uses public infrastructure and maintains privacy through the use of an encrypted tunnel. Many organizations now use a VPN in conjunction with their wireless network. This is often accomplished by allowing no access to internal or external resources from the WLAN until a VPN tunnel is established. When configured and deployed correctly, a VPN can be a very effective means of WLAN security. Unfortunately, in certain circumstances, VPNs in conjunction with wireless networks are deployed in a manner that can allow an attacker (or a penetration tester) to bypass the security mechanisms of the VPN.

Attacks

Although several different security mechanisms can be deployed with wireless networks, there are ways to attack many of them. Vulnerabilities associated with WEP, WPA, and LEAP are well known. Although there are tools to automate these attacks, to be a successful pen tester, it is important to understand the tools that perform these attacks, and how the attacks actually work.

Attacks Against WEP

There are two different methods of attacking WEP encrypted networks; one requires the collection of weak IVs, and the other requires collection of unique IVs. Regardless of the method used, a large number of WEP encrypted packets must be collected.

Attacking WEP Using Weak Initialization Vectors (FMS Attacks)

FMS attacks are based on a weakness in WEP's implementation of the RC4 encryption algorithm. Fluhrer, Mantin, and Shamir discovered that during transmission, about 9,000 of the possible 16 million IVs could be considered "weak," and if enough of these weak IVs were collected, the encryption key could be determined. To successfully crack the WEP key, at least 5 million encrypted packets have to be collected in order to capture around 3,000 weak IVs. Sometimes, the attack can be successful with as few as 1,500 weak IVs, and sometimes it will take more than 5,000 before the crack is successful.

After weak IVs are collected, they can be fed back into the Key Scheduling Algorithm (KSA) and Pseudo Random Number Generator (PRNG) and the first byte of the key is revealed. This process is then repeated for each byte until the WEP key is cracked.

Attacking WEP Using Unique Initialization Vectors (Chopping Attacks)

Relying on the collection of weak IVs is not the only way to crack WEP. Although chopping attacks also rely on the collection of a large number of encrypted packets, a method of chopping the last byte off the packet and manipulating enables the key to be determined by collecting unique IVs instead.

To successfully perform a chopping attack, the last byte from the WEP packet is removed, effectively breaking the Cyclic Redundancy Check/Integrity Check Value (CRC/ICV). If the last byte was zero, then xor a certain value with the last four bytes of the packet and the CRC will become valid again. This packet can then be retransmitted.

Commonalities in the Attacks Against WEP

The biggest problem with attacks against WEP is that collecting enough packets can take a considerable amount of time—weeks or sometimes months. Fortunately, whether you are trying to collect weak IVs, or just unique IVs, you can speed this process up. Traffic can be injected into the network, creating more packets. This is usually accomplished by collecting one or more Address Resolution Protocol (ARP) packets and retransmitting them to the access point. ARP packets are a good choice because they have a predictable size (28 bytes). The response will generate traffic and increase the speed that packets are collected.

Collecting the initial ARP packet for reinjection can be problematic. You could wait for a legitimate ARP packet to be generated on the network, but again, this can take a while, or you can force an ARP packet to be generated. Although there are several circumstances under which ARP packets are legitimately transmitted (see www.geocities.com/SiliconValley/Vista/8672/network/arp.html for an excellent ARP FAQ), one of the most common in regard to wireless networks is during the authentication process. Rather than wait for an authentication, if a client has already authenticated to the network, you can send a *deauthentication* frame, essentially knocking the client off the network and requiring reauthentication. This process will often generate an ARP packet. After one or more ARP packets have been collected, they can then be retransmitted or reinjected into the network repeatedly until enough packets have been generated to supply the required number of unique IVs.

Attacks Against WPA

Unlike attacks against WEP, attacks against WPA do not require a large amount of packets to be collected. In fact, most of the attack can actually be performed without even being in range of the target access point. It is also important to note that attacks against WPA can only be successful when WPA is used with a pre-shared key. WPA-RADIUS has no known vulnerabilities so if that is the WPA schema in use at a target site, a different entry vector should be investigated.

To successfully accomplish this attack against WPA-PSK, you have to capture the four-way Extensible Authentication Protocol Over LAN (EAPOL) handshake. You can wait for a legitimate authentication to capture this handshake, or you can force an association by sending *deauthentication* packets to clients connected to the access point. Upon reauthentication, the four-way EAPOL handshake is transmitted and can be captured. Then, each dictionary word must be hashed with 4,096 iterations of the Hashed Message Authentication Code-Secure Hash Algorithm 1 (HMAC-SHA1) and two *nonce* values, along with the MAC addresses of the supplicant and the authenticator. For this type of attack to have a reasonable chance of success, the pre-shared key (passphrase) should be shorter than 21 characters, and the attacker should have an extensive wordlist at his disposal. Some examples of good wordlists can be found at http://ftp.se.kde.org/pub/security/tools/net/Openwall/wordlists/ and www.securitytribe.com/~roamer/WORDS.TXT.

Attacks Against LEAP

LEAP is a Cisco proprietary authentication protocol designed to address many of the problems associated with wireless security. Unfortunately, LEAP is vulnerable to an offline dictionary attack, similar to the attack against WPA. LEAP uses a modified Microsoft Challenge Handshake Protocol version 2 (MS-CHAPv2) challenge and response that is sent across the network as clear text, which allows an offline dictionary attack. MS-CHAPv2 does not salt the hashes, uses weak Data Encryption Standard (DES) key selection for challenge and response, and sends the username in clear text. The third DES key in this challenge/response is weak, containing five NULL values. Therefore, a wordlist consisting of the dictionary word and the NT hash list must be generated. By capturing the LEAP challenge and response, the last two bytes of the hash can be determined, and then the hashes can be compared looking for the last two that are the same. Once a generated response and a captured response are determined to be the same, the user's password has been compromised.

Attacks Against VPN

Attacking wireless networks that use a VPN can be a much more difficult proposition than attacking the common encryption standards for wireless networks. An attack against a VPN is not a *wireless attack* per se, but rather an attack against network resources using the wireless network.

Faced with the many vulnerabilities associated with wireless networking, many organizations have implemented a solution that removes the WLAN vulnerabilities from the equation. To accomplish this, the access point is set up outside the internal network and has no access to any resources, internal or external, unless a VPN tunnel is established to the internal network. While this is a viable solution, often the WLAN, since it has no access, is configured with no security mechanisms. Essentially, it is an open WLAN, allowing anyone to connect, the thought being that if someone connects to it, he or she can't go anywhere.

Unfortunately, this process opens the internal network to attackers. To successfully accomplish this type of attack, you need to understand that most, if not all, of the systems that connect to the WLAN are laptop computers. You should also understand that laptop computers often fall outside the regular patch and configuration management processes the network may have in place. This is because updates of this type are often performed at night, when operations will not be impacted. This is an effective means for standardizing desktop workstations; however, laptop computers are generally taken home in the evenings and aren't connected to the network to receive the updates.

Knowing this, an attacker can connect to the WLAN, scan the attached clients for vulnerabilities, and if one is found, exploit it. Once this has been accomplished, keystroke loggers can be installed that allow an attacker to glean the VPN authentication information, which can be used to authenticate to the network at a later time. This attack can only be successful if two-factor authentication is not being used. For instance, if a Cisco VPN is in use, often only a group password, username, and user password are required in conjunction with a profile file that can either be stolen from the client or created by the attacker. This type of attack can also be performed against any secondary authentication mechanism that does not require two-factor authentication or one-time-use passwords.

Open Source Tools

Are you tired of theory and background information yet? Ready to actually put some of these tools to use? Now is the time to figure out how we use the open source tools available to us to perform a penetration test against a wireless network.

Footprinting Tools

To successfully penetrate a wireless network, we need to understand the physical footprint of the network. How far outside the target's facility does the wireless network reach? The easiest way to accomplish this is by using Kismet in conjunction with GPSMap's "circle map" functionality (see Figure 12.3).

To do this, use Kismet to locate the target WLAN. Once you have identified the target, you should drive around it a few times to get good signal data and four strong GPS coordinates. Using GPSMap, you can then plot the signal strength of the access points that have been discovered. There are several valuable options for GPSMap. The command line to generate circle maps is:

```
gpsmap -r -S2 -P0 -e *.gps
```

- *-r* indicates that gpsmap should create maps showing the range of the networks that have been detected.

- *-S2* indicates that the map should be downloaded from TerraServer. This provides satellite image maps, but there are other map servers you can use.

- *-P0* indicates the opacity, or the amount of background you can "see" through the map.

- *-e* indicates that a point should be plotted denoting the center of the network's range.

Figure 12.3 A GPSMap Circle Map Identifies the Network Range

Intelligence Gathering Tools

Unlike wired penetration tests, customers often want pen testers to locate and identify their wireless networks, especially if they have taken steps to obfuscate the name of their network. This is particularly common with red team penetration testing, where the pen tester, in theory, has no knowledge of the target other than the information he can find through his own intelligence gathering methods.

USENET Newsgroups

As Internet search engines have become more powerful, one tool available to penetration testers for intelligence gathering is often overlooked—USENET. As with all types of networks, wireless networks have connectivity and configuration issues from time to time. Administrators are likely to turn to other administrators of similar equipment to see if the problem has been experienced by others, and if so, is there a known solution. Searching USENET for our target's e-mail domain

(XXX@ourtaget.com) will often lead to messages posted by administrators looking for help. This can be a goldmine of information for a pen tester, revealing the manufacturer and model of access points in use (which can help exclude a network from or narrow our potential target list), the type of encryption standard in use, if any wireless intrusion detection mechanisms are in place, and many other essential pieces of information that will make the pen test easier as you proceed.

Google (Internet Search Engines)

Google is obviously one of the most powerful tools for performing this type of intelligence gathering. Assume that your target is in a large building or office complex where several other organizations are located and multiple WLANs are deployed. At this point, you want to take all of the SSIDs of the networks you discovered and perform a search of the SSID and the name of the target organization. If an organization has chosen not to use the company name as the SSID (many don't), they often will use a project name or other information that is linked to the organization. A search for the SSID and the organization name can often help identify these types of relationships and the target WLAN. With regard to Internet search engines, your imagination is your only barrier when performing searches; the more creative and specific your search, the more likely you are to come across information that will lead to identifying the target network.

Scanning Tools

There are several WLAN scanners available to penetration testers, both active and passive. Auditor includes two of these tools, Wellenreiter and Kismet. Both of these tools can be effective; however, there are certain circumstances where one may be more beneficial than the other. In any case, having multiple tools available to compare and verify results is always beneficial to a pen tester.

Wellenreiter

To start Wellenreiter, right-click on the **Auditor** desktop and select **Auditor -
>Wireless -> Scanner/Analyzer -> Wellenreiter (Wireless Scanner)**. A window opens prompting you for a data directory where your Wellenreiter results will be saved. Select a location, click **OK**, and then confirm the directory by clicking **Yes**. Next, you are prompted to provide a prefix that will be prepended to the Wellenreiter files as they are saved. This is useful for differentiating between multiple scans or sessions; for example, the date or target name can be prepended to the data files. After you have entered your prefix, click **OK** and Wellenreiter opens as shown in Figure 12.4.

Figure 12.4 The Wellenreiter Interface

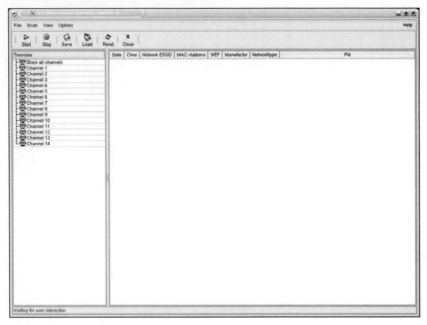

Wellenreiter does not start scanning for WLANs as soon as it is opened. You need to manually start the scan by clicking the **Start** icon located in the upper-right corner of the Wellenreiter interface. Wellenreither scans for WLANs and displays them by channel. By default, the Show all channels view is selected. By clicking on a channel listed in the left pane of the interface, WLANs transmitting on specific channels can be displayed. Wellenreiter also displays the state, channel number, SSID (Network ESSID), MAC Address, WEP status, Manufacturer, and Network Type and allows you to sort based on each of these fields by clicking on the field name. If the SSID is broadcast or has been determined due to an association, it is displayed in the Network ESSID field. If the SSID is not broadcast, "Non-broadcasting" is displayed in that field as demonstrated in Figure 12.5.

Figure 12.5 Wellenreiter Detects WLANs

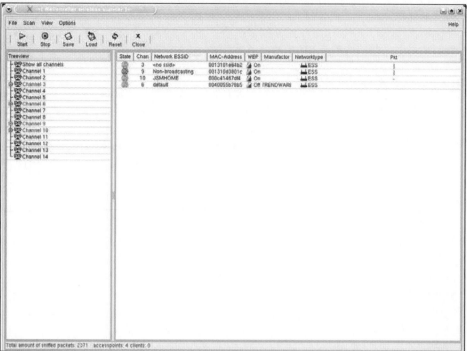

One drawback of using Wellenreiter is that it can only detect whether encryption is in use, but can't determine the type of encryption (WEP or WPA). WPA encrypted networks are displayed as WEP when using Wellenreiter and requires that further investigation is done using a different tool to determine the true type of encryption in use.

Wellenreiter saves two types of data files by default: a complete packet capture dump (.dump) that can be opened with a packet sniffer, and a text file detailing the results of the scan (.save) that can be opened with a text editor as shown in Figure 12.6.

Kismet

Probably the most versatile and comprehensive WLAN scanner is Kismet. Like Wellenreiter, Kismet is a passive WLAN scanner, detecting networks that are broadcasting the SSID and those that aren't. Kismet is started in much the same way as Wellenreiter. Select **Auditor | Wireless | Scanner/Analyzer | Kismet Tools | Kismet (Wireless Scanner)**. A window opens prompting you for a data directory where your Kismet results will be saved. Select a location, click **OK**, and then confirm the directory by clicking **Yes**. Next, you are prompted to provide a prefix that will be prepended to the Kismet files as they are saved. After entering the prefix, click **OK** and Kismet starts. Unlike Wellenreiter, Kismet is a text-based application, and begins collecting data as soon as it is started, as shown in Figure 12.7.

Figure 12.6 The Wellenreiter .save File

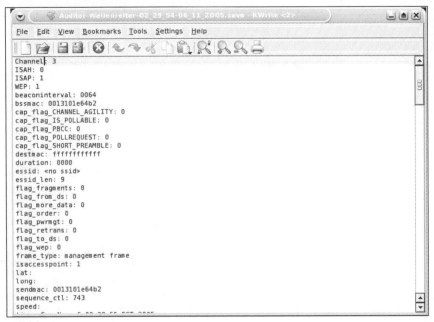

Figure 12.7 The Kismet Interface

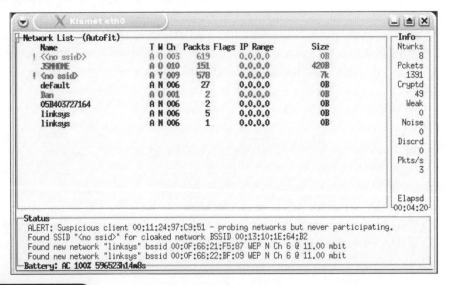

Kismet has a wide range of sorting and view options that allow you to learn view information that is not displayed in the main screen. Sort options can be selected by pressing the **s** key as shown in Figure 12.8.

Figure 12.8 The Kismet Sort Options

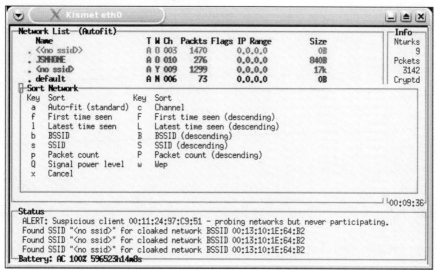

The default sorting view is Auto-Fit. To change the sort view, type **s** to bring up the sort options. Networks can be sorted by:

- The time they were discovered (first to last or last to first)
- The MAC address (BSSID)
- The network name (SSID)
- The number of packets that have been discovered
- Signal strength
- The channel on which they are broadcasting
- The encryption type (WEP or No WEP)

After you choose a sort view, information on specific access points can be viewed. Use the arrow keys to highlight a network, and then press **Enter** to get information on the network as shown in Figure 12.9.

Figure 12.9 Information on a Specific Network

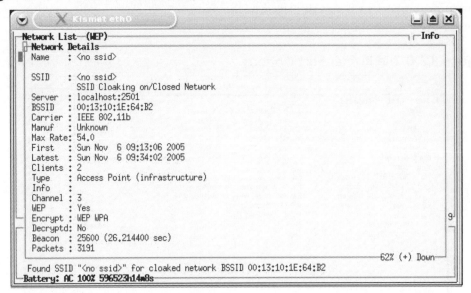

Kismet creates seven log files by default:

- Cisco (.cisco)

- Comma Separated Value (.csv)

- Packet Dump (.dump)

- Global Positioning System Coordinates (.gps)

- Network (.network)

- Weak IVs (.weak)

- Extensible Mark Up Language (.xml)

The range of log files created by Kismet allows pen testers to manipulate the data in many different ways (scripts, importing to other applications, and so forth).

Enumeration Tools

Once the target network has been located and the type of encryption identified, more information needs to be gathered to determine what needs to be done to compromise the network. Kismet is a valuable tool for performing this type of enumeration. It is important to determine the MAC addresses of allowed clients in case the target is filtering by MAC addresses. It is also important to determine the IP address range in use so the tester's cards can be configured accordingly (that is, if DHCP addresses are not being served).

Determining allowed client MAC addresses is fairly simple. Highlight a network and type **c** to bring up the client list, as shown in Figure 12.10. Clients in this list are associated with the network and obviously are allowed to connect to the network. Later, after successfully bypassing the encryption in use, spoofing one of these addresses will increase your likelihood of successfully associating. The client view also displays the IP range in use; however, this information can take some time to determine and may require an extended period of sniffing network traffic in order to capture.

Figure 12.10 The Kismet Client View Used for Enumeration

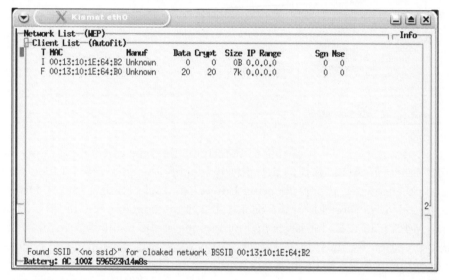

Vulnerability Assessment Tools

Vulnerability scans do not have to necessarily be performed on wireless networks, although once a wireless network has been compromised, a vulnerability scan can certainly be conducted on wireless or wire-side hosts. WLAN-specific vulnerabilities are usually based on the type of encryption in use. If the encryption is vulnerable, the network is vulnerable. There are two primary tools pen testers can use to test implementations of wireless encryption: Kismet and Ethereal

Using Kismet to determine the type of encryption in use is very simple, but not always effective. Use the arrow keys to select a network, and press **Enter**. The

"Encrypt" line displays the type of encryption in use. However, Kismet cannot always determine with certainty if WEP or WPA is in use, as shown in Figure 12.11.

Figure 12.11 Kismet Cannot Determine if WEP or WPA Is Used

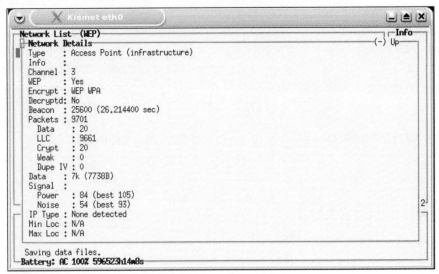

Luckily, even if Kismet is unable to determine the type of encryption on the network, Ethereal can be used to definitively identify the encryption. Open your Kismet or Wellenreiter .dump file using Ethereal and select a data packet. Drill down to the *Tag Interpretation* fields of the packet. If a frame contains ASCII ".P…." this indicates WPA is in use. This is verified by looking at the frame information. The Tag Interpretation for these bytes shows "WPA IE, type 1, version1" and conclusively identifies this as a WPA network as shown in Figure 12.12. An encrypted packet that does not contain this frame is indicative of a WEP encrypted network.

Exploitation Tools

The meat of any penetration test is the actual exploitation of the target network. Because there are so many vulnerabilities associated with wireless networks, there are many tools available to pen testers for exploiting them. It is important for a pen tester to be familiar with the tools used to spoof MAC addresses, deauthenticate clients from the network, capture traffic, reinject traffic, and crack WEP or WPA. Proper use of these tools will help an auditor perform an effective WLAN pen test.

Figure 12.12 WPA Is Positively Identified with Ethereal

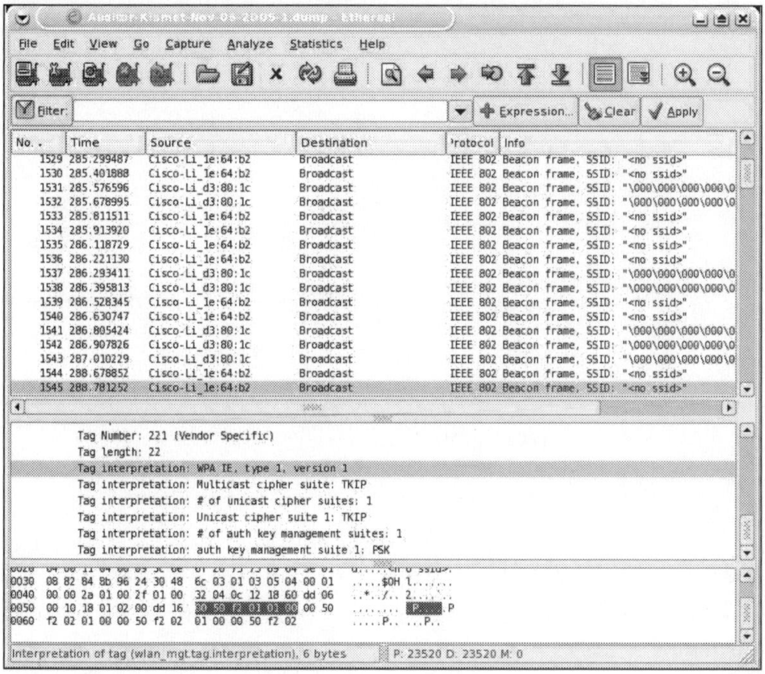

MAC Address Spoofing

Whether MAC address filtering is used as an ineffective, stand-alone security mechanism or in conjunction with encryption and other security mechanisms, pen testers need to be able to spoof MAC addresses. Auditor provides a mechanism to accomplish this called Change-Mac.

After determine an allowed MAC address, changing your MAC to appear to be allowed is simple with Change-Mac. Right-click on the **Auditor** desktop and choose **Auditor | Wireless-Change-Mac (MAC address changer)**. This opens a terminal window and prompts you to select the adapter for which you want to change the MAC address. Next, you are prompted for the method of generating the new MAC address:

- Set a MAC address with identical media type
- Set a MAC address of any valid media type
- Set a complete random MAC address
- Set your desired MAC address manually

While it is nice to have this many choices, the option that is most valuable to a pen tester is the last one, setting the desired MAC manually. Enter the MAC address you want to use and click **OK**. When the change is successful, a window pops up informing you of the change as shown in Figure 12.13.

Figure 12.13 Change-Mac Was Successful

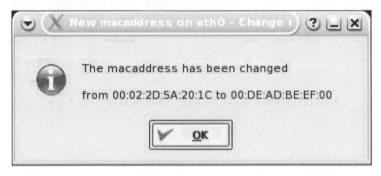

Deauthentication with Void11

To cause clients to reauthenticate to the access point to capture ARP packets or EAPOL handshakes, it is often necessary to deauthenticate clients that are associated to the network. Void11 is an excellent tool to accomplish this task.

To deauthenticate clients, you first need to prepare the card to work with Void11. The following commands need to be issued:

```
switch-to-hostap
cardctl eject
cardctl insert
iwconfig wlan0 channel CHANNEL_NUMBER
iwpriv wlan0 hostapd 1
iwconfig wlan0 mode master
```

The deauthentication attack is executed with:

```
void11_penetration -D -s CLIENT_MAC_ADDRESS -B AP_MAC_ADDRESS wlan0
```

which executes the deauthentication attack (demonstrated in Figure 12.14) until the tool is manually stopped.

Figure 12.14 Deauthentication with Void11

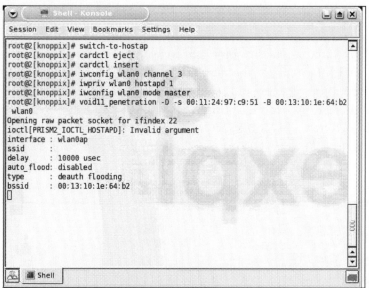

Cracking WEP with the Aircrack Suite

No wireless penetration test kit is complete without the ability to crack WEP. The Aircrack Suite of tools provides all of the functionality necessary to successfully crack WEP. The Aircrack Suite consists of three tools:

- **Airodump** Used to capture packets
- **Aireplay** Used to perform injection attacks
- **Aircrack** Used to actually crack the WEP key

The Aircrack Suite can be started from the command line, or using the Auditor menu system. To use the menu system, right-click on the desktop, navigate to **Auditor | Wireless-WEP cracker | Aircrack suite**, and select the tool you want to use.

The first thing you need to do is capture and reinject an ARP packet with Aireplay. The following commands configure the card correctly to capture an ARP packet:

```
switch-to-wlanng
cardctl eject
cardctl insert
monitor.wlan wlan0 CHANNEL_NUMBER
```

```
cd /ramdisk
aireplay -i wlan0 -b MAC_ADDRESS_OF_AP -m 68 -n 68 -d ff:ff:ff:ff:ff:ff
```

First, you need to tell Auditor to use the wlan-ng driver. The *switch-to-wlanng* command is an Auditor-specific command to accomplish this. Then, the card must be "ejected" and "inserted" for the new driver to load. The *cardctl* command coupled with the eject and insert switches accomplish this. Next, the *monitor.wlan* command puts the wireless card (wlan0) into rfmon or monitor mode, listening on the specific channel indicated by CHANNEL_NUMBER.

Finally, we start Aireplay. Here we are looking for a packet of size 68 bytes. Once Aireplay has collected what it thinks is an ARP packet, you will be given information and asked to decide if this is an acceptable packet for injection. To use the packet, certain criteria must be met:

- FromDS must be 0
- ToDS must be 1
- BSSID must be the MAC address of the target access point
- Source MAC must be the MAC address of the target computer
- Destination MAC must be FF:FF:FF:FF:FF:FF

You are prompted to use this packet. If it does not meet these criteria, type **n** for no. If, it does meet these criteria, type **y** and the injection attack will begin.

Aircrack, the program that actually performs the WEP cracking, takes input in pcap format. Airodump is an excellent choice, as it is included in the Aircrack Suite; however, any packet analyzer capable of writing in pcap format (Ethereal, Kismet, and so forth) will also work. To use Airodump, you must first configure your card to use it:

```
switch-to-wlanng
cardctl eject
cardctl insert
monitor.wlan wlan0 CHANNEL_NUMBER
cd /ramdisk
airodump wlan0 FILE_TO_WRITE_DUMP_TO
```

Airodump's display shows the number of packets and IVs that have been collected as shown in Figure 12.15.

Figure 12.15 Airodump Captures Packets

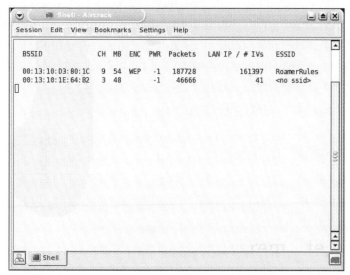

Once some IVs have been collected, Aircrack can be run while Airodump is capturing. To use Aircrack issue the following commands:

```
aircrack -f FUDGE_FACTOR -m TARGET_MAC -n WEP_STRENGTH -q 3 CAPTURE_FILE
```

Aircrack gathers the unique IVs from the capture file and attempts to crack the key. The fudge factor can be changed to increase the likelihood and speed of the crack. The default fudge factor is 2, but this can be adjusted from 1 to 4. A higher fudge factor cracks the key faster, but more "guesses" are made by the program so the results aren't as reliable. Conversely, a lower fudge factor may take longer, but the results are more reliable. The WEP strength should be set to 64, 128, 256, or 512 depending on the WEP strength used by the target access point. A good rule is that it takes around 500,000 unique IVs to crack the WEP key. This number will vary, and can range from as low as 100,000 to perhaps more than 500,000.

Cracking WPA with the CoWPAtty

CoWPAtty by Joshua Wright is a tool to automate the offline dictionary attack to which WPA-PSK networks are vulnerable. CoWPAtty is included on the Auditor CD and is very easy to use. Just as with WEP cracking, an ARP packet needs to be captured. Unlike WEP, you don't need to capture a large amount of traffic; you only need to capture one complete four-way EAPOL handshake and have a dictionary file that includes the WPA-PSK passphrase.

Once you have captured the four-way EAPOL handshake, right-click on the desktop and select **Auditor | Wireless | WPA cracker- | CoWPAtty (WPA PSK bruteforcer)**. This opens a terminal window with the CoWPAtty options.

Using CoWPAtty is fairly straightforward. You must provide the path to your wordlist, the dump file where you captured the EAPOL handshake, and the SSID of the target network (see Figure 12.16).

```
cowpatty -f WORDLIST -r DUMPFILE -s SSID
```

Figure 12.16 CoWPAtty in Action

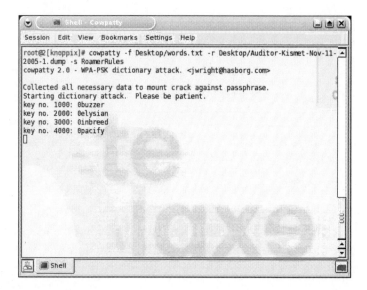

Case Studies

Now that you have an understanding of the vulnerabilities associated with wireless networks and the tools available to exploit those vulnerabilities it's time to pull it all together and look at how an actual penetration test against a wireless network might take place. First, we'll focus on a network using WEP encryption, and then turn our attention to WPA-PSK protected network.

Case Study—Cracking WEP

We have been assigned to perform a red team penetration test against Roamer Industries. We have been given no information about the wireless network, or the internal network. We have to use publicly available sources to gather information about Roamer Industries. We do know that Roamer Industries has deployed a wireless network, but that is all the information we have.

Before we do anything else, we'll investigate the company by performing searches on Google and other available search engines, as well as the USENET newsgroups. We'll also go to the Roamer Industries public Web site to look for information, and we'll perform an ARIN WHOIS lookup on the IP address of their Web site. Quite a bit of important information is gleaned from these searches. The address of their office complex is listed on their Web site. The WHOIS lookup reveals the name and e-mail address of an individual who we discover is a system administrator, judging from the posts he has made on USENET. Additionally, we discover that they are using Microsoft SQL Server on at least one system, because that administrator had described a configuration issue he was having while setting the server up on an MSSQL newsgroup.

Since we have specifically been tasked to test the WLAN, we note the address of the office complex, where the WLAN is almost certainly located, and head to that area. Upon arrival, we fire up Kismet and drive around the building several times. We find 23 access points in the area of our target. Fifteen of these are broadcasting the SSID, but none is named Roamer Industries. This means that we have to gather the SSIDs of the other eight (obviously cloaked) networks. Since we don't want to inadvertently attack a network that does not belong to our target, and thus violate our Rules of Engagement, we have to be patient and wait for a user to authenticate so we can capture the SSIDs. It takes us most of a day to gather the SSIDs of the eight cloaked networks, but once we have them all, we can try to determine which network belongs to our target. None of the SSIDs is easily identifiable as belonging to them, so we go back to Google and perform searches for each SSID we discovered. About halfway through the list of SSIDs we see something interesting. One of the SSIDs is InfoDrive. Our search for *InfoDrive Roamer Industries* locates a page on the Roamer Industries Web site describing a research and development project named InfoDrive. While it is almost certain that this is our target's network, before proceeding, we contact our white cell to ensure that this is, indeed, their network. Once we have confirmation we are ready to continue with our pen test.

Opening the Kismet dumps with Ethereal, we discover that WEP encryption is in use on the InfoDrive network. Now we are ready to start our attack against the WLAN. First, we fire up Aireplay and configure it to capture an ARP packet that we can inject into the network and generate the traffic necessary to capture enough unique IVs to crack the WEP key. Once Aireplay is ready, we start Void11 and perform a deauthentication flood. After a few minutes of our flood, Aireplay has captured a packet that it believes is suitable for injection, as shown in Figure 12.17.

Figure 12.17 Aireplay Searches for a Suitable Packet for Injection

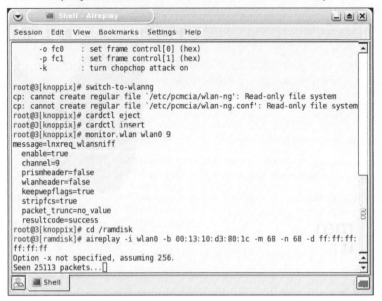

Based on our criteria, we decide that this packet is probably going to work, and we begin the injection attack. Now that Aireplay is injecting traffic, we start Airodump to collect the packets and determine the number of unique IVs we have captured. Aireplay works pretty quickly, and after about 20 minutes, we have collected over 200,000 unique IVs. We decide it is worth checking to see if we have gathered enough IVs for Aircrack to successfully crack the WEP key. Once we have fired up Aircrack and provided our Airodump capture file as input, we find that we have not collected enough IVs. We continue our injection and packet collection for another 15 minutes, at the end of which we have collected over 370,000 unique IVs. We try Aircrack again. This time, we are rewarded with the 64-bit WEP key "2df6ef3736."

Armed with our target's WEP key, we configure our wireless adapter to associate with the target network:

```
iwconfig wlan0 essid "InfoDrive" key:2df6ef3736
```

Issuing the `iwconfig` command with no switches returns the information about the access point with which we are currently associated. Our association was successful, as revealed in Figure 12.18.

Figure 12.18 A Successful Association to the Target WLAN

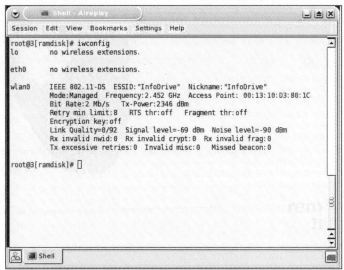

Now that we have associated, we need to see if we can get an IP address and connect to the network resources. First, we try running `dhclient wlan0` to see if they are serving DHCP addresses. This doesn't work, so we go back to Kismet and look at the IP range that Kismet discovered. Kismet shows that the network is using the 10.0.0.0/24 range. We have to be careful here because we don't want to take an IP address that is already in use. We look at the client list in Kismet and determine that 10.0.0.69 is available. Now, we have to make some educated guesses as to how the network is set up. First, we try configuring our adapter with a default subnet mask of 255.255.255.0 and 10.0.0.1 as the default gateway:

```
ifconfig wlan0 10.0.0.69 netmask 255.255.255.0
route add default gw 10.0.0.1
```

Next, we ping the router to see if we have connectivity. Sure enough, we do. At this point, we have successfully established a foothold on the wireless network. Now we can probe the network for vulnerabilities and continue our red team engagement. Our first avenue to explore would likely be the MS SQL server since we know that this service is often configured in an insecure manner, especially by administrators who aren't very experienced in setting up and configuring them. Since our target's administrator was asking for configuration help on a public newsgroup, chances are that he is not an extremely experienced MS SQL administrator, so our chances are good. From here, we continue our penetration test following our known methodologies. The WLAN was the entry vector we needed.

Case Study—Cracking WPA-PSK

Thanks to our success with our penetration test of Roamer Industries, we have been contracted to perform a similar penetration test on the Law Offices of Jack Meoffer. Again, before beginning, we do our information gathering and find valuable information about our target. This time in addition to the address of our target's offices, we are able to harvest 12 different e-mail addresses from our Google and USENET searches.

When we arrive at the target, we again drive around the perimeter of the building where our target's office is located. Using Kismet, we discover 15 WLANs in the area. Ten of these are broadcasting the SSID, including one called Meoffer. We open our Kismet dump with Ethereal and discover that this network is using WPA. Since we have CoWPAtty in our arsenal, we are ready to try to crack the WPA passphrase. First, we look at the client list using Kismet and see that three clients are associated to the network. This is going to make our job a bit easier since we can send a deauthentication flood and force these clients to reassociate to the network, allowing us to capture the four-way EAPOL handshake. To accomplish this, we again fire up Void11 and send deauthentication packets for a couple of minutes. Once we feel like we are likely to have captured the EAPOL handshake, we end our deauthentication.

Since Kismet saves all of the packets collected in the .dump file, we use this as our input file for CoWPAtty. We provide CoWPAtty with the path to our dictionary file, the SSID of our target, and the path to our Kismet .dump file. CoWPAtty immediately lets us know that we have, in fact, successfully captured the four-way handshake, and begins the dictionary attack. We have an extensive wordlist, so we sit back and wait a while. After about 20 minutes, CoWPAtty determines the passphrase is "Syngress" and we are ready to proceed with our intrusion (see Figure 12.19).

Now that we have cracked the passphrase, we edit our *wpa_supplicant.conf,* file, the file where WPA network information and configuration is stored, to reflect the correct SSID and PSK.

Figure 12.19 CoWPAtty Cracks the WPA Passphrase

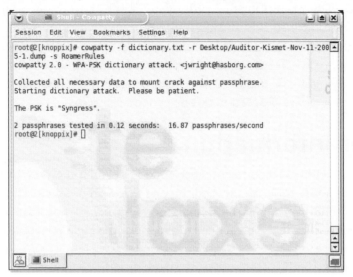

```
network={
ssid="Meoffer"
psk="Syngress"
}
```

After editing the conf file, we restart the wpa_supplicant and check for association with the Meoffer network by issuing the iwconfig command with no parameters. An association was not made. It would appear that our target has taken a step to restrict access. We make an educated guess that they are using MAC address filtering to accomplish this. Again, we look at the client list using Kismet and copy the MAC addresses of the three clients associated with the network. We don't want to use these while the clients are on the network, so we have to sit back and wait for one of them to drop off. After a couple of hours, one of the clients does drop off, and we change our MAC address using the Change-Mac utility that is included with Auditor to the MAC of the client that just left the network.

Now that our MAC has been changed, we again try to associate to the network by restarting the supplicant. This time, we are successful. Now, we try issuing the dhclient wlan0 command to see if a DHCP server is connected to the network. Luckily for us, one is. We are assigned an address, subnet mask, and default gateway. We are also assigned DNS servers.

Now that we have our foothold on the network, it's time to propagate. Since our information gathering didn't turn up much useful information about specific servers and services that are on the network, we decide to use the information we were able to gather to our advantage. Our first path of attack is to take the usernames we gleaned from the collected e-mail addresses (for example, if an e-mail address is jack@meoffer.org, there is a good chance that "jack" is the network username) and try to find blank or weak, easily guessable passwords. Now that we have our initial foothold into the network and are armed with possible usernames, we have many options open to us as we proceed with our penetration test.

Further Information

The tools discussed here to perform penetration tests aren't the only ones available. In fact, there are more tools on the Auditor CD that weren't discussed in this chapter. Those tools have much of the same functionality as tools that were discussed, or functionality that isn't generally beneficial during a penetration test of wireless networks.

In addition to Auditor, some other outstanding tools to be aware of when pen testing are NetStumbler (for Windows) and KisMAC (for Mac OS X). NetStumbler is an active scanner, so its application is limited, but it can be an outstanding resource, particularly for use with direction finding due to its excellent Signal to Noise Ratio (SNR) display. KisMAC is a fantastic tool for penetration testers that provides the ability to perform both active and passive scanning and has a strong graphical signal display. Additionally, the functionality of many of the tools discussed in this chapter is built in to KisMAC, including deauthentication, packet injection, WEP cracking, and WPA cracking.

If you want a quick tool to change MAC addresses, SirMACsAlot (www.securitytribe.com/~roamer/SirMACsAlot.tar.gz) provides a simple, command-line interface for changing MAC addresses.

This list is still not complete, and more tools are released every day, so it is important to stay current and understand the tools you need and what tools are available. One advantage of Auditor for penetration testers is that it incorporates a large selection of tools, and with each update, more are added, bringing even more functionality to an already outstanding resource.

Additional GPSMap Map Servers

TerraServer satellite maps (such as those shown in Figure 12.3) are not the only types of maps available. GPSMap allows you to generate maps from a number of different sources and types. The following list shows the map server options and types available for GPSMap.

- **-S-1** Creates a representation of the networks with no background map

- **-S0** Uses Mapblast

- **-S1** Uses MapPoint (this functionality does not work as of the time of this writing)

- **-S2** Uses TerraServer satellite maps

- **-S3** Uses vector maps from the U.S. Census

- **-S4** Uses vector maps from EarthaMaps

- **-S5** Uses TerraServer topographical maps

Appendix A

Solutions Fast Track

This Appendix will provide you with a quick, yet comprehensive review of the most important concepts covered in this book.

Chapter 1

Introduction to Wireless: From Past to Present

Exploring Past Discoveries That Led to Wireless

☑ *Wireless technology* is the method of delivering data from one point to another without using physical wires, and includes radio, cellular, infrared, and satellite.

☑ The discovery of electromagnetism, induction, and conduction provided the basis for developing communication techniques that manipulated the flow of electric current through the mediums of air and water.

☑ Guglielmo Marconi was the first person to prove that electricity traveled in waves through the air, when he was able to transmit a message beyond the horizon line.

☑ The limitations on frequency usage that hindered demand for mobile telephone service were relieved by the development of the geographically structured cellular system.

Exploring Present Applications for Wireless

☑ Vertical markets are beginning to realize the use of wireless networks. Wireless technology can be used for business travelers needing airport and hotel access, gaming and video, for delivery services, public safety, finance, retail, and monitoring.

☑ Horizontal applications for wireless include new technology for messaging services, mapping (GPS) and location-based tracking systems, and Internet browsing.

Chapter 2

Wireless Security

Enabling Security Features on a Linksys WRT54G, a D-Link DI-624 AirPlus Xtreme G, a Apple Airport Extreme, and a Cisco 1100 Series Access Point

These have been consolidated because they are the recommendations for securing any AP/router and are not specific to a particular hardware:

☑ Assigning a unique SSID to your wireless network is the first security measure that you should take. Any attacker with a "default" configuration profile is able to associate with an access point that has a default SSID. Assigning a unique SSID in and of itself doesn't offer much protection, but it is one layer in your wireless defense.

☑ Many attackers use active wireless scanners to discover target wireless networks. Active scanners rely on the access point beacon to locate it. This beacon broadcasts the SSID to any device that requests it. Disabling SSID broadcast makes your access point "invisible" to active scanners. Because your access point can still be discovered by passive wireless scanners, this step should be used in conjunction with other security measures.

☑ Wired Equivalent Privacy (WEP) encryption, at a minimum, should be used on your home wireless network. Although there are tools available that make it possible to crack WEP, the fact that encryption is enabled on the access point may be the difference between an attack on your AP or your neighbor's. Adequate security for these networks is provided by 128-bit WEP.

☑ Enabling Wi-Fi Protected Access (WPA) on your home network is the most secure solution in use today. WPA uses enhanced encryption and dynamically changing keys that make the process of cracking your encryption key more difficult. Only a dictionary attack is possible at this time, so ensure that your passkey/passphrase is robust and not a common dictionary word.

☑ Filtering by Media Access Control (MAC) address allows only wireless cards that you specifically designate to access your wireless network. Again, it is possible to spoof MAC addresses, therefore you shouldn't rely on MAC address filtering exclusively. It should be part of your overall security posture.

☑ Each of the four security steps presented in this chapter can be defeated. Fortunately, for most home users they do provide adequate security for a wireless network. By enacting a four-layer security posture on your wireless network, you have made it more difficult for an attacker to gain access to your network. Because the likelihood of a strong "return" on the attacker's time investment would be low, he is likely to move on to an easier target. Don't allow your wireless network to be a target of convenience.

Configuring Security Features on Wireless Clients

☑ Windows XP clients are configured using the Wireless Connection Properties and the Windows XP Wireless Client Manager. To associate with your access point once the security features have been enabled, your access point must be added as a Preferred Network. You need to enter the SSID and the WEP key during the configuration process. On the same token, you can also enable WPA during this process, including your passkey/passphrase for connection.

☑ Windows 2000 does not have a built-in wireless client manager like Windows XP. You need to enter the SSID and WEP key into a profile in the client manager software that shipped with your wireless card. Remember that Microsoft does not natively support WPA in Windows 2000. You must obtain client software from your network card vendor in order to use WPA with Windows 2000.

☑ Apple makes wireless connections seem trivial in their 10.x versions of their operating system. By simply adding the SSID and encryption key, in either WEP or WPA mode, you are able to gain access to the network in a small amount of time.

☑ Linux users now have the ability to install and use the Wireless Tools package for their distribution. This package includes the *iwconfig* binary that makes quick configuration of connecting to a WEP encrypted network. WPA can be easily implemented using the *wpa_supplicant* application, with supported wireless network cards and configuration files. There is plenty of information on the Internet for configuring wireless clients to use WEP and WPA in Linux.

Understanding and Configuring 802.1X RADIUS Authentication

☑ RADIUS provides for centralized authentication and accounting.

☑ 802.1X provides for a method of port-based authentication to LAN ports in a switched network environment.

☑ For 802.1X authentication to work on a wireless network, the AP must be able to securely identify traffic from a particular wireless client. This identification is accomplished using authentication keys that are sent to the AP and the wireless client from the RADIUS server.

Chapter 3

Dangers of Wireless Devices in the Workplace

Intruders Accessing Legitimate Access Points

- ☑ Disable SSID broadcasts
- ☑ Use an obscure SSID
- ☑ Enable encryption
- ☑ Filter MAC addresses
- ☑ Control RF signal strength
- ☑ Implement a wireless DMZ
- ☑ Implement wireless IDS

Intruders Connecting to Rogue Access Points

- ☑ Implement clear organizational policy
- ☑ Conduct user awareness training
- ☑ Control the procurement process
- ☑ Conduct periodic wireless assessments
- ☑ Scan your network from the wired side

Intruders Connecting to WLAN Cards

- ☑ Implement clear organizational policy
- ☑ Conduct user awareness training
- ☑ Utilize a host-based firewall
- ☑ Restrict administrator privileges
- ☑ Manage procurement
- ☑ Disable wireless networking
- ☑ Enforce wireless network policies

Chapter 4

WLAN Rogue
Access Point Detection and Mitigation

The Problem with Rogue Access Points

- ☑ A rogue access point is an unauthorized access point installed by an employee without permission from the IT or Security departments.

- ☑ One rogue access point can dismiss an entire security architecture.

- ☑ Employees install rogue access points for their own benefit without realizing that they have created a back door to the corporate LAN.

Preventing and Detecting Rogue Access Points

- ☑ The first step in protecting against rogue access points is having a security policy. A security policy should outline the rules against unauthorized wireless devices and employees must be educated about the policy.

- ☑ A wireless sniffer can aid in the detection of wireless access points throughout an area that can then be compared against a list of authorized access points.

- ☑ Cisco offers a centralized solution with a WLSE engine where all Cisco-aware wireless devices work together to detect possible rogue access points and report them to the central management station.

- ☑ Rogue access points can be detected from the wired network by using a network port scanner. Unlike a user's workstation, rogue access points usually have port 80 (HTTP) and 23 (Telnet) open for administration purposes.

- ☑ A port scanner can trigger false alarms and extra traffic on already congested traffic by scanning every device. Coordinated scanning should be performed to avoid confusion.

IEEE 802.1x Port-based Security to Prevent Rogue Access Points

☑ The 802.1x protocol allows mutual authentication where the access point authenticates the user and the user authenticates the access point, to ensure that the user is connecting to a valid, not a rogue, access point.

☑ In 802.1x protocol, users are prompted for authentication credentials as soon as they plug their workstation into the switch port. Devices such as rogue access points that do not support such authentication will not be allowed to connect to the wired port.

☑ A third-party authentication server that supports RADIUS protocol is required to store all user credentials and perform the actual authentication. The access point or the catalyst switch is used as a proxy server between the authenticating client and the RADIUS server.

Using Catalyst Switch Filters to Limit MAC Addresses per Port

☑ Port security in catalyst switches allows you to restrict devices that can physically connect to the port by their MAC addresses.

☑ The three types of MAC addresses in port security feature are static, dynamic, and sticky.

☑ When an unauthorized device connects to a secured port, a violation occurs. The three configurable reactions to a violation are protect, restrict, and shutdown modes.

☑ In shutdown violation mode the port is shut down and requires the administrator to manually bring it back up.

Chapter 5

Wireless LAN VLANs

Understanding VLANs

- ☑ A VLAN is used to define the logical separation of a LAN network into multiple broadcast domains.

- ☑ Two configured VLANs cannot interact with each other unless they are routed with a Layer 3-aware device such as router.

- ☑ A trunk port is a configured interface port that allows for multiple VLAN communications. A trunk port is used between the access point and the switch to transfer multiple VLANs using the 802.1q encapsulation standard.

VLANs in a Wireless Environment

- ☑ SSID is used to bind a wireless user to the proper VLAN.

- ☑ Each VLAN can have unique characteristics such as the authentication method, IP filters, and the encryption method. This allows one access point or bridge to support multiple groups of users and devices.

- ☑ A native VLAN is used to tag traffic originating and directed to the IP address of the access point or bridge, such as SSH and HTTP administration.

Wireless VLAN Deployment

- ☑ Currently you can configure up to 16 VLANs. You can only configure up to 16 SSIDs on Cisco's wireless devices.

- ☑ VLANs are supported in VxWorks 12.00T release and IOS 12.2.4-JA release and later.

- ☑ Av 802.1q trunk port must be configured between two bridges supporting multiple VLAN communications.

Configuring Wireless VLANs in IOS

☑ Multiple SSID configurations using the **ssid** command are configured under interface configuration mode.

☑ Radio and Ethernet interfaces are split into logical sub-interfaces to represent each VLAN configuration.

☑ You should always copy the running configuration and startup configuration to save your configuration in case the device reboots.

Broadcast Domain Segmentation

☑ A broadcast domain segmentation prevents broadcast-directed traffic from one VLAN reaching other VLANs that are considered to be in a separate broadcast domain.

☑ Unlike in wired broadcast segmentation, in 802.11 all broadcasts are seen and processed by every wireless user, even if they are in a different VLAN.

☑ To overcome the differences between 802.11 and a wired network, a broadcast WEP key configuration is required per VLAN. This still does not prevent broadcasts from reaching every wireless user, but it allows only specific VLAN users who know the broadcast key to read its content.

Primary (Guest) and Secondary SSIDs

☑ A guest mode SSID allows users without any SSID to associate to the access point.

☑ The access point sends out a guest SSID in its broadcast beacon to announce its presence.

☑ Only the primary (Guest) SSID can be used in beacons.

Using RADIUS for VLAN Access Control

☑ RADIUS can be used to verify user VLAN mapping and prevent VLAN hopping using unauthorized SSIDs.

☑ RADIUS can either send a list of SSIDs to the user that they are allowed to use, or statically assign a user to a specific VLAN without the need for an SSID.

☑ You can only use RADIUS in a per-user authentication environment such as EAP.

Chapter 6

Designing a Wireless Network

Exploring the Design Process

☑ The design process consists of six major phases: preliminary investigation, analysis, preliminary design, detailed design, implementation, and documentation.

☑ In the early phases of the design process, the goal is to determine the cause or impetus for change. As a result, you'll want to understand the existing network as well as the applications and processes that the network is supporting.

☑ Because access to your wireless network takes place "over the air" between the client PC and the wireless access point, the point of entry for a wireless network segment is critical in order to maintain the integrity of the overall network.

☑ PC mobility should be factored into your design as well as your network costs. Unlike a wired network, users may require network access from multiple locations or continuous presence on the network between locations.

Identifying the Design Methodology

☑ Lucent Worldwide Services has created a network lifecycle methodology, called the Network Engagement Methodology (NEM), for its consultants to use when working on network design projects. The design methodology contains the best-of-the-best samples, templates, procedures, tools, and practices from their most successful projects.

☑ The NEM is broken down into several categories and stages; the category presented in this chapter is based on the execution and control category, for a service provider methodology. The execution and control category is broken down into planning, architecture, design, implementation, and operations.

☑ The planning phase contains several steps that are responsible for gathering all information and documenting initial ideas regarding the design. The plan consists mostly of documenting and conducting research about the needs of the client, which produces documents outlining competitive practices, gap analysis, and risk analysis.

☑ The architecture phase is responsible for taking the results of the planning phase and marrying them with the business objectives or client goals. The architecture is a high-level conceptual design. At the conclusion of the architecture phase, a high-level topology, a high-level physical design, a high-level operating model, and a collocation architecture will be documented for the client.

☑ The design phase takes the architecture and makes it reality. It identifies specific details necessary to implement the new design and is intended to provide all information necessary to create the new network, in the form of a detailed topology, detailed physical design, detailed operations design, and maintenance plan.

Understanding Wireless Network Attributes from a Design Perspective

☑ It is important to take into account signal characteristics unique to wireless technologies from several design perspectives. For example, power consumption and operating system efficiency are two attributes that should be considered when planning applications and services over wireless LAN technologies.

☑ Spatial density is a key wireless attribute to focus on when planning your network due to network congestion and bandwidth contention.

Chapter 7

Wireless Network Architecture and Design

Fixed Wireless Technologies

☑ In a fixed wireless network, both transmitter and receiver are at fixed locations, as opposed to mobile. The network uses utility power (AC). It can be point-to-point or point-to-multipoint, and may use licensed or unlicensed spectrums.

☑ Fixed wireless usually involves line-of-sight technology, which can be a disadvantage.

☑ The *fresnel* zone of a signal is the zone around the signal path that must be clear of reflective surfaces and clear from obstruction, to avoid absorption and reduction of the signal energy. *Multipath reflection* or interference happens when radio signals reflect off surfaces such as water or buildings in the fresnel zone, creating a condition where the same signal arrives at different times.

☑ Fixed wireless includes Wireless Local Loop technologies, Multichannel Multipoint Distribution Service (MMDS) and Local Multipoint Distribution Service (LMDS), and also Point-to-Point Microwave.

Developing WLANs through the 802.11 Architecture

☑ The North American wireless local area network (WLAN) standard is 802.11, set by the Institute of Electrical and Electronics Engineers (IEEE); HiperLAN is the European WLAN standard.

☑ The three physical layer options for 802.11 are infrared (IR) baseband PHY and two radio frequency (RF) PHYs. The RF physical layer is comprised of Frequency Hopping Spread Spectrum (FHSS) and Direct Sequence Spread Spectrum (DSSS) in the 2.4 GHz band.

☑ WLAN technologies are not line-of-sight technologies.

☑ The standard has evolved through various initiatives from 802.11b, to 802.11a, which provides up to five times the bandwidth capacity of

802.11b. 802.11g also provides the same bandwidth as 802.11a but on the same 2.4 GHz as 802.11b, and provides interoperability for both. 802.11i provides enhanced security features. Now, accompanying the ever-growing demand for multimedia services is the development of 802.11e.

☑ 802.11b provides 11 Mbps raw data rate in the 2.4 GHz transmission spectrum.

☑ 802.11a provides 25 to 54 Mbps raw data rate in the 5 GHz transmission spectrum.

☑ 802.11g provides up to 54 Mbps raw data rate in the 2.4 GHz transmission spectrum, and provides backward compatibility for 802.11a/b devices.

☑ 802.11i or WPA2 provides additional security features and utilized AES encryption.

☑ HiperLAN type 1 provides up to 20 Mbps raw data rate in the 5 GHz transmission spectrum.

☑ HiperLAN type 2 provides up to 54 Mbps raw data rate and QOS in the 5 GHz spectrum.

☑ The IEEE 802.11 standard provides three ways to provide a greater amount of security for the data that travels over the WLAN: use of the 802.11 Service Set Identifier (SSID); authentication by the Access Point (AP) against a list of MAC addresses; and the use of encryption technologies.

Developing WPANs through the 802.15 Architecture

☑ Wireless personal area networks (WPANs) are networks that occupy the space surrounding an individual or device, typically involving a 10m radius. This is referred to as a personal operating space (POS). WPANs relate to the 802.15 standard.

☑ WPANs are characterized by short transmission ranges.

☑ Bluetooth is a WPAN technology that operates in the 2.4 GHz spectrum with a raw bit rate of 1 Mbps at a range of 10 meters. It is not a line-of-sight technology. Bluetooth may interfere with existing 802.11 technologies in that spectrum.

☑ HomeRF is similar to Bluetooth but targeted exclusively at the home market. HomeRF provides up to 10 Mbps raw data rate with SWAP 2.0.

Mobile Wireless Technologies

☑ Mobile wireless technology is basic cell phone technology; it is not a line-of-sight technology. The United States has generally progressed along the Code Division Multiple Access (CDMA) path, with Europe following the Global System for Mobile Communications (GSM) path.

☑ Emerging technologies are known in terms of *generations*: 1G refers to analog transmission of voice; 2G refers to digital transmission of voice; 2.5G refers to digital transmission of voice and limited bandwidth data; 3G refers to digital transmission of multimedia at broadband speeds (voice, video, and data).

☑ The Wireless Application Protocol (WAP) has been implemented by many of the carriers today as the specification for wireless content delivery. WAP is a nonproprietary specification that offers a standard method to access Internet-based content and services from wireless devices such as mobile phones and PDAs.

☑ The Global System for Mobile Communications (GSM) is an international standard for voice and data transmission over a wireless phone. A user can place an identification card called a Subscriber Identity Module (SIM) in the wireless device, and the device will take on the personal configurations and information of that user (telephone number, home system, and billing information).

Optical Wireless Technologies

☑ Optical wireless is a line-of-sight technology in the infrared (optical) portion of the spread spectrum. It is also referred to as free space optics (FSO), open air photonics, or infrared broadband.

☑ Optical wireless data rates and maximum distance capabilities are affected by visibility conditions, and by weather conditions such as fog and rain.

☑ Optical wireless has very high data rates over short distances (1.25 Gbps to 350 meters). Full duplex transmission provides additional bandwidth capabilities. The raw data rate available is up to a 3.75 kilometer distance with 10 Mbps.

☑ There are no interference or licensing issues with optical wireless, and its data rate and distance capabilities are continuously expanding with technology advances.

Chapter 8

Monitoring and Intrusion Detection

Designing for Detection

- ☑ Get the right equipment from the start. Make sure all of the features you need, or will need, are available from the start.

- ☑ Know your environment. Identify potential physical barriers and possible sources of interference.

- ☑ If possible, integrate security monitoring and intrusion detection in your network from its inception.

Defensive Monitoring Considerations

- ☑ Define your wireless network boundaries, and monitor to know if they're being exceeded.

- ☑ Limit signal strength to contain your network.

- ☑ Make a list of all authorized wireless Access Points (APs) in your environment. Knowing what's there can help you immediately identify rogue APs.

Intrusion Detection Strategies

- ☑ Watch for unauthorized traffic on your network. Odd traffic can be a warning sign.

- ☑ Choose an intrusion detection software that best suits the needs of your environment. Make sure it supports customizable and updateable signatures.

- ☑ Keep your signature files current. Whether modifying them yourself, or downloading updates from the manufacturer, make sure this step isn't forgotten.

Conducting Vulnerability Assessments

- ☑ Use tools like NetStumbler and various client software to measure the strength of your 802.11b signal.

☑ Identify weaknesses in your wireless and wired security infrastructure.

☑ Use the findings to know where to fortify your defenses.

☑ Increase monitoring of potential trouble spots.

Incident Response and Handling

☑ If you already have a standard incident response policy, make updates to it to reflect new potential wireless incidents.

☑ Great incident response policy templates can be found on the Internet.

☑ While updating the policy for wireless activity, take the opportunity to review the policy in its entirety, and make changes where necessary to stay current. An out-of-date incident response policy can be as damaging as not having one at all.

Conducting Site Surveys for Rogue Access Points

☑ The threat is real, so be prepared. Have a notebook computer handy to use specifically for scanning networks.

☑ Conduct walkthroughs of your premises regularly, even if you don't have a wireless network.

☑ Keep a list of all authorized APs. Remember, Rogue APs aren't necessarily only placed by attackers. A well-meaning employee can install APs as well.

Chapter 9

Designing a Wireless Enterprise Network: Hospital Case Study

Introducing the Enterprise Case Study

- ☑ Hospital requires wireless access for laptops in satellite buildings.
- ☑ Hospital requires wireless access in conference room.
- ☑ Hospital requires building-to-building wireless links.

Examining Network Requirements

- ☑ The area in the satellite buildings has rooms that will cause interference to the wireless buildings.
- ☑ The area in the conference room is small.
- ☑ There is clear line-of-sight from the main hospital buildings to provide a wireless solution.
- ☑ The distance between buildings permits wireless links.

Designing a Wireless Solution

- ☑ Use two access point bridges per floor in the satellite buildings with omnidirectional antennas.
- ☑ Add an access point bridge in the conference of the main hospital building.
- ☑ Use access-point outdoor routers with directional antennas for hospital to satellite building wireless connectivity.
- ☑ Add wireless links between building pairs for redundancy.
- ☑ Use encryption for security.

Implementing and Testing the Wireless Solution

☑ Install, configure, and test the access points and antennas in the satellite buildings. Test that laptops can access the LAN from all locations in the floor.

☑ Install, configure, and test the access point in the hospital conference room.

☑ Install, configure, and test the outdoor routers and wireless links. Then install the redundant wireless links.

Lessons Learned

☑ Using multiple access point devices on a floor will provide additional access range and redundancy.

☑ Using an access point with a directional antenna in the conference room will provide wireless access for those attending meetings.

☑ Using encryption will provide data security for the wireless network.

☑ Using IEEE 802.11b outdoor routers with wireless directional antennas provides increased bandwidth to 11 Mbps between buildings.

Chapter 10

Designing a Wireless Industrial Network: Retail Case Study

Introducing the Industrial Case Study

- ☑ Wireless technology addresses the emerging mobility needs in the industrial setting. Recent coupling of 802.11b technology with handheld devices promotes widespread uses, from mobile inventory to network administration, to increase employee productivity and customer service.

- ☑ In the case study, the store owner wants to make his existing wired network more efficient and address customer needs. Handheld devices must be implemented to provide mobility.

- ☑ By streamlining the network, the store owner provides employees and customers easy access to store data, such as pricing and inventory.

Designing and Implementing the Wireless Network

- ☑ The network consultants approach the design by categorizing the physical store into three subdomains: the first floor, the warehouse, and the second floor.

- ☑ The consultants obtained a physical map and reviewed the existing network.

- ☑ The store owner provided estimates of the maximum number of customers and employees on each subdomain.

- ☑ The store owner also provided the constraint that all network elements must be hidden for aesthetics.

- ☑ Planning for the RF patterns took place. The consultants planned the placement of the network elements. IP addresses were established.

Planning the Equipment Placement

☑ The following hardware was selected: the Orinoco AP-1000 access point, the Orinoco PCI card, the Orinoco Range Extender, the Orinoco PCI/MCA card, and the SPT1700 handheld device.

☑ The consultants set up the IP addresses, installed the access points, and installed the related software. They installed the radios in the access points and handheld devices and installed the PCI/MCA card in the shipping/receiving PC. All of the hardware and software underwent testing to ensure functionality.

Lessons Learned

☑ You learned how a consulting company can apply the design principles described in previous chapters.

☑ The planning phase contains the details you must be aware of when implementing a similar type of wireless network.

☑ The implementation section of this chapter walks you through the process of integrating the existing wired network with the proposed wireless network.

☑ The most important lesson is to adequately evaluate software development.

Chapter 11

Designing a Wireless Home Network: Home Office Case Study

Introducing the Wireless Home Network Case Study

☑ New technologies and services include online music services, video-on-demand, and Internet-ready entertainment devices. Companies are also networking kitchen appliances, heating and cooling appliances, lighting, and security functions.

☑ This case study explores the user's opportunities in designing and implementing a home network with high-speed access to the company LAN and to a printer. The user in the case study has an existing PC with broadband access that is for family use only, and is located far from the home office.

Designing the Wireless Home Network

☑ The functional requirements include the security of files passed to the company's LAN, the family budget, and the accessibility of the home office.

☑ The user identifies her options, investigates costs, and weighs the results.

☑ In the preliminary design, the user checks for interference, opts for a DHCP server, and reviews the access points within her budget.

Implementing the Wireless Home Network

☑ In the detailed design, the user assembles the network components, determines the Broadband configuration, installs the hardware, installs and configures the software, and tests the resulting network.

Designing a Robust Wireless Home Network for Data, Voice, and Beyond

☑ A user in a related case study is excited by the convenience and mobility of a wireless home network. He wants to create an audio/video server on his home PC.

☑ Currently there are three wireless standards competing for the wireless home-network space: IEEE 802.11b, Bluetooth, and HomeRF. The technologies are, for the most part, incompatible, and it is still unclear which will eventually emerge as the technology (or technologies) of choice.

☑ The major drawbacks of 802.11b products are their interference with 2.4 GHz phones and the fact that they support data only (no native voice integration). New products based on the IEEE 802.11a standard will be emerging in the near future. These products will support an even higher bandwidth and will not interfere with the cordless phones.

☑ Because the voice and data signals are integrated, products using the HomeRF standard should find voice recognition and automation applications easier to develop and support.

Lessons Learned

☑ Building a simple home network is a relatively easy and inexpensive task that can be completed with technology available today. However, as with any network, you should complete a thorough process of investigation, analysis, and design before purchasing any wireless networking solutions.

☑ The advantages of wireless are that it offers you mobility and is available in locations where a wired outlet doesn't exist.

☑ The disadvantages of wireless include some minor interference problems and a quickly evolving and competitive group of standards.

Index

broadcast domain segmentation, 171–174
configuration with IOS, 165–171
guest, secondary SSIDs, 174–175
logical segmentation, 153–155
numbers, 156
overview of, 152–153, 178–179
RADIUS for access control, 175–177
trunk ports, 158–159
VTP in wired network, 156–158
in wireless environment, 159–162
wireless VLAN deployment, 162–165
Virtual Private Network (VPN)
attacks against, 392–393
overview of, 389
VLAN hopping, 169, 175–176
VLAN Trunking Protocol (VTP)
modes, 157
in wired network, 156–158
in wireless network, 161
VLANs. See virtual local area networks
voice, 372–375
Void11, 382, 404–408
vulnerability assessment
conducting, 279–282
overview of, 293
security professional for, 294
tools, 401–402

W

walk-through, 325, 333
WAN (Wide Area Network), 8
WAP (Wireless Access Protocol), 248
WAP (Wireless Application Protocol), 249–250
WAPs. See Wireless Access Points
WarDriving, 103
water, 5
Weaknesses in the Key Scheduling Algorithm of
    RC4" (Fluhrer, Mantin, and Shamir), 281
weather
FSO and, 252
PTP Microwave and, 224
web clipping, 207
web site resources
for home network standards, 217–218
for NMAP, 138
for pen tools, 382
web surfing, 14
WECA (Wireless Ethernet Compatibility Alliance),
    229–230
Wellenreiter, 395–397
WEP. See Wired Equivalent Privacy
WEPCrack, 281
Wide Area Network (WAN), 8
WIDS (wireless intrusion detection system), 106
Wi-Fi Protected Access 2 (WPA2), 241
. See also 802.11i
Wi-Fi Protected Access (WPA)
attacks against, 391
CoWPAtty, cracking with, 407–408
enabling on 1100 Series AP, 52–53
enabling on Airport AP, 41–42

enabling on DI-624 wireless router, 33–34
enabling on WRT54G AP, 24–26
for intrusion prevention, 103, 105
overview of, 98, 388–389
WEP vs., 100
for Windows XP clients, 57–59
for WLAN data transmission security, 241–242
Windows 2000, Microsoft, 59–61, 99
Windows XP, Microsoft
802.1X RADIUS authentication, 93–96
security features, configuring, 98
wireless clients, configuring, 57–59
wireless clients, enabling security on, 56–57
wire logs, 188
Wired Equivalent Privacy (WEP)
Aircrack Suite, cracking with, 405–408
attacks against, 389–391
for broadcast domain, 173
cracking, 408–411
enabling on 1100 Series AP, 49–51
enabling on Airport AP, 41
enabling on DI-624 wireless router, 31–33
enabling on WRT54G AP, 22–24
for intrusion prevention, 103, 105
overview of, 98, 388
protection with, 230
snooping and, 281–282
for VLAN configuration, 168–169
for WLAN data transmission security, 241–242
WPA vs., 100
wired network, VTP in, 156–158
Wireless Access Points (WAPs)
disabling SSID on, 105
intruders accessing legitimate, 102–108
rogue, intruders connecting to, 108–111
Wireless Access Protocol (WAP), 248
Wireless Application Protocol (WAP), 249–250
wireless device, distribution services for, 237–238
wireless devices in workplace
legitimate APs, intruders accessing, 102–108
overview of, 115–116
questions/answers about, 117–118
rogue APs, intruders connecting to, 108–111
WLAN cards, intruders connecting to, 111–114
wireless drop box case study, 106–108
wireless enterprise network, hospital case study
applying wireless in enterprise network, 298–299
conference room networking landscape, 303–304
current network, evaluation, 303
design of wireless solution, 304–310
implementation/testing, 310–314
lessons learned, 314
network requirements, evaluation of, 300–301
opportunity assessment, 299–300
overview of, 298, 315–317
physical landscape, evaluation of, 301–302
satellite buildings' physical landscape, 301
Wireless Ethernet Compatibility Alliance (WECA),
    229–230
wireless intrusion detection system (WIDS), 106
Wireless LAN Solution Engine (WLSE), 128–131
wireless local area networks (WLANs)
802 Standards Committee initiatives, 227–229

X

Syngress: *The Definition of a Serious Security Library*

Syn·gress (sin‑gres): *noun, sing.* Freedom from risk or danger; safety. See *security.*